Communications Wiring and Interconnection

Other McGraw-Hill Communications Books of Interest

AZEVEDO • *ISPF*

BERSON • *APPC: A Guide to LU6.2*

BLACK • *Network Management Standards*

BLACK • *TCP/IP and Related Protocols*

BLACK • *The V Series Recommendations*

BLACK • *The X Series Recommendations*

CHORAFAS • *The Complete LAN Reference*

COOPER • *Computer and Communications Security*

DAYTON • *Integrating Digital Services*

FOLTS • *The McGraw-Hill Compilation of Open Systems Standards*

FORTIER • *Handbook of LAN Technology*

HA • *Digital Satellite Communications*

HELD, SARCH • *Data Communications*

HUGHES • *Data Communications*

INGLIS • *Electronic Communications Handbook*

KESSLER • *ISDN*

KESSLER, TRAIN • *Metropolitan Area Networks*

KIMBERLY • *Electronic Data Interchange*

KNIGHTSON • *Standards for Open Systems Interconnection*

LEE • *Mobile Cellular Telecommunications Systems*

LEE • *Mobile Communications Engineering*

NEMZOW • *Ethernet Management Guide*

PELTON • *Voice Processing*

RADICATI • *Electronic Mail*

RANADE • *Advanced SNA Networking*

RANADE • *Introduction to SNA Networking*

RANADE • *VSAM: Performance, Design, and Fine Tuning*

RANADE, RANADE • *VSAM: Concepts, Programming, and Design*

RHEE • *Error Correction Coding Theory*

ROHDE, BUCHER • *Communication Receivers*

SABIN • *Single-Sideband Systems and Circuits*

SCHLAR • *Inside X.25*

TUGAL, TUGAL • *Data Transmission*

Communications Wiring and Interconnection

Fred J. McClimans

McGraw-Hill, Inc.
New York St. Louis San Francisco Auckland Bogotá
Caracas Lisbon London Madrid Mexico Milan
Montreal New Delhi Paris San Juan São Paulo
Singapore Sydney Tokyo Toronto

Library of Congress Cataloging-in-Publication Data

McClimans, Fred J.
Communications wiring and interconnection / Fred J. McClimans.
p. cm.
Includes index.
ISBN 0-07-044847-7
1. Telecommunication—Equipment and supplies. 2. Electric wiring.
I. Title.
TK5103.M38 1992
621382—dc20 91-29964
CIP

For Shannon, Chloe, and my parents, who gave me both time and energy. Also, JC, TK, JM, Remo, and Chiun for general ideas, commentary, and inspiration.

And in memory of June, who started this whole thing a long, long time ago, on a mountain far, far away. Wish you were here.

1 2 3 4 5 6 7 8 9 0 DOC/DOC 9 7 6 5 4 3 2 1

ISBN 0-07-044847-7

The sponsoring editor for this book was Daniel A. Gonneau, the editing supervisor was Nancy Young, and the production supervisor was Pamela A. Pelton. This book was set in Century Schoolbook by McGraw-Hill's Professional Book Group composition unit.

Printed and bound by R. R. Donnelley & Sons Company.

Contents

Preface xiii

Chapter 1. A Case for Building Wiring 1

1.1 Introduction 1

1.2 The Intelligent Building 3

1.2.1 The Intelligent Building Concept 3

1.2.2 Building Resource Integration 5

1.3 Purpose of This Book 6

1.3.1 Book Structure 7

1.3.2 Book Usage 9

1.4 Before We Begin 9

Chapter 2. Twisted-Pair Cable 11

2.1 What Is Twisted-Pair? 11

2.2 What Twisted-Pair Is Not 15

2.3 Different Types of Twisted-Pair 16

2.4 Twisted-Pair Plugs and Jacks 20

2.4.1 Standard Phone Jacks 20

2.4.2 Pin Connectors 21

2.4.3 Genderless Connectors 23

2.5 Twisted-Pair Cross-Connects 23

2.5.1 Types of Cross-Connect Blocks 23

2.5.2 66-Blocks 25

2.5.3 110-Blocks 27

2.5.4 BIX-Blocks 30

2.5.5 Modular Cross-Connect Products 31

2.6 Common Twisted-Pair Applications 33

2.7 Cost of Twisted-Pair Applications 34

2.8 Advantages of Twisted-Pair 36

2.9 Disadvantages of Twisted-Pair 37

2.10 Summary 38

Chapter 3. Coaxial Cable 41
3.1 What Is Coaxial Cable? 41
3.2 Types of Coaxial Cable 42
3.3 Coaxial Plugs and Jacks 44
3.4 Coaxial Cross-Connect Systems 46
3.5 Common Coaxial Applications 47
3.6 Cost of Coaxial Installations 48
3.7 Advantages of Coaxial Cable 49
3.8 Disadvantages of Coaxial Cable 50
3.9 Summary 51

Chapter 4. Fiber-Optic Cable 53
4.1 What Is Fiber-Optic Cable? 53
4.1.1 Fiber-Optic Cable Components 54
4.1.2 Characteristics of Fiber-Optic Cable 56
4.1.3 Types of Fiber-Optic Cable 58
4.2 Fiber-Optic Connectors 61
4.2.1 Difficulties with Fiber-Optic Connections 62
4.2.2 Types of Fiber-Optic Connectors 63
4.2.3 Fiber-Optic Splicing 64
4.3 Common Fiber-Optic Applications 66
4.4 Cost of Fiber-Optic Systems 67
4.5 Advantages of Fiber-Optic Systems 68
4.6 Disadvantages of Fiber-Optic Systems 69
4.7 Summary 71

Chapter 5. Secondary Transmission Systems 73
5.1 Microwave/Radio/Transmission Systems 73
5.1.1 What Is Microwave? 73
5.1.2 Microwave Connectors 75
5.1.3 Common Microwave Applications 76
5.1.4 Cost of Microwave Installations 77
5.1.5 Advantages of Microwave 77
5.1.6 Disadvantages of Microwave 78
5.2 Infrared Transmission Systems 79
5.2.1 What Is Infrared? 79
5.2.2 Infrared Connectors 79
5.2.3 Common Infrared Applications 80
5.2.4 Cost of Infrared Installations 81
5.2.5 Advantages of Infrared 81
5.2.6 Disadvantages of Infrared 82
5.3 Transmission Support Systems 83
5.3.1 Patch Panels 83
5.3.2 Cross-Connect Adapters 84
5.3.3 Baluns 85

5.4 Telecommunications and Electrical Systems 87
5.4.1 Electrical Noise 87
5.4.2 Crosstalk 88
5.4.3 Electrical Placement 88
5.5 Summary 89

Chapter 6. Vendor Plans 91
6.1 Vendor Plan Capabilities 92
6.2 Vendor List 93
6.3 Selecting a Vendor Plan 94
6.4 AT&T 96
6.4.1 Components 97
6.4.2 Applications 97
6.5 DEC 98
6.5.1 Components 99
6.5.2 Applications 100
6.6 IBM 101
6.6.1 Components 101
6.6.2 Applications 103
6.7 Modular Equipment Vendors 103
6.7.1 Components 103
6.7.2 Applications 104
6.8 Summary 105

Chapter 7. Standard Information Outlet 107
7.1 SIO Selection 107
7.1.1 Type of Jacks 108
7.1.2 Number of Jacks per SIO 114
7.2 Outlet Mounting 115
7.2.1 Flush-Mount Outlets 116
7.2.2 Surface-Mount Outlets 118
7.2.3 Pedestal-Mount Outlets 119
7.3 Station Outlet Cables 119
7.3.1 Twisted-Pair Outlet Cables 119
7.3.2 Coaxial Outlet Cables 121
7.3.3 Fiber-Optic Outlet Cables 121
7.4 Design Parameters 121
7.4.1 Station Sizing 121
7.4.2 Station Placement 123
7.4.3 Special Cases 127
7.4.4 Surge Protection 130
7.5 Installation Guidelines 131
7.5.1 Installation Scheduling 131
7.5.2 Configuration Testing 132
7.6 Summary 132

Chapter 8. Horizontal Distribution System 133

8.1 Cable Selection 133
8.1.1 Type of Media 134
8.1.2 Shielding Requirements 134
8.1.3 Plenum Requirements 135

8.2 Distribution Systems 136
8.2.1 Underfloor Distribution 136
8.2.2 Ceiling Distribution 141
8.2.3 Molding Distribution 150
8.2.4 Undercarpet Distribution 154
8.2.5 Raised Floor Distribution 155
8.2.6 Power Pole Distribution 156
8.2.7 Poke-Through Distribution 157

8.3 Design Criteria 157
8.3.1 Distribution Hardware 158
8.3.2 Cable Selection 158
8.3.3 Power Integration 158
8.3.4 Sizing 160
8.3.5 Building Segmentation 162

8.4 Installation Guidelines 163
8.4.1 Installation Scheduling 164
8.4.2 Fire-Stopping 166
8.4.3 Testing Activities 168

8.5 Summary 170

Chapter 9. The Distribution Frame 171

9.1 Distribution Frame Characteristics 172
9.1.1 Distribution Frame Functionality 172
9.1.2 Distribution Frame Equipment 173
9.1.3 Distribution Frame Room 174
9.1.4 Grounding 176
9.1.5 Multiuse Facilities 176
9.1.6 Secondary Distribution Frame 176

9.2 Cross-Connect Blocks 178
9.2.1 Block Placement 179
9.2.2 Cross-Connect Cable 181

9.3 Network Access Point 182
9.3.1 Building Access 182
9.3.2 Riser Access 183
9.3.3 Horizontal Access 185
9.3.4 Surge Protection 185

9.4 Campus Distribution Frame 187
9.4.1 CDF Placement 188
9.4.2 CDF Room Characteristics 188

9.5 Building Distribution Frame 189
9.5.1 BDF Placement 190
9.5.2 BDF Room Characteristics 191

9.6 Intermediate Distribution Frame 191
9.6.1 IDF Placement 192
9.6.2 IDF Equipment Placement 193
9.7 Satellite Distribution Frame 194
9.7.1 SDF Placement 194
9.7.2 SDF Room Characteristics 194
9.8 Design Criteria 195
9.8.1 Block Block Selection 196
9.8.2 Sizing 197
9.9 Installation Parameters 197
9.9.1 Installation Scheduling 198
9.9.2 Fire-Stopping 199
9.9.3 Configuration Testing 199
9.10 Summary 200

Chapter 10. Frame Interconnection System 201
10.1 Cabling Selection 202
10.1.1 Media Selection 203
10.1.2 Cable Sizing 204
10.1.3 Plenum Rating 205
10.1.4 Shielding 206
10.1.5 Protective Jackets 207
10.2 Distribution Hardware 208
10.2.1 Horizontal Building Distribution 208
10.2.2 Vertical Building Distribution 208
10.2.3 Campus Distribution 209
10.3 Design Criteria 213
10.3.1 Aesthetics 214
10.3.2 Distribution Frame Configuration 214
10.3.3 Flexibility 217
10.3.4 Installation and Management 217
10.4 Installation Guidelines 218
10.4.1 Installation Scheduling 218
10.4.2 Danger Points 221
10.4.3 Configuration Testing 224
10.4.4 System Marking and Identification 225
10.5 Summary 226

Chapter 11. Cable Management 227
11.1 The Entropy Factor 227
11.2 Native Record Systems 229
11.2.1 Numbering Systems 230
11.2.2 Color Coding 232
11.3 Hard-Copy Records 234
11.3.1 The Role of Hard-Copy Records 234
11.3.2 Information Content 234
11.3.3 Administration 235

11.4 Automated Record Systems 236
11.4.1 The Role of Automated Records 236
11.4.2 System Design 237
11.4.3 Administration 241
11.5 Troubleshooting 243
11.5.1 Fault Recognition 243
11.5.2 Fault Isolation 245
11.5.3 Fault Resolution 246
11.6 Organizational Requirements 248
11.7 Expert Systems 249
11.7.1 Expert System Definition 250
11.7.2 Advantages 251
11.7.3 Implementation Issues 252
11.8 Summary 253

Chapter 12. Design Principles 255
12.1 Design Principle Goals 255
12.2 Requirements Driven Design 257
12.2.1 Developing a Requirements Specification 257
12.2.2 Building Wiring Consultants 260
12.3 Wiring Codes 262
12.4 Project Planning 263
12.4.1 The Initial Concept 264
12.4.2 Space Planning 264
12.4.3 Installation Duration 265
12.4.4 Estimating Costs 266
12.5 Design Programming 267
12.5.1 Design Programming Goals 267
12.5.2 Monitoring the Design 268
12.6 Procurement Issues 269
12.6.1 Competitive Bidding 269
12.6.2 System Installation 270
12.6.3 Testing and Acceptance Procedures 272
12.7 Summary 273

Chapter 13. System Integration 275
13.1 The System Integration Process 275
13.2 OSI Model 278
13.2.1 OSI Layer 1—Physical 279
13.2.2 OSI Layer 2—Data Link 280
13.2.3 OSI Layer 3—Network 280
13.2.4 OSI Layer 4—Transport 281
13.2.5 OSI Layers 5, 6, and 7 281
13.3 Voice Systems 282
13.3.1 Physical Requirements 282
13.3.2 Implementation Issues 283

13.4 Local Area Networks 284
13.4.1 Physical Media 286
13.4.2 Topology 287
13.4.3 Access Methods 288
13.4.4 Ethernet Background 290
13.4.5 Token-Ring Background 294
13.4.6 Appletalk and Localtalk Background 294
13.4.7 FDDI Background 295
13.4.8 LAN Internetworking 297
13.4.9 LAN System Design Options 299
13.5 Host and Terminal Systems 300
13.6 Summary 301

Chapter 14. Universal Wiring Standard 303
14.1 Standard Goals 303
14.1.1 Standard versus Plan 305
14.1.2 Standardization Process 308
14.2 Standards Limitations 311
14.3 Existing Standards 312
14.3.1 EIA Wiring Standard 313
14.3.2 BICSI Wiring Standard 314
14.3.3 State Standards 314
14.4 Summary 317

Chapter 15. Looking to the Future 319
15.1 Telecommunications Trends 319
15.2 Computing Trends 322
15.3 Global Networking 323
15.4 Wiring Trends 324

Appendix: Abbreviations and Acronyms 327

Glossary 331

Index 349

Preface

During the past decade, there have been two major events that have struck fear into the MIS world: the rise of local communications systems beyond PBXs and the deregulation of inside wiring from the telephone company (telco). In effect, in 1984 the end user became the telco. Inside wiring was now the responsibility of the customer organization. Had communications maintained a steady state, this may not have been much of a problem. However, as with everything else in this era of information technology, the requirement for advanced forms of local communications systems has overloaded the original capability of most building wiring systems. Systems originally designed to support analog voice and low-speed modem-oriented data communications were quickly overwhelmed by the advent of high-speed local area networks (LANs) operating at multiple megabits per second.

Almost overnight, a new ad-hoc approach to communications wiring was implemented. But since, inside the building, the user is the telco, management of these new LANs and cabling systems was a disaster for many years. For many users, salvation was found in the structured wiring system, an approach to building wiring that creates a flexible, modular system capable of adapting easily to new requirements. This type of standardized building wiring system has now become a cornerstone of most building environments. It allows support for a wide range of communications systems (voice, data, video, control, etc.), which simplifies the job of being the telco. This is very important, since most organizations would much rather focus on their mission-critical business than on pulling cables from office to office.

The building wiring system should be thought of as another building resource, just like power or water facilities. They should all be kept fairly simple and easy to manage, yet provide ubiquitous reliable service for all. This is becoming very important as we move toward the use of high-speed enterprise networks where interconnectivity of our communications resources requires the wiring designer to consider the holistic viewpoint.

The use of a structured building wiring plan can not only save time but money as well through simplified management and the ability to incorporate new technology as it develops. This approach is even extended to new nonwire or wireless technologies being developed today.

Many people mistakenly feel that the advent of wireless communications systems spells the immediate demise of in-house wiring. Actually, the use of wireless systems is made much easier when there is a structured wiring support system as its foundation.

The biggest surge in wireless communications is taking place within the PBX and LAN supplement areas, which use the application of wireless technology to enhance the capabilities of existing systems. For example, wireless PBX systems are becoming available that offer a small hand-held handset to allow a person to roam within a building while having constant phone access. However, there is still a wire-based handset that resides on the person's desk. This makes sense given the size constraints and limited power capabilities of the wireless units (most have to be recharged fairly frequently and are still in the $500-and-up price range).

Likewise, the use of wireless LAN technology is seeing the most potential in enhancing existing wire-based LANs. For example, wireless systems (spread spectrum radio, cellular, and infrared) are ideal for adding several temporary workstations or for workstations that are constantly on the move, such as are used by a building technician (i.e., palmtop or laptop PCs). Again, the extreme cost of wireless LAN systems ($1500 and up in addition to the normal cards and hub devices) and their under development status mean that industry standardization and widespread public use for common desktop applications is probably at least 5 years away. Additionally, as the bandwidth required for many new applications (such as CAD/CAM and desktop imaging) continues to rise above the 100 megabits per second range, it is likely that fiber-optic cables are the only currently foreseeable technology available; wireless systems currently have difficulty in exceeding 10 megabits per second. Even those few wireless systems in existence today acknowledge that wire is the best transport means between wireless concentrators.

Building wiring is as important as ever. Understanding the broad scope of wiring systems can help you to better design and manage your building wiring resource, letting you focus on your mission-critical functions, such as how to better manufacture or sell your organization's products or services. As you work your way through your own business' wiring systems, remember that innovation and flexibility are often the keys to any successful action. Although there are many wrong answers in building wiring, there are also many more right answers. Don't be afraid to lead the way or be inventive, provided you have a standardized base from which to work.

Fred J. McClimans

Chapter

1

A Case for Building Wiring

Definition: *Telecommunications wiring system* Any transmission media used to pass communications signals between two or more computers or telephone systems. May include twisted-pair, coaxial cable, fiber-optic cable, microwave, infrared, radio, etc. See headache, nightmare, heartburn....

1.1 INTRODUCTION

Telecommunications wiring is perhaps the most overlooked and least planned for resource within the building environment, often not even considered until after the building has been occupied by a tenant. For many building developers and tenants alike, this can be a catastrophic error. The cost of installing a wiring system after a building has been occupied can be monumental. Installation costs may include structural changes to the building (for cable distribution systems) as well as potential costs incurred by tenants because of office disruption while cable installers invade like the wild hoards from the north.

Building wiring systems installed today have much greater demands placed upon them than their earlier predecessors. Looking back 5 or 10 years, the main form of communications within a building (including from a building to external communications) was the phone system. The phone system was usually either Centrex (phone service offered by the local telephone company) or a small private branch exchange (PBX) and required little more than plain old telephone wire. The wire used was typical "telephone wire" that used between 4 and 50 conductors, depending upon the phone system in use. The wiring system's main function was to support voice-based communications. Most data communications, typically over coaxial cable, were considered a unique case and not part of the building wiring system.

Today, the idea of data communications being labeled "unique" has faded from all but a few organizations. Most office environments are inundated with data devices. These devices include a wide variety of hardware systems, including terminals, personal computers, intelligent engineering stations, and minicomputers. Even nontraditional data systems, such as facsimile machines, are being connected to data devices directly and can be considered "data" oriented.

Along with this proliferation of data equipment came the closely related invasion of wire and cable required to connect the data devices together. In most buildings, this meant the ad hoc installation of bulky coaxial cable between devices. Or perhaps it meant the installation of twisted-pair cable installed point to point to support RS-232 applications. These installations were generally very costly, not only for resources and labor costs but also for the amount of productive office time lost while the installation was taking place. In a typical existing office, a new cable installation to support a single user could easily cost thousands of dollars.

Since a failure to plan for a solid and structured wiring system can be so costly, it would make sense for basic economic principles to dictate proper planning. However, this is not always the case. In search of the answer as to why building wiring systems are not always adequately planned for, it is necessary to determine how people perceive building wiring. Is it a high-tech nightmare or is it a benefit of technological progress? Most people who must deal with wiring systems on a regular basis are fairly opinionated about wiring systems and support their stance with extreme, almost religious, vigor. Interestingly, this is usually a love/hate relationship.

Those individuals who are willing to deal with the issues of planning, implementation, and management usually think that wiring is certainly a benefit. On the other hand, most people who do not take the time to understand and properly design a building wiring system tend to describe building wiring as a nightmare.

It is easy to see why telecommunications building wiring is thought of as a confusing morass or nightmare. Just look at the growing number of cabling vendors and installers. Look at the number of proprietary wiring "standards" that are being pushed on the business community today. Look at the lack of management systems that handle voice as well as data wire. This complexity has led many people to feel that telecommunications wiring is overly complicated and perhaps the most troublesome aspect of their telecommunications system:

> My communications room is a mess: cables everywhere. Everytime we add a new device I need to pull new wire. Most of my LAN failures are traced to bad cable connections. My technicians spend too much time trying to find the right cables.

These are legitimate complaints that are being experienced by many telecommunications managers around the world. There are, however, some people who have the other opinion. They feel that telecommunications wiring has been a tremendous advantage:

> My new shielded cable lets me run my token-ring in my machine shop that is full of EMI [electro-magnetic interference]. The new fiber-optic cable will connect my entire campus. And even if there are several disadvantages, the benefits outweigh the extra time it takes to manage such a system.

Which user accurately described building wiring? To be honest, telecommunications wiring is often both a nightmare and a benefit. However, this does not always have to be the case. A properly designed telecommunications wiring system can provide positive benefit to users and management while not taking 3 years off the telecommunications manager's life.

1.2 THE INTELLIGENT BUILDING

One of the driving forces behind the evolution of building wiring has been the development of the intelligent building. The advent of the intelligent building, where a building has "intelligent" value-added abilities, has required that building wiring systems be capable of supporting the abilities. In the current slow-growth real estate market, the addition of intelligent features into a building can make the difference between making a profit and taking a loss. It is important that the basic underlying concept of the intelligent building, as well as the requirements that will be placed upon a building wiring system, be understood.

1.2.1 The Intelligent Building Concept

When most people hear the phrase "intelligent building," an image of futuristic science fiction comes to mind. They point to the written works by authors such as Isaac Asimov and Arthur Clark as examples of worlds where buildings have been taken over by artificial intelligence (AI) in computers and robots. When discussing the capabilities of intelligent buildings today, they think of examples such as talking elevators, automated security systems, and intelligent environmental control. Perhaps they point to examples of automated fire-suppression systems or even of a large computer that "controls" everything within a single building. However, while these components may be present in an intelligent building, they are only the end result of the intelligent building concept.

The real intelligence in any building lies in communications. The ability to create and rely upon information has become an overwhelming cornerstone of our society. Part of this has been our requirement to share information with others. Information that we create ourselves is relatively useless unless we can communicate it to others. What good are the complete works of Shakespeare if you cannot get to the library to read the books. This integral component of our society has manifested itself in the rapidly growing telecommunications and computer industries. The number of computer and communications networks being installed today, such as Ethernet and Token-Ring, is rivaled only by the number of high-performance computers appearing on our office desks.

The current wave of high-technology has been very active in shaping the areas of building control. Many areas of the building control and maintenance cycle are now automated. Using computer technology we have elevators that "speak." The environmental control of many buildings is now based upon a microcomputer with hundreds of remote sensors throughout the building feeding information to the main computer on temperatures, airflow, air quality, and outside temperatures. Security systems are available that recognize security codes and pass keys at building entrances and can automatically deny or allow building access. Even building functions, such as turning lights on or heating water, can now be controlled and automated. The one area these features have in common is their reliance upon computer technology, in effect, giving the building control system the intelligence to react to certain predetermined situations.

For every computerized control system within a building, there must be some form of communications system that provides information or commands for remote devices (an environmental control system cannot turn up the air conditioning if it does not know that it is too warm on the fourth floor). This is where the functions of the intelligent building and building wiring come together. The intelligent building provides the automation of the command and control system while the building wiring provides the physical means by which these control systems input and output commands to remote sensor and functional devices.

The building wiring support for an intelligent building comprises universal access, where the communications wiring system can reach all locations within a building, and universal support, where the communications wiring can support the interfaces and communications throughput requirements of these systems. This is what makes a "smart" building.

1.2.2 Building Resource Integration

Another aspect of an intelligent, or smart, building involves the support of end-user communications. Just as the building command and control systems must communicate between various devices, so too must office occupants communicate with each other. These communications will take place through many different types of systems, such as telephones and computer networks. These end-user communications systems help to form the local communications system (LCS), which comprises all of the components that provide communications support for end users. This includes voice, data, and video systems. The foundation of these systems is in the building wiring. The building wiring system must be a universal resource that can be utilized to support information technologies. Just as with the command and control systems, the building wiring system must provide both universal access and universal support.

An interesting task for the building wiring designer is the integration of the command and control and LCS systems. Traditionally these systems have been isolated from each other. For years, the LCS consisted of coaxial data networking systems and twisted-pair voice devices. Prior to the automation of the command and control process, most building support systems were either stand-alone systems or operated over non-LCS-compatible wiring systems (often using electrical signals that would disrupt normal LCS communications). This dichotomy has recently begun to converge into a single building wiring requirement. See Fig. 1.1.

The current trend in LCS devices is toward twisted-pair and fiber-optic cable for all types of end-user devices. Likewise, the trend in command and control systems is toward the use of twisted-pair for device connectivity. This convergence of wiring requirements has been difficult in some areas. While the command and control devices now use computers for automation and twisted-pair for communications, they often require building wiring in locations not commonly associated with end-user systems. There are also many of these systems available today that do not support, or require, building wiring for all control functions (i.e., pneumatic air control systems that use air pressure to control the flow of air through ducts). In general, however, the accepted rule is that the LCS can be expanded to include building

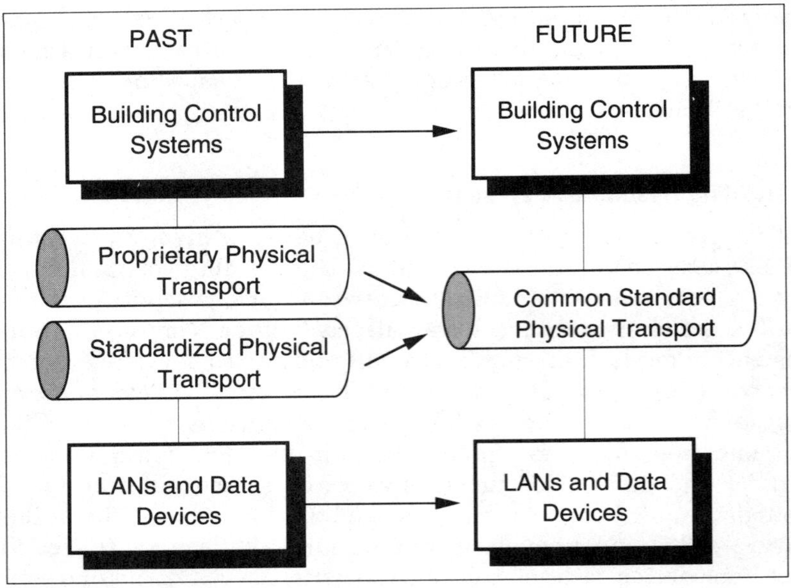

Figure 1.1 Building integration.

command and control functions. Understanding this, and its place in providing an intelligent building where communications between multiple types of devices is available through resource elements, is important to understanding the overall design and planning of a building wiring system.

1.3 PURPOSE OF THIS BOOK

The purpose of this book is to define and discuss the technology and methodologies that can be applied to the development of a successful telecommunications wiring system. It is structured such that a wiring novice can first learn about the components and then address the issues of design and management. This format is ideal for any individual involved in the process of building wiring design but perhaps unschooled in all of the basics or unfamiliar with the various tasks involved in the wiring process. It provides a broad picture of the necessary components and tasks that must take place during the planning, design, implementation, and management of a successful telecommunications building wiring system.

This book will not teach you how to pull cable faster through a round conduit or how to install a component from vendor X. It will, however, instruct you in the benefits of proper cable installation and the value of designing a wiring system to support, rather than compete with, vendor communications equipment. Once an understanding of wiring systems is achieved, the process of dealing with low-level issues, such as how hard to pull a cable through a conduit, can be much easier to handle.

1.3.1 Book Structure

This book has 15 chapters, each dealing with a particular aspect of building telecommunications wiring systems, plus an appendix. They are as follows:

- Chapter 1: Introduction—An overview of the demand behind the development of a standardized building wiring system.
- Chapter 2: Twisted-Pair Cable—Twisted-pair cable has become the dominant transmission media for both voice and data communications applications. This chapter discusses the role of twisted-pair cable in today's telecommunications wiring system.
- Chapter 3: Coaxial Cable—Once the dominant transmission media for data networking, coaxial cable has become a dinosaur in the eyes of many users. Nonetheless, this cable is still being installed in a large number of sites. This chapter discusses the background and role of coaxial cable in today's networking environment.
- Chapter 4: Fiber-Optic Systems—The use of fiber-optic cable for telecommunications systems has risen rapidly during the past few years. This chapter takes a look at characteristics of optical fiber that make it well suited for high-speed networking applications.
- Chapter 5: Secondary Transmission Systems—Often overlooked when the topic turns to building wiring, there are a multitude of components that can be used to support and enhance the performance of a building wiring system. This chapter discusses some of the more popular secondary transmission components.
- Chapter 6: Vendor Wiring Plans—Many equipment and cable vendors offer "prepackaged" wiring plans for implementation in end-user facilities. This chapter presents a brief discussion of the motives behind vendor-offered plans and what users can expect to get with an off-the-shelf building wiring plan.
- Chapter 7: Standard Information Outlet (SIO)—The SIO (i.e., telephone jack) is the main interface between the human and machine worlds. This chapter discusses the uses, options, and configuration details of the SIO unit.
- Chapter 8: Horizontal Distribution—If getting there is half the bat-

tle, this chapter could win the war. Included here are discussions on the design, selection, and implementation of horizontal distribution systems designed to transport telecommunications wire through the floor of a building.

- Chapter 9: Distribution Frame—The core of any wiring plan. Also known as the telephone closet, the distribution frame is the location at which cables serving different locations all come together.
- Chapter 10: Frame Interconnection—This chapter discusses the methods available to interconnect multiple distribution frames to form a single and cohesive building resource.
- Chapter 11: Cable Management—Proper cable management techniques can allow a building wiring system to be properly maintained. This chapter discusses several options for establishing and automating such a system.
- Chapter 12: Design Principles—A building wiring plan can only be as strong as the design process which was responsible for the plan. This chapter discusses some of the important aspects of designing and implementing a building wiring system.
- Chapter 13: System Integration—Knowing what types of voice and data applications are available today can help with the planning and design of a building wiring system. This chapter presents some background information on networking systems available today, along with some hints on how they can affect, and be affected by, a telecommunications wiring system.
- Chapter 14: Universal Wiring Standards—Developing a site-specific wiring plan may not be enough to fully support the requirements of a large organization. This chapter focuses on the advantages of developing a standardized wiring approach that can be used as the basis for telecommunications wiring plans at multiple site locations.
- Chapter 15: Future Issues—Looking at only today's user requirements is not sufficient when trying to set performance metrics for a building wiring plan. It is also important to know that issues and developments within the telecommunications industry as a whole may affect the usefulness of a telecommunications wiring system in the future. This chapter closes out the book by presenting some of the major trends in telecommunications networking and pointing out some of the possible effects these trends could have on building wiring issues.
- Appendix and Glossary—A key to understanding the function of a building wiring system is to understand the language that describes

it, as well as the terms used by systems that must be supported by a building wiring system. Included in this book is a complete glossary of terms and a list of abbreviations and acronyms commonly used within the wiring and telecommunications industry. They can serve as quick reference guides to help the reader understand where particular components of telecommunications systems are used.

Although these chapters are designed as independent discussions of a single topic, the combination of the information provided in them will provide the reader with a complete understanding of wiring systems. In effect, the sum of the book is greater than the individual chapters. This premise is key to the understanding of both this book and telecommunications wiring systems.

As in many activities, the short-sighted, or narrow, view very rarely succeeds in wiring system design. A well-rounded understanding of wiring systems can be just as advantageous as a well-rounded education. As such, this book is designed to provide a complete understanding of building telecommunications wiring systems, not just what they are or how much they will cost. It addresses how a wiring system is driven by communications requirements, standards, growth projections, and technology.

1.3.2 Book Usage

Each of the chapters within this book is intended to present to the reader bits and pieces of information relative to the telecommunications and building wiring systems. While there is a great deal of information included in all chapters, there are bound to be some readers who are more interested in certain chapters than others. For the seasoned telecommunications staffer, it is possible to seek out those chapters that are of most interest first. However, for the reader wishing to gain a broad understanding of both the technology and the application of this technology, it is suggested that the book be read in the order of each chapter.

1.4 BEFORE WE BEGIN

In addition to the guidelines presented in this book, there are many guidelines and rules set forth by local, state, and federal government agencies concerning the design and installation of building wiring, power, and communications systems. In no way should any guideline discussed in this book be construed as the final authority or in any way supersede any government regulation. Any building construction or wiring installation should involve personnel familiar with any relevant local code that may have jurisdiction over the design or implementation process in that area. When in doubt about a code or regulation, it always pays to check first with the proper local authorities to determine what regulations may apply to the particular installation site.

Chapter

2

Twisted-Pair Cable

Definition: *Twisted-pair cable* A type of cable consisting of two copper conductors each surrounded by an insulating material. The most common form of telecommunications support wire, used for both voice and data applications.

One of the most common types of telecommunications cable in use today is twisted-pair cable. Installed in almost every commercial building built within the past half-century, twisted-pair has evolved from an analog voice provider to a high-speed digital system capable of supporting both voice and data transmissions. This transformation has allowed twisted-pair to support modern voice applications (almost every telephone system in use today uses twisted-pair in some form), data applications, such as local area networks (LANs) and data private branch exchanges (PBXs), and emerging integrated voice/data systems, such as Integrated Services Digital Network (ISDN). This chapter presents a discussion of twisted-pair as it is used within the building wiring system.

2.1 WHAT IS TWISTED-PAIR?

There are many types of cable known as twisted-pair, many of which do not conform to the basic criteria of twisted-pair. True twisted-pair consists of the following:

- Two solid copper conductors, each encased in a polyvinylchloride (PVC) sheath
- A copper conductor diameter of between 20AWG and 26AWG
- A twist rate for each pair of 2 to 12 twists per foot

- A color coding scheme involving set colors representing the tip and ring (or + and −) for each pair
- A characteristic impedance of between 90 ohms (Ω) and 110 Ω

Individual pairs of conductors may be combined with other pairs to form larger multipair cables. Typical cables have 25, 50, 100, or 300 pairs. These larger-pair cables are generally used for bulk distribution of cable within a building riser or campus backbone system. See Fig. 2.1.

The inner copper conductors are used to carry the communications signal (either analog or digital), which consists of an electrical signal. The insulation material, consisting of a dielectric material, is not capable of conducting an electrical signal and thus serves to isolate the inner copper core from potential outside conductive electrical signals. The dielectric material is made of a nonconductive material such as PVC or some other synthetic plastic casing.

Twisted-pair cable is, however, susceptible to nonconductive electrical currents (i.e., magnetic fields). For this reason, the pair of copper conductors encased individually in an insulating material are twisted together. The twisting of the conductors, typically between 2 and 12 twists per linear foot, helps to make the cable immune to electrical noise and crosstalk.

The banded color coding scheme of twisted-pair cable uses a color pattern that identifies not only what conductors make up a pair but also what pair in sequence it is relative to other pairs within a

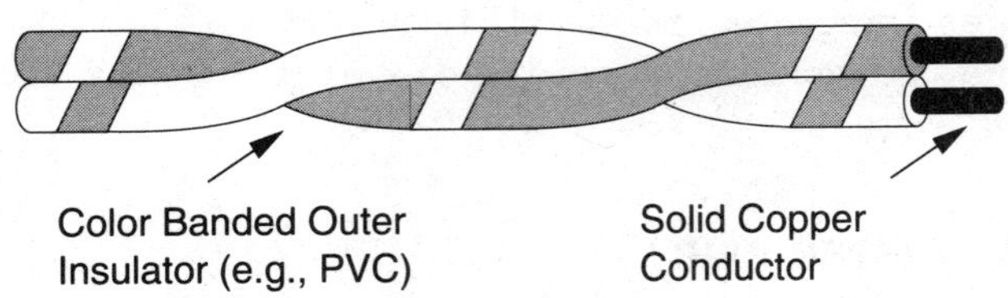

Common Pair Counts	Common Applications
2 Twisted-Pairs →	Patch Cables, Special Data
3 Twisted-Pairs →	Station/Horizontal Distribution
4 Twisted-Pairs →	Station/Horizontal Distribution
25-300 Bulk Pairs →	Distribution Between Telephone Closets

Figure 2.1 Twisted-pair.

multipair sheath. This is also used to determine which conductor in a pair is the tip and which is the ring (or + and − in the data world). The banding scheme uses two opposing colors to represent a single pair. One color is considered the primary while the other color is considered the secondary. For example, given the primary color of white and the secondary color of blue, a single twisted-pair would consist of one cable that is white with blue bands on it and a second cable that is blue with white bands on it. The five primary colors are white, red, black, yellow, and violet. See Fig. 2.2.

In multipair cables the primary color is responsible for an entire group of pairs (five pairs total). For example, as shown in the figure, the first five pairs all have the primary color of white. Each of the secondary colors, blue, orange, green, brown, and slate, are paired in a banded fashion with white. This continues through the entire primary color scheme for all four primary colors (comprising 25 individual pairs). In larger cables (50 pairs and up), each 25-pair group is wrapped in a pair of ribbons, again representing the groups of primary colors matched with their respective secondary colors. These color-coded band markings help cable technicians to quickly identify and properly terminate cable pairs.

The impedance (Ω) of a cable is the total electrical resistance that a

Pair #	TIP	RING
1	W	BLUE
2	H	ORANGE
3	I	GREEN
4	T	BROWN
5	E	SLATE
6	R	BLUE
7		ORANGE
8	E	GREEN
9		BROWN
10	D	SLATE
11	B	BLUE
12	L	ORANGE
13	A	GREEN
14	C	BROWN
15	K	SLATE
16	Y	BLUE
17	E	ORANGE
18	L	GREEN
19	L	BROWN
20	O	SLATE
	W	
21	V	BLUE
22	I	ORANGE
23	O	GREEN
24	L	BROWN
25	E	SLATE
	T	

Figure 2.2 Primary color band marked code.

cable presents to an electrical signal (in this case a communications signal) being passed through the cable. We can define the basic electrical characteristics of a cable as follows:

Impedance: The total electrical resistance of a metallic cable to an electrical signal. In the communications environment, the electrical impedance of a cable is very important. An impedance level too high or too low could affect the distance an electrical signal could travel while still maintaining a useful wave pattern. Additionally, an impedance mismatch between two cable segments spliced together (i.e., a horizontal distribution cable connected to a vertical riser cable) could cause signal reflections and/or decrease the strength of the original signal. When an electrical signal encounters a large change in the characteristic impedance of a cable, usually $\geq 15\ \Omega$, the signal can "bounce," or reflect, off the new cable segment (leaving a weakened original signal still traveling in the original direction). This is similar to sound waves bouncing off a door in a long hallway. The reflected signal, in turn, can interfere with other signals (messages) being sent down the cable. This is very important since many advanced networking systems (such as Ethernet and Token-Ring) have specific distance ratings that are based upon specific electrical cable characteristics. Any deviation from these preset parameters can cause significant disruption in the networking systems' capabilities. Typical twisted-pair cable has an impedance measurement of approximately $100\ \Omega$.

Attenuation: A measurement (in decibels per thousand feet, dB/kft) representing the loss of an electrical signal as it travels down a cable. This loss of signal strength can negatively affect the ability of a destination on a network to properly recognize and interpret a message.

Capacitance: The amount of electrical signal that passes from one cable to another as a result of difference between the electrical currents present in the two cables. This is measured in nanofarads per thousand feet (nF/kft). The capacitance of a cable is directly related to electrical crosstalk, where the signal on one twisted-pair overpowers or corrupts the signal on another, adjacent pair.

Velocity of propagation: The velocity of propagation (Vp) is a factor representing the time required for an electrical signal to propagate through a segment of cable. The Vp is used predominantly for the calibration of cable test equipment.

Resistance: The opposing force a cable exerts on an electrical signal resulting in resistance to the signal. The resistance (measured in Ω) can be thought of as a cable's ability to "steal" energy from an electrical signal. This causes the cable to release heat and weakens the original signal.

Inductance: A measurement (in henrys, H) that represents the creation of an external electromagnetic field when current is passed through it. The creation of the field (EMI) can affect the signal on adjacent cable pairs or circuits.

Although these various measurements of cable characteristics exist, they are not commonly used in the day-to-day building wiring environment. The important cable characteristic to monitor is the cable impedance, which represents the best overall measure of a cable's ability to properly pass a communications signal.

2.2 WHAT TWISTED-PAIR IS NOT

To fully understand what twisted-pair cable is, it is necessary to discuss what twisted-pair cable is not. Many older buildings (read this as most buildings built before the mid-1970s) have telephone wire installed that is either technically not twisted-pair or fails to match the high electrical standards required of today's building wiring for supporting high-speed LAN and data applications. Some common types of cable that inappropriately fall under the generic heading of twisted-pair include:

Stranded cable: This type of cable uses a stranded conductor instead of a solid conductor. This stranded conductor is made up of a dozen or more strands of cable with each strand being considerably thinner than a standard solid conductor. Some examples of stranded cable that many people might be familiar with include most "silver/satin" telephone cables, stereo speaker cable, and many household electrical cables. Although stranded cables are typically available in the same sizes and pair counts as solid twisted-pair, they cannot always be used in the same manner; both electrical and physical characteristics may differ considerably. One of the most common problems associated with stranded cable is the difficulty involved in punching-down or terminating the cable in a jack or cross-connect block. Stranded cables tend to splay or spread out when pressure is applied to them. Unfortunately, most jacks and cross-connect blocks rely upon pressure to split the outer insulation material and establish a solid contact between the cable and the jack and block component. For this reason, stranded cables are used only in limited instances. The most common application is patch cords, such as silver/satin telephone cables and limited-distance data cables, such as used by the IBM Cabling System.

Quad cable: In the early days of telecommunications (more specifically the early days of telephones), a type of cable known as *quad* cable was installed in many buildings for use by the in-house telephone system. It consists of four solid conductors (occasionally this is stranded and not solid) sealed within a single PVC sheath. The four inner conductors are insulated from each other by an individual PVC insulator and the overall PVC sheath which encases all conductors. The four conductors are not twisted in any pair-oriented pattern, although they may be all twisted together in a common rotation pattern. The color code of quad cable differs from the standard twisted-pair color scheme. Each conductor has a solid color, typically red, black, green, and yellow.

Untwisted twisted-pair: The most difficult type of imitation twisted-pair cable to spot is the notorious untwisted twisted-pair. This type of cable usually follows most color and physical characteristics of twisted-pair, such as the color-coded band marking and the use of solid inner conductors. The major difference between the real and imitation twisted-pair is the twisting (or lack thereof) of individual pairs. The best way to identify this type of cable is to physically inspect the cable for the twist rate. Fortunately, most "twisted-pair" cable installed within the past decade or so conforms to today's accepted specifications for twisted-pair. Common cables that may not provide adequate twist rates include most silver/satin patch cables as well as most multipair apparatus cables (such as printer or modem cables). Patch cables are rarely twisted and should be limited in use wherever digital data systems are being used (special twisted patch cables, while not common, are available from most modular wiring vendors).

In addition to the above-mentioned cables, there are other, less common types of cable installed within buildings that may or may not meet the physical and electrical characteristics of acceptable twisted-pair (such as the NYC Class E twisted-pair for use with fire-alarm systems). If there is any question about the integrity of twisted-pair, testing for electrical characteristics can usually be performed fairly easily with standardized test gear available from many vendors.

2.3 DIFFERENT TYPES OF TWISTED-PAIR

Despite, or perhaps because of, the pervasiveness of twisted-pair cable in the building environment, there are many different types of cable that come under the twisted-pair moniker. These cables typically differ in a standardized (or common) fashion that does not significantly alter the electrical characteristics of the cable. The best way to understand these different types of cable is to think of these differences as options that enhance the standard characteristics of twisted-pair cable. They are:

AWG (American wire gauge): The diameter of a cable (the inner conductor) is measured in AWG units (i.e., 24AWG, or 24 gauge). Common cable AWG measurements for twisted-pair include 22, 24, and 26. The AWG is an inverse measurement where the higher the AWG, the smaller the cable diameter. As cable diameter increases, the distance over which a signal can be transmitted without significant loss increases. For this reason, 22AWG cable is usually reserved for high-speed data applications that must travel extended distances. 24AWG is typically used for most telephone and data applications. 26AWG cable, with its reduced transmission distance, is usually used for patch-panel cables. See Fig. 2.3.

Shielding: Any copper communications cable is susceptible to electromagnetic interference (EMI) or crosstalk. Such interference can disrupt a communications signal causing bit errors to occur. To protect against this type of interference, a shield, consisting of a braided mesh or solid metal conductor, is wrapped around the cable. At one end of the cable (typically the cross-connect end), the cable shield is connected to the telecommunications ground through either the mesh shield or a drain wire (a solid conductor in the mesh shield). While shielding can protect the integrity of a communications cable, it can also have adverse effects. Shielding slightly alters the electrical characteristics of the cable. This can reduce the allowable transmission distance with systems that are not intended for use with shielded cables (i.e., systems designed for use with unshielded twisted-pair).

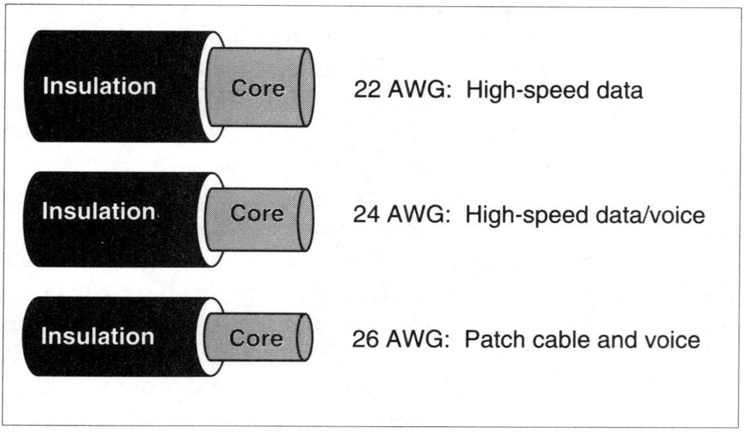

Figure 2.3 American wire gauge.

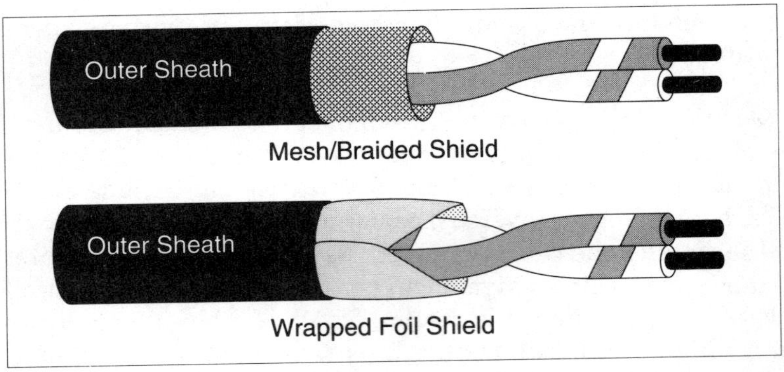

Figure 2.4 Cable shielding.

See Fig. 2.4. Many cables, such as riser or campus distribution cables, have a protective solid metal shield surrounding the twisted-pair conductors. This shield is generally intended to add strength to the cable and prevent cables from being severed (by rodents or backhoes). This protective shield is not intended for use as a "shielding" provider and should not be grounded for transmission purposes.

Plenum rating: First, a word on the term *plenum*. A plenum is an area or duct where air is passed for circulation in a building. This includes air ductwork as well as the area found above most suspended ceilings. Air passage areas are commonly referred to as *air plenums* or *plenum-rated areas*. In the typical office building, air is continually routed throughout an entire building (in other words, each office does not have a dedicated air supply). For example, in a large office, if a person in one end of a building lights up a cigarette, the odor can be detected almost immediately anywhere else in the building. The moral of this little story is that smoke travels fast throughout a building; especially if the source is in an air plenum. Most twisted-pair is encased in PVC or similar sheath. PVC has been used for many years as a cable casing because of its low cost and physical characteristics. Unfortunately, PVC (and similar plastic-based cable casings) can give off extremely toxic fumes when heated or burned. This can pose quite a dilemma since the majority of communications cables installed today (and for the past few decades) have been installed in air plenums, such as above drop-tile ceilings. If a fire were to occur within an office building, toxic fumes from burning or smoldering PVC cables could be transported throughout the building in a matter of minutes (remember here that the leading cause of death and injury in fires is from smoke inhalation). Enter the plenum-rated cable, which is a cable that is encased in a material that has been certified to not give off toxic fumes when heated or burned. Cables that are plenum rated are referred to as *plenum cables*. A plenum casing on a cable does not alter the electrical characteristics of the inner conductor. What it does do, however, is decrease the flexibility of the cable (because of increased overall diameter) and increase the cost.

Specialty cables: This category includes any cable that has been manufactured to meet a specific requirement while still retaining the characteristics of twisted-pair. An example of a specialty cable would be undercarpet twisted-pair. Many users today rely upon cables placed under floor carpets for connection to communications devices or wiring systems. This typically occurs in open office environments where modular furniture is used or where other distribution facilities, such as underfloor ducts or power poles, are not available. Most twisted-pair is encased in a round sheath. Placing such a cable under a carpet would cause a noticeable bump in the carpet which someone could trip over. Twisted-pair cable also tends to be very sensitive when it comes to people stepping on or rolling chairs over it. The undercarpet cable was designed to be placed under a carpet without a significant disruption and to provide additional protection for the encased twisted-pairs.

Another type of specialty cable is the composite cable, which is cable that contains multiple types of media within a single sheath. An example of this is the IBM Type 2 cable which houses both four unshielded twisted-pairs (24AWG) and two shielded twisted-pairs (22AWG). Another example, which actually mixes media materials, is the 1090/2090 composite cable offered by AT&T, which includes two sets of four twisted-pairs (24AWG) and two fiber-optic strands (62.5 μ/125 μ) within a single sheath. These types of specialty cables have become increasingly popular since they can simplify many aspects of cable installation and can usually offer a reduced cost compared to individual cables. More information on this type of cable, and its installation merits, is found in Chap. 8.

Configuration options for twisted-pair can be selected to best suit your specific application. Influencing factors may include:

- The distance of the cable run
- The amount of EMI within the path of the cable run
- The plenum rating, or lack of one, for the cable distribution path
- Physical constraints that may be placed on a cable run

In addition to basic differences in type of physical characteristics of a twisted-pair cable, there can be diverse electrical characteristics shared by the same type of physical cable. For example, vendors such as AT&T and Northern Telecom provide high-performance "data" versions of unshielded 24AWG twisted-pair cable that greatly increases the distance allowable for LAN transmissions over conventional wire with the same physical characteristics (i.e., standard unshielded, 24AWG). It is these special cables that allow high-speed LANs, such as 16-Mbps Token-Ring and potentially 100-Mbps Fiber Distributed Data Interface (FDDI), to operate over unshielded twisted-pair.

2.4 TWISTED-PAIR PLUGS AND JACKS

One of the more important factors regarding twisted-pair implementations is the cable jack or cross-connect block. These items are vital since without the proper interface, any twisted-pair cable would be relatively useless. In the twisted-pair arena, there are three major types of twisted-pair jacks: the RJ-type (a.k.a. the phone plug), the pin-connector, and the genderless connector (i.e., the IBM sexless data connector). The *RJ-type* (registered jack) name generally refers to the standard format used for most telephone jacks. The term *pin-connector* refers to twisted-pair connectors, such as the RS-232 connector, which provide connection through male and female pin receptacles. Genderless connectors are connectors in which there is no separate male or female component; each component can plug into any other similar component.

2.4.1 Standard Phone Jacks

The standard phone jack is specified by a variety of different nomenclatures, such as RJ and RG, which refer to the physical and electrical characteristics of the unit. These jacks consist of a male and a female component. The male component snaps into the female receptacle. The important point to note, however, is the number of conductors each type of jack can support. The actual pin configuration or electrical characteristics of the jack are likely to be application-specific and do not necessarily affect the physical connector.

Common configurations for phone jacks include support for four, six, or eight conductors. A typical example of a four-conductor jack, supporting two twisted-pairs, would be the one used for connecting most telephone handsets to their receivers. See Fig. 2.5.

A common six-conductor jack, supporting three twisted-pairs, is the RJ-11 jack used to connect most telephones to the telco or PBX facilities. An example of an eight-conductor jack is the RJ-45 jack (also covered by the RG-45/48 heading), which is intended for use by ISDN as the user-site interface for ISDN terminals. In the building wiring

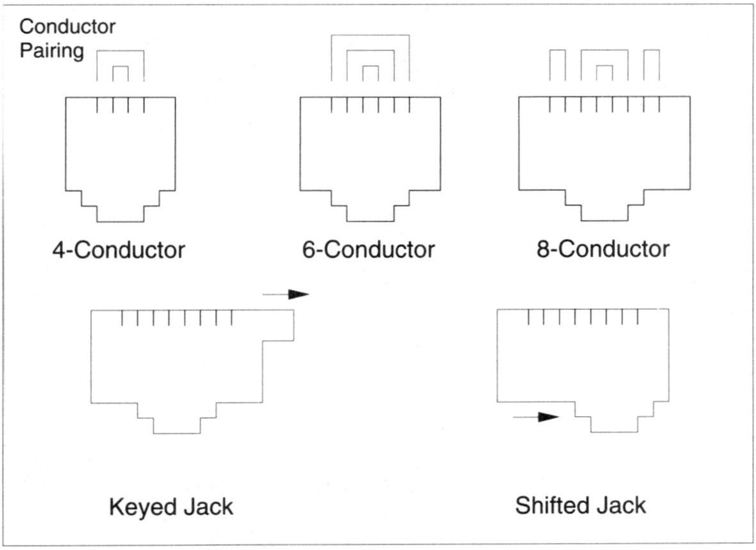

Figure 2.5 Standard phone jacks.

arena, the six-conductor jack is by far the most prevalent. The eight-conductor jack, however, is rapidly becoming the most widely installed jack as more corporations are pulling twisted-pair in four-pair bundles for both voice and data. The eight-conductor jack, in addition to being used for ISDN, is also specified by several other popular applications, such as the new IEEE 802.3 10BaseT standard for Ethernet over twisted-pair.

These types of jacks are often keyed to deny the insertion of a plug into the wrong type of receptacle. There are two types of keying: side keying and shift keying. Side keying relies upon an additional piece of plug placed on the side of the plug. This type of keying is commonly used in outlets where multiple jacks (typically for voice *and* data) are present. Shift keying, instead of using an addition to the plug, actually shifts the snap connector on the plug to the left or right. This type of keying is not as common as side keying in the building wiring arena; it is more often used in the world of data connectors for computer or communications equipment.

2.4.2 Pin Connectors

There are many different types of pin connectors, each designed to serve a specific application. Two common types are the DB-type connectors and the 50-pin "champ" connectors. See Fig. 2.6.

The pin connector is available in many different configurations,

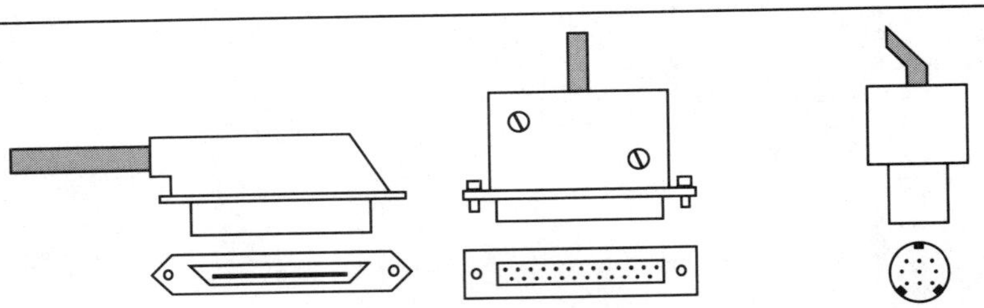

Figure 2.6 Twisted-pair pin connectors.

supporting anywhere from 5 to 50+ conductor leads. The most common use for this type of connector is for computer peripheral connection, such as printers and modems. These connectors consist of a male and female pair which often have locking mechanisms (screws or clasps) to prevent the plugs from coming apart. The 50-pin champ connector (a.k.a. the 50-pin amphenol connector, ANSI X3.131) has been in use for years as a standard wiring system interface. Perhaps the most common use was for the old key-systems telephones that relied upon the 50-pin connector for interface to telco Centrex systems. Today the 50-pin interface is commonly used by many communications products to connect multiple twisted-pairs to cross-connect equipment, patch panels, and communications gear (such as networking systems). One of the more common applications for the X3.131 connector is as the small computer serial interface (SCSI) I connector used by computers to attach peripheral devices. This connector is very popular in the Apple Macintosh environment. Some other common examples of pin connectors include the following:

- *DB-15:* Used for the Ethernet AUI transceiver cable
- *DB-25:* Used for the RS-232 interface commonly used for serial interface connections on computer equipment
- *SCSI II:* A high-density version of the SCSI I interface that provides 50 connectors in a unit the size of a DB-25 connector
- *M34:* Used to support the physical interface for the V.35 standard

In addition to the above pin configurations, there are many other pin-oriented interfaces available for connecting twisted-pair systems. Most of these product offerings are tailored to specific equipment product lines or applications and do not commonly appear as part of a building wiring system.

2.4.3 Genderless Connectors

Genderless connectors are predominantly used in the building wiring environment for one application only—the IEEE 802.5 Token-Ring network. This type of connector does not rely upon two different male and female components. Instead, a single connector that is capable of connecting into a similar connector is used. Often supporting only two shielded twisted-pairs, these connectors (referred to as data connectors in the IBM world) are generally not used for standard twisted-pair outlets except where IBM data systems are installed.

2.5 TWISTED-PAIR CROSS-CONNECTS

There are several major types of cross-connect blocks available in the market today. These different types of blocks have been developed to suit different purposes and as such have different strengths and weaknesses, as well as different configuration options.

The twisted-pair cross-connect block is the device or unit which connects multiple segments of twisted-pair together. Cross-connect blocks are designed to bridge multiple different distribution systems, such as the vertical and horizontal distribution systems, together into a single cohesive distribution plan.

Cross-connect systems (multiple blocks) typically consist of three main components. In generic terms, these are the inside cross-connect block, outside cross-connect block, and cross-connect (or jumper) cable. The terms *inside* and *outside* cross-connect refer to the cross-connect block's relative position in the total building wiring system (this is different from inside and outside wiring that refers to wires in or out of a building). As an example, let us suppose that a building wiring system consists of a single primary distribution frame (DF) feeding several lesser secondary DFs. Each secondary DF in turn feeds end-user station outlets.

Focusing on the secondary DF, the inside cross-connect blocks would attach to the cables originating from the primary DF. The outside cross-connect blocks would be attached to the distribution cables feeding the individual end-user station outlets.

The cross-connect cable, consisting of a short segment of twisted-pair that matches the electrical characteristics of the distribution twisted-pair, is used to link the inside and outside cross-connect blocks together. In this manner, a complete connection from an end-user station outlet can be established to other locations accessible through the entire building wiring system. In some cross-connect systems, a patch cable, using a modular plug and jack, is used.

2.5.1 Types of Cross-Connect Blocks

There are several popular types of cross-connect blocks in use today. These include the 66- and the 110-block, which are by far the most popular in use today and are installed in the majority of building wiring installations. In addition to these systems, there are also numerous cross-connect block types available from modular equipment vendors. While these cross-connect systems may vary considerably in design and configuration, they all employ a method of installation displacement to establish an electrical connection.

Each conductor of a twisted-pair consists of an inner conductor and an outer insulation material. To establish an electrical connection with a cross-connect block (and in turn another segment of cable), the insulation must be removed from the inner conductor. The process of connecting a cross-connect cable to a cross-connect block is known as *punching down* a cable, which refers to physical force that is used to force the twisted-pair into the cross-connect block. This force is required to displace the insulation surrounding the inner conductor. Insulation displacement is a process in which the insulation on a cable is displaced from a portion of the cable to allow an electrical connection. There are two primary methods of insulation displacement, pressure and cutting:

Pressure displacement: In pressure displacement systems, extreme force is applied to either side of a cable to force the outer insulation material away from the inner copper conductor. This force can, in some cases, stress the inner conductor to the point at which the inner conductor can break or not establish a solid contact with the cross-connect block. Additionally, since pressure displacement systems require a substantial piece of metal, a much larger segment of the cable is stressed than with other methods. Since stress on a cable can alter the electrical characteristics of the cable, this method can increase the line loss caused by a cross-connect punch-down.

Cutting displacement: Cutting displacement systems rely upon a sharp metallic material which cuts through the outer insulation, establishing a connection with the inner solid conductor. While the cutting displacement method does not apply extreme pressure which can damage the inner conductor, improper cross-connect implementation can lead to a cable being severed. The implications of this are quite apparent (to overcome this, some cutting displacement cross-connect blocks use modular adapters to ensure that undue force or improper connections are not made). Despite this shortcoming, the cutting displacement method does in most cases offer a better cross-connect connection and is the preferred method of the better cross-connect blocks.

2.5.2 66-Blocks

The 66-block is the oldest type of cross-connect block currently in use. Developed as the basic cross-connect system supporting traditional telephony applications (analog voice), the 66-block is by far the largest installed cross-connect block today. It is questionable, however, whether or not the 66-block is the most commonly installed block today. Many new building applications use newer cross-connect products that provide better electrical and/or physical characteristics.

The 66-block (and there are several minor variations of this product) comprises a series of dual pressure displacement units (made of metal) arranged in two vertical columns of 50 units each. Thus, the standard 66-block can terminate 50 twisted-pairs in a single unit. The dual pressure displacement unit is used to establish a connection between the twisted-pair conductor and the cross-connect cable conductor. See Fig. 2.7. The twisted-pair cable being connected to the cross-connect block is placed in one of the "Vs" of each unit. The cable is

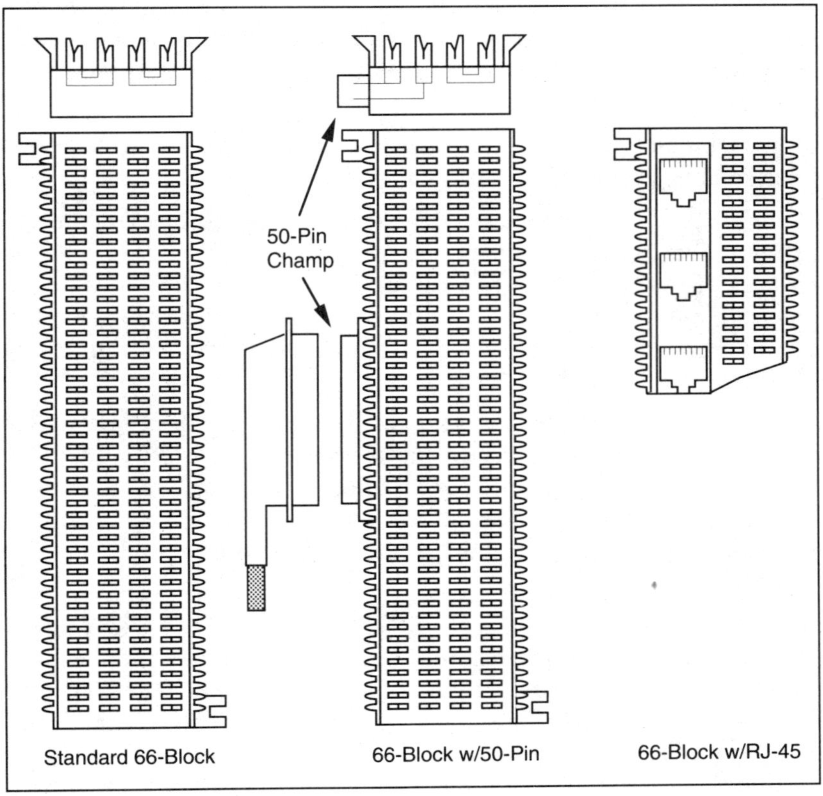

Figure 2.7 66-block.

then forced into the small of the V, which exerts the pressure on the outer insulation material. The cross-connect cable is then connected to the second "V" to provide connection to another cross-connect block and twisted-pair cable.

There are several advantages and disadvantages to the use of 66-blocks as compared to other cross-connect blocks available today.

2.5.2.1 Advantages of 66-blocks

Ease of installation: The 66-block is very simple to install and configure. It typically mounts to a wall with only two screws and provides a solid, if basic, unit. Also, since the 66-block was designed for and used in most basic telephony (voice) systems installed over the past half-century, there is quite a large number of qualified installation personnel who are familiar with the 66-block and its installation.

Ease of crossconnection: It is physically very simple to punch down or remove a cable with a 66-block. There are no clamps or physical components securing a cable onto the block (this becomes, as we will later see, a liability as well as an asset). This ease of crossconnection is very appealing to installers and maintainers who must frequently reconfigure cross-connect systems.

Large installed base: The 66-block has been installed in the vast majority of commercial buildings in use today.

Availability: Most wire and cable vendors offer 66-blocks, in some configuration, for sale in conjunction with their standard cabling products. The 66-block is not directly associated with any single vendor or manufacturer. This contrasts sharply with other vendor-specific cross-connect blocks that are only available through a single vendor or manufacturer.

Low cost: The 66-block has become a commodity item because of the vast number of vendors offering the product. It is difficult today to look through any networking or telecommunications equipment catalog without seeing some reference to the 66-block architecture. This has created a very low price not only for the product but also for installation.

Configuration options: As a result of the pervasiveness and age of the 66-block, many manufacturers have developed alternate configurations for the 66-block which enhance its product capabilities. These options, or enhancements, typically involve the integration of some type of twisted-pair connector with the standard 66-block insulation displacement units. For example, one configuration might have a 66-block teamed with a 50-pin amphenol connector. Another configuration might promote the use of patch cables in place of cross-connect cable through the integration of RJ-type jacks. In this type of configuration, the insulation displacement unit is hard-wired (directly connected) to a corresponding pin on a corresponding RJ-type jack. Such a configuration is ideal for data networking installations where the use of a patch-panel cable would allow for frequent system reconfiguration.

2.5.2.2 Disadvantages of 66-blocks

Poor electrical characteristics: While the pressure insulation displacement method is fairly simple to work with, it does require a significant amount of metal to ensure that the cable is not pinched and thus severed. Such an amount of metal, easily 2 or 3 times the amount used in new cross-connect systems, can disrupt a modern digital communications signal.

Poor size configuration: When the 66-block was originally designed, office space was not at a premium and data systems were not yet widely in use. As such, the 66-block was not designed to put a large number of cross-connects into limited area, as are many of today's newer cross-connects, and thus is lacking in the density of connects.

Lack of color coordination for twisted-pairs: The standard 66-block has no specific color coding to indicate what twisted-pairs in a large group are placed where. Special color-coded attachments are available from some vendors. These are, however, usually a temporary attachment that must be removed in order to work on the 66-block.

Limited space for distribution cables: The design of the 66-block does not adequately provide for the location of bulk cables underneath the system. This requires that the cables be placed to the side of the unit. Such a configuration can waste valuable space. To overcome this limitation, special 66-block frame holders are available that provide some additional cable space by raising the 66-block from the surface of the wall. However, this configuration does not provide the amount of extra cable space that is available with other types of cross-connect blocks.

2.5.3 110-Blocks

The 110-block is a newer type of cross-connect block that has gained wide acceptance within the building wiring market. The 110-block was designed by AT&T as the successor to the 66-block. Using a more modular approach to wiring connections, the 110-block is the cross-connect of choice in many new installations, especially where high-speed data applications are anticipated.

The 110-block configuration is substantially different from the 66-block. It consists of four rows of cutting displacement connectors. Mounted in a horizontal fashion, the 110-block can support 25 twisted-pairs per row for a total of 100 twisted-pairs per unit. Each 110-block includes built-in supports that place the unit approximately 1¾ inch (in) above the wall. See Fig. 2.8.

To establish a connection between an incoming feeder cable and a cross-connect cable, the feeder cable is first punched down into the color-coded slots on the main 110-block unit. A second unit, the insulation displacement clip, is then placed over the cable, providing the real punch down and the electrical connection. The insulation displacement clip is available in three-, four-, and five-pair configura-

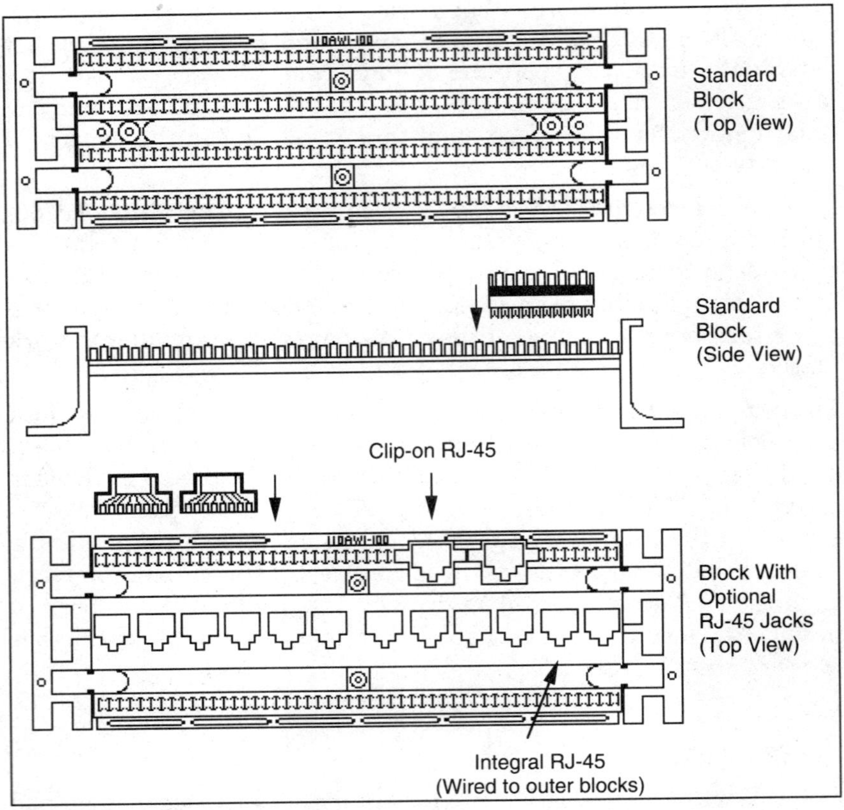

Figure 2.8 AT&T 110-block.

tions. A cross-connect cable, or a modular patch cable, is then punched down onto the top of the insulation displacement clip. The modular patch cables provide a multipair option for cross-connecting pairs. The patch cables are typically used in data applications where ease of reconfiguration and multipair support are desirable.

There are several different configurations of the 110-block. The more common options are as follows:

- *110A:* This is the standard block used for voice and/or data applications.
- *110P:* This version of the 110-block includes additional supports for multipair patch cables (used in place of individual twisted-pair cross-connect cable). This option is typically used for data applications only.
- *110(A/P)-100:* This is the standard 100-pair configuration of the 110-block.
- *110(A/P)-300:* This configuration of the 110-block comprises three 110 units (attached together) to provide support for up to 300 individual twisted-pairs in a single block unit. This configuration is very economical from a wall-space perspective and is used in many larger installations or where wall space in a telephone closet is at a premium.

There are several advantages and disadvantages to the use of 110-blocks as compared to other cross-connect blocks available today.

2.5.3.1 Advantages of 110-blocks

Improved electrical characteristics: The 110-block is based on the cutting displacement method. This allows a better electrical contact with less stress on the individual twisted-pair cable. Typical decibel loss levels associated with a single 110-block punch down are ≤ 75 percent of a standard 66-block cross-connect decibel loss. Additionally, the composition of the metal used for 110-blocks is very stringently controlled. For example, a minimum amount of gold (for its high conductive powers) must be present in each metal component used.

Pair density: The 110-block affords a higher pair-per-foot installation count than many other types of traditional cross-connect equipment.

Cable support: Given the raised architecture and horizontal alignment of the 110-block, greater pass-through of cables under the unit is supported. This allows for a cleaner and properly organized cross-connect configuration supporting a higher number of bulk cable pairs within a single cross-connect.

Color-coded units: The 110-block has been designed with built-in color-coded components. Each twisted-pair row of the base unit is marked in five-pair increments with the primary colors of the standard band-marking system for twisted-pair (i.e., white, red, blue, yellow, and purple).

Patch cable options: The 110-block provides a system-designed configuration option for supporting patch cables in place of cross-connect cable. The patch cables allow for multipair modularity and relatively easy cross-connect reconfiguration, which can simplify the maintenance of the cross-connect system.

2.5.3.2 Disadvantages of the 110-block

Fixed-punch architecture: The insulation displacement clips of the 110-block clip onto the base unit. Once the unit has been clipped into place, it is very difficult to remove (removal is possible, but reterminating the unit can be difficult) and requires a special insertion tool.

Limited source: The 110-block, unlike the 66-block, is typically not readily available from multiple sources. Since the supply for the 110-block, developed by AT&T, is limited (telco-related companies and distributors), prices tend to be higher and less flexible than for other commercially available products.

Limited installation support: Given that the 110-block is fairly new to the market (compared to the 66-block), there is less of an installed base of reputable installation personnel. This should change as the 110-block becomes more established in the new-building arena.

2.5.4 BIX-Blocks

The BIX-block is a cross-connect block designed by Northern Telecom. This product uses a cutting displacement method and provides many of the advantages typical of modular units. The BIX-block uses a removable two-sided bar that connects distribution cables on one side to cross-connect wires on the other side. The configuration is horizontal like the 110-block and there is space provided behind the unit to allow cables to pass through.

There are several advantages and disadvantages to the use of a BIX-block system as the main component of a building wiring system.

2.5.4.1 Advantages of the BIX-block

Small footprint: The BIX-block, like the 110-block, is a fairly compact unit providing a high-density twisted-pair to square inch ratio.

Color coding: Each cross-connect strip on the BIX-block can be color coded with the standard cable color markings.

Electrical characteristics: The BIX-block has been designed with high-speed data applications in mind. By using a cutting displacement system and carefully planning the metal content of the unit, the BIX-block offers a high-quality electrical characteristic that provides a minimal disruption to a cross-connected cable.

2.5.4.2 Disadvantages of the BIX-block

Limited availability: Being a proprietary product, there is limited availability in the commercial market. Many equipment vendors do not routinely stock the product.

Limited installation support: Third-party installation support for this product is limited. The insertion tool (punch-down tool) required for the BIX-block uses a blade (tool head) that is not easily interchangeable with other products. For example, although the 66-block and the 110-block require different blades, the blade units themselves can easily be inserted for use into the other product's handle (this makes good business sense since the blade is typically only $15 while a good handle unit can cost in excess of $50).

Poor design: The removable-bar architecture of the BIX-block can be difficult to install or remove. This is important since it may be necessary to remove the bar to test the connections on the inside strip. When removing or installing the bar, there can be a great deal of stress placed on the physical connections. This stress can loosen or break the connection between the cross-connect cutting displacement unit and the twisted-pair cable.

2.5.5 Modular Cross-Connect Products

In addition to the more established 66- and 110-blocks, there are numerous modular cross-connect products available today. These units are typically designed to provide highly modular support for high-speed data applications. A form of cutting displacement is usually used to provide physical and electrical connection with some components consisting of high-density jacks, such as RJ-11 and RJ-45 components. While they do offer many advantages, there are several things to consider and reasons to be wary of such products.

2.5.5.1 Advantages of modular cross-connect products

Superior electrical characteristics: Most modular cross-connect products are designed to satisfy the tough electrical characteristics of emerging high-speed networks that cannot tolerate significant cross-connect signal disruption. Some products go so far as to use combinations of modular plugs and jacks connected with printed circuit boards (PCBs). While this extreme is typically an issue of overkill, it does demonstrate the high-performance commitment of the designers.

High cross-connect density: Given the use of modular components, cross-connect products are often able to support a high number of connections within a limited space. Additionally, many modular cross-connect units can function as a combination cross-connect–patch-panel unit. This combination reduces the requirement for multiple separate components.

Product flexibility: The modularity inherent in modular cross-connect systems can allow a vast array of component combinations. Single units supporting RJ-jacks, 50-pin amphenol connectors, and standard wire termination components are not uncommon. This flexibility can provide an ideal solution for users who require a custom configuration to suit a specific application.

2.5.5.2 Disadvantages of modular cross-connect products

Untested performance: As with any product new to the market, there is a certain risk involving the integrity of the component. While products like the 110-block and 66-block have been field tested for years (even decades), many newer modular products have only been on the market for a few years at best. This is often not enough time to measure true product durability.

Single source: Most modular products are designed and offered by a single vendor. This puts some severe limitations on the number of system integrators that can support the product, which can result in high prices and/or limited supply.

Limited installation support: It is difficult for the typical cable and wire vendor to maintain a trained staff who can support all of the different modular products currently available.

Short life span: Many new communications products are offered by vendors each year. Given the competitive nature of this industry, it is not uncommon for products to be dropped or for vendors to disappear from the market completely. While this is not a major problem with most reputable vendors, it has been known to happen, often leaving an end user with an installed base of equipment that cannot be properly supported.

When selecting any cross-connect system, it is important to make the right choice for your specific requirements. Not all advantages or disadvantages may be significant to your application. For most installations, the appropriate choice will probably involve an established cross-connect with reliable electrical characteristics. However, for smaller application-specific installations which tend to be more limited in scope (i.e., a cross-connect system serving a specific type of hardware), a newer product offering may be the right choice.

2.6 COMMON TWISTED-PAIR APPLICATIONS

The twisted-pair transmission media has become a fundamental part of any building wiring system. As such, it has become the media of choice for many different applications, ranging from PBX systems to LANs. Some common applications include the following:

Analog PBX systems: Analog PBX systems, usually installed to support analog voice applications, use twisted-pair cabling to connect both handsets to the PBX switch and to connect the PBX switch to the local telco office. These systems typically use one or two twisted-pairs. Analog PBX systems are generally restricted to voice and low-speed data applications using modems because of the loose requirements for twisted-pairs supporting analog applications (mainly since voice applications are less error sensitive than data applications—your ear and brain can tolerate moderate signal disturbance on a communications line and still interpret most voice signals). Typical configuration: star topology.

Analog key systems: Key systems, functioning as a user interface between the end user and a telco Centrex system, provide a function very similar to the analog PBX. The major difference between PBX and key systems is the lack of services offered by key systems. These systems typically use one or two twisted-pairs. Typical configuration: star topology.

Digital PBX systems: Digital PBX systems have become increasingly popular as end users have increased the number of data applications in use today. Digital PBX systems are ideally suited for both error-insensitive (voice) and error-sensitive (data) applications. The advent of digital PBX systems allows enhanced services such as ISDN to be offered. Typical configuration: star topology.

Data switches: Data switch systems are small digital PBXs designed expressly for use by data systems. Voice applications, supported by traditional PBX systems, are not supported by data PBXs. Data switches range in size from small eight-port systems to large systems supporting thousands of users. Typical configuration: star topology.

Local area network systems: LANs have become the most commonly installed form of computer networking system today. They offer a wide range of configuration options for connecting a wide variety of computing devices, such as personal computers, printers, plotters, file servers, and modems. Some common LAN systems that can function over twisted-pair include the following:

LAN*	Transmission media
IEEE 802.3 Ethernet	Unshielded twisted-pair
IEEE 802.5 Token-Ring	Unshielded or shielded twisted-pair
ARCnet	Unshielded twisted-pair
Appletalk	Unshielded or shielded twisted-pair
StarLAN	Unshielded twisted-pair

*Typical configuration: star or bus topology.

Intercom systems: Most intercom systems in operation today use some form of twisted-pair between the microphone system and the speakers (via amplifiers) to provide basic analog voice transmission. The twisted-pair used for intercom systems, like that of analog PBX systems, does not require a great deal of electrical integrity and can vary greatly from vendor to vendor. Typical configuration: star or bus topology.

Security systems: Many security systems use twisted-pair to connect sensing and communications equipment to a central control location. Although not usually considered part of a telecommunications wiring system, many security systems can use twisted-pair to satisfy some portion of the security application. Typical configuration: star or bus topology.

2.7 COST OF TWISTED-PAIR APPLICATIONS

Twisted-pair installations may vary greatly in cost because of the wide range of configuration and cable options that can be implemented. However, in relation to other types of cable installation (i.e., coaxial or fiber), twisted-pair is relatively inexpensive, for several reasons:

Low cost of production: Twisted-pair cabling is installed in almost every form of commercial or residential building in use today. In addition, the vast majority of telephone systems in use today around the world employ twisted-pair cable. This large demand allows suppliers to enjoy the economies of scale, lowering the cost of production. Money is made in mass production, not high prices.

Large number of sources: Given the almost ubiquitous demand for twisted-pair cabling, a multitude of vendors have offered to provide these services in exchange for a little slice of the cabling market.

Ease of installation: Twisted-pair cable is by far the easiest form of communications cable to install. Not only is the cable fairly flexible, but there is a very large supply of cable installers who are fully trained and have years of experience installing this type of cable.

The cost of twisted-pair cable is extremely low. The following chart represents a sample of cable and connector prices. These prices reflect standard 24AWG cable and are intended for reference purposes only. Pricing may vary between different vendors (especially if the cable is high-quality data cable).

	Price ($)*	
Product	PVC	Plenum
Unshielded twisted-pair (2 pair, price per foot)	0.10	0.30
Unshielded twisted-pair (3 pair, price per foot)	0.15	0.35
Unshielded twisted-pair (4 pair, price per foot)	0.15	0.40
Unshielded twisted-pair (25 pair, price per foot)	0.35	1.10
Unshielded twisted-pair (50 pair, price per foot)	0.50	2.00
Unshielded twisted-pair (100 pair, price per foot)	0.80	3.75
RJ-11 jack (plug)	0.50	
RJ-45 jack (plug)	0.80	
RS-232 connector	3.00	
Genderless token-ring connector	8.00	

*Prices represent 1991 prices averaged from various sources. The addition of shielding can result in a two- or threefold increase in price.

When pricing a complete wiring system, such as that installed to support a PBX or LAN (star topology), a price range of $150 to $450 per cable run can be used as a good working estimate. Note that this price includes all physical components (plugs and receptacles) as well as all installation labor costs but not necessarily hardware costs. The cost per installation may vary greatly depending upon the type of twisted-pair selected, the quantity installed, the type of installation environment (i.e., new construction, renovation, age of building), and the location of the installation site.

2.8 ADVANTAGES OF TWISTED-PAIR

There are numerous advantages to the selection and installation of twisted-pair cabling in a building wiring system. These advantages range from pricing factors to availability of product and ease of installation. Together they have combined to make twisted-pair the number 1 installed transmission media in new and renovated buildings. The following is a review of the advantages of twisted-pair cabling:

Multiple sources of product: There is currently an extremely large demand for twisted-pair cabling. With very few exceptions, every building, residential or commercial, has some quantity of twisted-pair cabling installed. In response to this universal product demand, a multitude of vendors are currently available to provide twisted-pair products. These vendors range from the very small, employing only a few installation persons, to the very large, employing hundreds of trained installation personnel.

Low cost: Given the large number of vendors providing wire and cable services, this has become a very commodity-oriented market. Prices for cable and wire products are fairly universal and extremely competitive in most market areas. The biggest fluctuation in the pricing arena involves system installation pricing, which can increase markedly in most large city and urban areas because of varying labor costs.

Simplified configuration planning: Most twisted-pair networking systems, voice or data, are configured in a star topology with a central hub location. This topology is very easy to model and plan for. Unlike other systems, such as coaxial-based networks, which are typically serial bus (point-to-point) oriented, twisted-pair offers a consistent approach to network configuration, which can simplify planning and installation procedures.

Qualified installation personnel: Given the large number of wire and cable vendors that support twisted-pair, and the large demand for the product, there is a large pool of qualified installation personnel available for twisted-pair installations. Telco organizations, such as local phone companies, have developed detailed procedures and instruction sets for training their installation crews.

Ease of installation: Twisted-pair cable is relatively easy to install in most situations, including new construction and renovation. This is primarily because of the small cable diameter and extreme flexibility of twisted-pair cable.

Simplified problem resolution: The star and hub configuration of twisted-pair installations is very predictable and helps to simplify fault location and resolution. As an example, in a star-shaped twisted-pair LAN, if a single cable link were to fail, service to only one user would typically be lost. In the traditional coaxial bus-based network, a break in the network cable would disrupt service to all users on the entire network. Locating such a break can be very difficult if every user on the network loses service: Where would you begin to look? In a star and hub configuration, a cable break is very easy to isolate since only a single user has had service disrupted.

Equipment vendor support: As recently as the mid-1980s, most data applications were supported by coaxial cable systems. Only voice systems, Centrex, PBX, or key system, were universally supported by twisted-pair. The most popular data systems, such as the IBM 3270 terminal and the Ethernet networking system, all relied heavily on some form of thick coaxial cable. This trend of all-coaxial support for data systems has changed drastically in recent years. Predominantly as a result of consumer demand, many vendors have modified their networking systems to support either shielded or unshielded twisted-pair. Products such as media adapters (i.e., baluns) allow twisted-pair to replace segments of coaxial cable in IBM 3270 environments. New networking standards, such as the IEEE 10BaseT standard, allow Ethernet systems to support unshielded twisted-pair in environments that only a few short years ago would have required thick coaxial cable. Increasingly, vendors are recognizing the universal appeal of twisted-pair cabling and designing new systems that support this media. This vendor support has bolstered the number of cabling vendors that support data-grade twisted-pair, which in turn has helped to lower the price and increase the level of support in the market today.

Each of these factors alone might not be enough to warrant the popularity of twisted-pair. However, when combined, they form a broad spectrum of benefits that cannot be overlooked in today's competitive business world.

2.9 DISADVANTAGES OF TWISTED-PAIR

Now that we have discussed the wondrous benefits of twisted-pair cabling, it is time to mention those areas in which twisted-pair is slightly deficient. The deficiencies, in a certain respect, can be considered highly subjective. To paraphrase a popular historical quotation, "Not every user will need all the capabilities all the time. But some of the users will want these capabilities some of the time." Now that I have succeeded in destroying the original quotation (possibly beyond repair), I shall explain the point of this discussion.

In the world of telecommunications there is a distinct tendency for the mean, or average, to rule the implementation of telecommunications systems. This is very reasonable since the goal of most telecommunications directors is to satisfy as many people as possible as much of the time as possible. There are situations, however, when those left out of the standard organization will demand satisfaction from a telecommunications system. Perhaps a user will require 5 gigabits per second (Gbps) to pass data between two communications systems. Or perhaps you will add additional floorspace to your office, 3 miles away. Perhaps it can even be as simple as networking an IBM PC, an Apple Macintosh, and a printer on the same desk at the same time.

Any of the above-mentioned events can stress the limitations of a twisted-pair wiring system. Specific areas of deficiency (again, this is a subjective list) are as follows:

Data throughput limitations: Twisted-pair cabling, while offering many advantages, does have several disadvantages. One of the most important shortcomings involves the data throughput limitation of twisted-pair cabling relative to other transmission media. While systems such as coaxial and fiber-optic cables can offer transmission rates as high as 45 to 600 megabits per second (Mbps) (or higher), twisted-pair is limited to 16 to 20 Mbps peak.

Distance limitations: In the analog voice system environment, distance limitations of twisted-pair transmission media are typically not a problem. Most analog PBX systems can support hundreds, if not thousands, of feet between a telephone handset and the main PBX processor. However, most twisted-pair data systems are restricted to several hundred feet between the end-user workstation and the central processor hub. This restriction can often impose severe configuration limitations on a data system if the building wiring system was not originally designed with the cable limitations of twisted-pair data systems in mind.

Susceptibility to interference: Twisted-pair cable is very susceptible to EMI. EMI signals can be caused by a variety of sources including power lines, unshielded machinery, or other twisted-pairs. The result of this susceptibility can be increased errors or even disruption of a communications signal. To overcome this, shielded twisted-pair cables are available. However, these systems still do not provide the stability and immunity to EMI that coaxial and fiber-based systems offer today.

2.10 SUMMARY

The following are the top key points of this chapter:

- Twisted-pair is the most common telecommunications wiring in use today.
- Twisted-pair is ideally suited for a broad range of applications and equipment configurations.
- Relative to other media, twisted-pair is cost effective and easy to install and maintain.
- Twisted-pair has shortcomings in the areas of data throughput, distance, and EMI susceptibility.

Chapter

3

Coaxial Cable

Definition: *Coaxial cable* A cable consisting of a single copper conductor surrounded by an insulating material and a mesh ground. Commonly used for terminal/host connection, data networking, and security systems.

Coaxial cable has been, over the past 10 to 20 years, the basic workhorse of the data communications environment. Providing reliable transmission support for a multitude of products has made coaxial cable a favorite among many data networking managers.

Despite the current popularity of twisted-pair and fiber-optic systems, coaxial cable is still installed in the majority of existing local networking systems because for many years coaxial cable was the only viable option for local networking systems. As a result of this installed base, and occasional applications for which it is the best alternative, coaxial cable is still installed on a fairly regular basis. The current level of installation, however, will almost certainly decline through the early 1990s. This chapter takes a look at coaxial cable, defining what it is and what its future will be in the building wiring system.

3.1 WHAT IS COAXIAL CABLE?

Coaxial cable comprises two (or more, as we will shortly discuss) conductive cables contained within a single sheath. Unlike twisted-pair, where both conductors are solid and of equal electrical characteristics, only one of the conductors is solid. This solid (inner) conductor is encased in a nonconductive dielectric insulation material which is then encased in a braided mesh conductor. Being of different construction, the two conductors have different electrical characteristics. This dif-

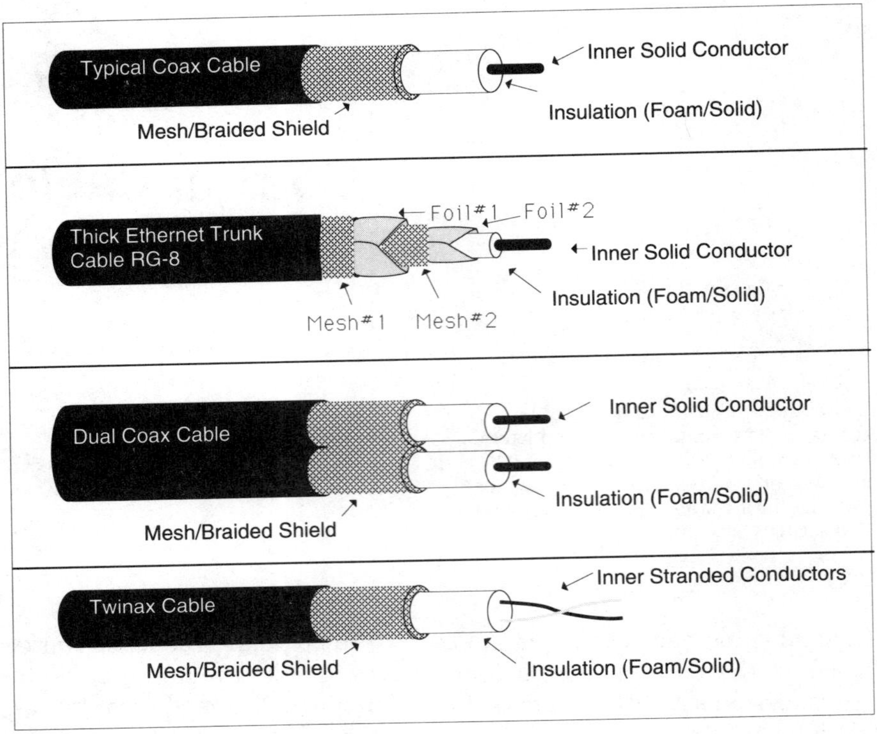

Figure 3.1 Coaxial cable.

ference between the two conductors results in an unbalanced circuit, which helps to prevent EMI or crosstalk interference. Both the inner and outer conductors of coaxial cable are usually copper, tinned copper, or copper coated steel. See Fig. 3.1.

The dielectric insulation that separates the inner and outer conductors can be made of either a hardened foam or a solid material such as polyethylene or fluorinated ethylene propylene. These differences can alter the velocity of propagation (Vp) slightly and as such are designed for specific cables only.

Common impedance levels for coaxial cable include 50 Ω, 75 Ω, and 93 Ω. These different levels are designed for specific applications only, which is generally reflected in the fact that most coaxial-based applications specify specific cable types for usage.

3.2 TYPES OF COAXIAL CABLE

There are many different types of coaxial cable since it is usually designed to support specific applications. Unfortunately, many applications developers decided to specify different types of coaxial cable for each application they developed. Unlike the twisted-pair market, there was not a great deal of standardization amongst coaxial developers. This is primarily because during the big coaxial development period (the 60s and 70s), there were not many standards organizations. Most networking systems were in their infancy and were developed as proprietary systems for sale by a single vendor only.

Looking at two of the biggest coaxial systems in the data networking world, the IBM 3270 terminal and the traditional Xerox Ethernet, the coaxial cable systems used by each system are completely incompatible. When these systems were developed, the concept of vendor interoperability (even to be able to reuse a common transmission media) did not exist. Only through user demand has this occurred within the twisted-pair and fiber-optic markets.

Looking at the various types of coaxial cable available today can result in a confusing and sometimes contradicting appraisal. Differences can be found in almost every category: impedance levels, jack and plug type, cable diameter (AWG), number of shields, plenum rating, and even number of inner conductors. A change in any of these parameters can drastically alter the applicability of a particular cable for your installation.

The most common differentiator between coaxial cable characteristics is the *RG* rating of a cable (RG: Radio Government—MIL-C-17, also RG/U: Universal). This rating indicates a specific set of physical characteristics including impedance, inner conductor AWG, dielectric insulation thickness, shield configuration and material, and outer sheath diameter. Many people mistakenly assume that the RG rating can generally be used as a rough equivalent to the AWG of twisted-pair when determining the size of a cable. For example, the RG-8 cable (used for some Ethernet networks) is approximately twice the diameter of the RG-58 cable (used for some other Ethernet networks). However, the only true cable diameter measurement is the AWG of the inner conductor.

Here is a look at some of the different types of coaxial cable available today. The listing is based not on specific technical characteristics but more on the common names associated with coaxial cable today.

Standard coaxial cable: In the data networking environment, this refers to a cable with a single inner conductor and one or two outer sheaths. An example of standard coaxial cable is the RG-62 cable used for IBM 3270 and ARCnet local area network (LAN) systems.

Thick coax: A term applied to the original coaxial cable specified for use by the Xerox Ethernet (IEEE 802.3 10Base5) networking system. Thick coax is RG-8 (nonplenum) or RG-9 (plenum) consisting of a single standard coaxial cable surrounded by a dual layer of aluminum and/or polyester tape and metal shielding.

Thin coax: This term applies to a version of coaxial cable specified for the Xerox Ethernet (IEEE 802.3 10Base2). Thin coax is RG-58 (nonplenum) or RG-59 (plenum). Thin coax is a standard form of coaxial cable consisting of a single inner conductor and a single outer mesh conductor. This type of cable is also known as *thin-net* cable since it was developed as a "thinner" cable option for Ethernet.

Dual coax: Dual coax is a cable consisting of two individual coaxial cables housed together in a common sheath (occasionally the two cables are two independent cables connected by a common seam linking the outer sheaths). There are no major changes caused to the electrical characteristics of such a cable as a result of this physical arrangement. The main reason for this cable arrangement is to accommodate networking systems that require two independent cable connections.

Twinax: Twinax cable is a major variant of coaxial cable in which there are two stranded inner conductors (twisted around each other) instead of the typical single solid conductor. Twinax cable is most commonly associated with IBM 5250 and Series/1 networking products and is often referred to as *data cable.*

Triaxial cable: Triaxial cable is a form of coaxial cable that uses two braided shields surrounding a single solid inner conductor. The two braided shields, unlike some forms of coaxial cable, are separated by a layer of polyethylene or similar dielectric material.

Trunk cable: Trunk cable is a general name associated with thick Ethernet (thick coax) or CATV cable.

CATV cable: CATV cable is the standard cable used for CATV networks. This 75-Ω cable (commonly RG-6) is also the same cable used in most broadband LANs. It is fairly inflexible and uses a single solid inner conductor.

In addition to the above cables, there are numerous other less common variants of coaxial cable available to support the multitude of coaxial-based networking systems produced today.

3.3 COAXIAL PLUGS AND JACKS

Just as there are numerous variants of coaxial cable, there are also numerous variants of coaxial cable connectors. Many different types of connectors are available with different locking mechanisms as well as physical configuration differences (such as right-angle plugs designed to connect to components as a 90° angle). Typical shield components are nickel-plated with inner conductor components typically gold-plated for enhanced electrical contact. The following is a list of some of the more common connectors and their commonly associated cables.

BNC connector: The BNC (bayonet) network connector, designed for standard coaxial cable, is commonly used for most networking applications today including Ethernet, ARCnet, and IBM 3270-terminal applications. The BNC connector has male and female components that attach through a spring-loaded twist-and-lock mechanism. See Fig. 3.2.

TNC connector: The TNC (threaded) network connector is designed for standard coaxial cable connection. The TNC connector uses a threaded mechanism for connecting the male and female components.

N-series connector: The N-series connector is commonly used as a connector for thick Ethernet cable (RG-8/9). The male component attaches to the female component with a threaded connector.

F-series connector: The F-series connector is the connector used for most CATV and broadband LAN systems. The F-series male component attaches to the female component with a threaded connector. The EIA 550 standard for F-series connectors (also known as the FD standard) is one of the more widely known standards for this type of connector (the F-series has traditionally been defined only by its general appearance and not by specific design and electrical parameters).

Twinax connector: As the name implies, the twinax connector is designed to terminate twinaxial cable. The two inner conductors are fitted to the stranded conductors of the cable and the outer shield is connected to the connector ground.

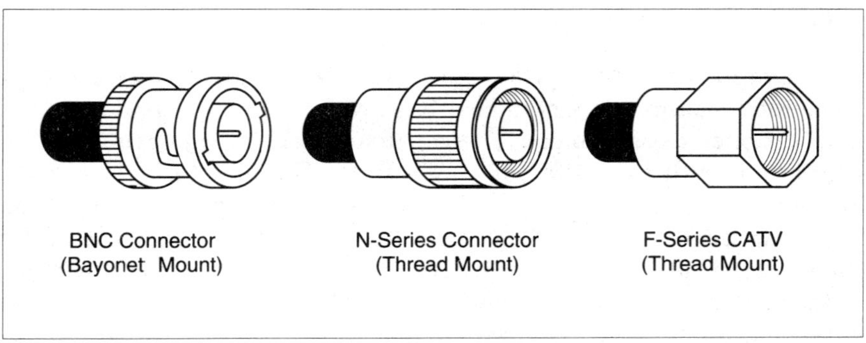

Figure 3.2 Coaxial connectors.

3.4 COAXIAL CROSS-CONNECT SYSTEMS

Coaxial systems are almost universally oriented toward the data application. Given that data applications development within the typical end-user organization was separate from the voice systems development, there are a few differences between coaxial (data) and twisted-pair (voice and data) cross-connect systems; there are no true coaxial cross-connect systems. This occurs for several reasons:

System size: Most coaxial installations were, and still are, minor in size compared to twisted-pair installations. Whereas every person in an office must have a telephone, only a small number must have data terminals (yes, this is changing rapidly, and in doing so, is pushing for conversion from coaxial cable to twisted-pair as the optimal data transmission media). Since relatively few users required coaxial connections, there was not much of a need for a well-organized, high-density approach to cross-connects.

Cable inflexibility: Coaxial cable is fairly rigid. Organizing such a cable into a tightly compacted area (as is done with twisted-pair) would be extremely difficult.

Ad hoc implementation: In almost every instance, coaxial cable was installed as an extra addition to an existing telecommunications system. Moreover, these ad hoc installations tended to be point to point in nature between a terminal and a server or controller. Since the data center was typically not in the same location as the phone system, it made little sense to operate in concert with the phone system wiring structure. With direct connections between a terminal and a server, there was little reason for a universal cross-connect system.

Diversity of media: The various types of coaxial cable can be extremely different. These differences, including number of inner conductors, diameter of inner conductors, and number of outer sheaths, have made it virtually impossible to develop a universal coaxial cable cross-connect system.

Many product vendors today offer patch panels specifically designed to be rack mounted in a standard 19-in rack and serve the function of a traditional cross-connect. These patch-panel systems are typically configured with two different panel units. The first unit terminates the incoming coaxial cable to the backside of the panel. The second unit also terminates the outgoing coaxial cable to the backside of the panel. Short intermediate coaxial patch cables are used to connect together the appropriate ports on each panel.

Since the interfaces required for each panel will differ according to the type of coaxial cable being used, there has not been a large move toward standardization as there has with twisted-pair. Standardization has been further hampered by the decreasing interest in coaxial solutions by end-user organizations.

3.5 COMMON COAXIAL APPLICATIONS

The use of coaxial cable for networking systems connection for the past few decades has allowed a multitude of manufacturers to develop numerous applications for coaxial cable. These applications, ranging from the low-speed (a few megabits per second) to the high-speed (45 Mbps), are almost exclusively concentrated in data applications. The only real system that supports voice, T-3, is based upon digital voice networking. The general classification of coaxial applications include LANs, terminal and server systems, CATV, security systems, and high-speed data (T-3) systems.

Baseband LANs: Many baseband LAN systems were originally designed to operate over coaxial cable. Examples of this include both the Ethernet and ARCnet networking systems. Baseband systems are available in both star and bus configurations. Distances supported in this configuration are typically in the few thousand feet range.

Broadband LANs: Broadband LAN systems, relying upon a distribution system similar to CATV systems, almost exclusively use coaxial cable for transmission media. The physical configuration for broadband networks is a star (referred in the broadband arena as a *tree* topology). Broadband LANs can support many thousands of feet between the top of the tree (called the *head-end*) and a remote drop through the use of signal amplification systems which are used to regenerate weakened signals.

Terminal and server systems: Many terminal and server systems rely upon coaxial cable point to point between the terminal and server (or cluster controller) unit. The IBM 3270-terminal and the Wang word processing system are the two most common examples of this application. Distances of 2000 ft between the terminal and server are not uncommon for this application.

CATV: Cable TV networks are configured much the same as broadband LANs (they use the same technology). There is a current trend toward the use of hybrid fiber and coaxial CATV networks where fiber provides the geographical servicing area distribution and coaxial is used to provide the tail circuit, or drop to each individual house.

Security systems: Many security systems, such as video surveillance equipment, rely upon coaxial cable to carry signals from remote cameras to a central monitoring station.

T-3 systems: The 45-Mbps T-3(DS3) system (28 T-1 signals at 1.544 Mbps) relies upon a coaxial cable with BNC connection to connect a T-3 unit to the telco carrier. While most telcos provide a fiber interface at the building demarcation point, coaxial cable is still the standard interface that must be supported in the building. This means running a coaxial cable between any customer T-3 equipment and the telco demarcation point.

Microwave systems: Most microwave systems in use today require a coaxial cable (typically 50 or 75 Ω) between the microwave radio unit and the microwave antenna. While these two components are usually in fairly close proximity, there are often situations where locating the radio equipment remotely from the antenna (for example, within a special communications room) is required.

3.6 COST OF COAXIAL INSTALLATIONS

Coaxial cable installations can be fairly expensive as compared to twisted-pair installations. In some cases, the cost can easily be twice as much as twisted-pair. There are two very straightforward reasons for this cost differential: increased materials and labor costs.

Materials costs: The product materials required for a coaxial installation can be fairly expensive. Cable costs, for instance, can run anywhere from $0.25/ft to well over $1.00/ft. Cable connectors for coaxial cable are typically manufactured from a combination of copper, brass, nickel, and gold, resulting in a product that usually starts in price at around $2.00 per connector. Sample pricing for coaxial cable, in price per foot, is as follows:

Product	PVC ($)	Plenum ($)
RG-58/59	0.25	0.75
RG-8/9	0.80	2.00
Twinaxial cable	0.30	1.00

Labor installation costs: Coaxial cable can be very rigid depending upon the diameter of the cable. Thin coax (used for 10Base2 Ethernet) is one of the more flexible coaxial cables with a bend radius of 2 to 3 in, while thick coax (used for 10Base5 Ethernet) has a bend radius closer to 1 ft. This lack of flexibility can make cable installation more difficult, thus requiring more time and increasing the labor costs associated with the project.

Coaxial connectors (i.e., BNC, TNC, N) range in price from $1.50 to $5.00 per unit depending upon the quality of the component. These prices reflect 1991 figures averaged from several sources and may vary depending upon source and quantity purchased (prices reflect minimum quantity).

3.7 ADVANTAGES OF COAXIAL CABLE

Although coaxial cable is no longer considered the transmission media of choice, there are some solid advantages of coaxial cable. This is especially true if your application either requires coaxial cable or cannot support twisted-pair because of environmental conditions. Some strong advantages of coaxial cable include immunity to EMI, support for a wide variety of networking systems, a large installed base, and ease of termination and splicing.

Immunity to EMI: The conductive configuration of coaxial cable, with the inner conductor completely surrounded by an outer mesh conductor, provides coaxial cable with a built-in resistance to EMI and other forms of noise interference.

Network support: Coaxial cable is currently the media of choice for broadband networks. Coaxial cable provides a high-bandwidth media capable of supporting the multiple signals required of broadband systems. Despite the popularity of twisted-pair, baseband network systems (i.e., Ethernet, ARCnet) are still supported on coaxial cable. Additionally, until fiber to the home becomes popular, CATV networks will be predominantly coaxial cable.

Installed base: The large installed base of coaxial cable has helped to maintain a continued support level for new coaxial installations. For example, many users routinely install coaxial cable for new portions of existing networks if the network is currently based on coaxial cable. Given that coaxial cable was the media of choice for Ethernet for many years, this installed base cannot be easily dismissed.

Simplified termination and splicing: Despite the bulkiness and lack of flexibility of coaxial cable, splicing and termination are fairly easy. For example, a twist-on BNC jack can be connected to a stripped coaxial cable in under 5 seconds (s). The size of coaxial cable is also an advantage when it comes to cable termination. The larger, and stiffer, leads can be easier to work with than thinner twisted-pair or stranded cables.

3.8 DISADVANTAGES OF COAXIAL CABLE

When local data communications systems were first being developed, coaxial cable was an ideal selection as the media of the masses. It offered reasonable distances, could support multiple types of networks, and provided a high level of noise (EMI) immunity. However, since the development of networking systems that support other types of media, the disadvantages of coaxial cable have become more apparent.

Lack of commonality: An important point to note concerning coaxial cable is that there are not a great deal of common media between different coaxial applications. For example, the cable used to support CATV cannot readily support Ethernet signals as specified in the original Ethernet specification. This means that previously installed cable cannot always be reused for new network systems.

Cable size: Compared to fiber-optic cable, and even twisted-pair, coaxial cables are typically much larger in diameter. This large size requires additional space in cable distribution systems, such as risers and conduit.

Installation costs: Coaxial cable installations can be fairly expensive because of the physical characteristics of the cable and the topology of coaxial LAN installations. The thickness of the coaxial cable decreases the flexibility and makes it difficult to install in conduit or raceways, especially if the distribution system is small or has a great deal of bends. The most common coaxial LAN installed today is Ethernet. The coaxial Ethernet network is configured in a bus topology. This means that installations tend to be point to point between different node sites (often located randomly around an office) instead of point to point between multiple node sites and a single centralized wiring closet. This random point-to-point topology can be difficult to implement since there are usually no established cable distribution systems between adjacent offices. Working with a hub-oriented network system allows you to install cable along established distribution paths between individual offices and a building wiring closet housing a wiring hub.

Product costs: Coaxial cable is composed of multiple layers of cable and insulation material. This layered design can cost more to manufacture than a simple twisted-pair made of a single inner conductor surrounded by a single insulating material. Additionally, the plugs and jacks used by coaxial cable are typically bulky and made of metal. This type of plug and jack system is much more expensive to produce than a small plastic RJ-jack such as is used with unshielded twisted-pair.

Limited configuration options: Given the size and inflexibility of coaxial cable, configuration options may be limited in some installations. For example, installations where space in conduit is limited may not allow sufficient room for new coaxial cable to be installed. This can be commonly seen in vertical riser systems where a single riser shaft may be choked with only a handful of coaxial cables. This contrasts to the high number of fiber and twisted-pair cables that could be placed in the same riser space.

Difficulty of maintenance: Coaxial systems, especially bus topology LANs, can be very difficult to maintain. For example, a cable break in a bus network could disable the entire network. Locating the break can be difficult because of the high number of cable taps present in the bus network (cable taps can alter or reflect a test signal). To realize the magnitude of this inability to locate cable faults, imagine that a single cable break could bring down your entire phone system. Certainly such a network system would not be tolerated very long. This is exactly what can happen in bus coaxial networks.

3.9 SUMMARY

The following are the key summary points of this chapter:

- Coaxial cable is predominantly associated with older data networking systems, such as terminal and mainframe connections and early LAN systems.
- Coaxial cable is more expensive than twisted-pair.
- Installation and maintenance of coaxial cable can be difficult in traditional bus network topologies.
- Coaxial cable does offer some advantages in certain environments, such as heavy EMI locations.

Chapter

4

Fiber-Optic Systems

Definition: *Fiber-optic cable* A type of cable that transmits light pulses, representing digital information, through an optical glass fiber. Usually associated with high-speed voice and data applications.

If there is one dominant trend in cabling that has, and will continue to, shape the future of networking in the building environment, it is the acceptance of fiber-optic cable as a media for local area and metropolitan area network (LAN and MAN) transport media. Fiber optics have enabled the development of extremely high-capacity networks, which has enabled modern computer applications (such as CAD/CAM) to evolve. The merits of fiber are many, while the drawbacks are diminishing rapidly.

4.1 WHAT IS FIBER-OPTIC CABLE?

To fully understand the applications and the pros and cons of fiber-optic cable, it is necessary to understand the fundamental differences between fiber optics and other forms of communications transmission media. The basic premise of fiber-based systems is the representation of information in light pulses. Digital 1s and 0s can be represented by sending pulses of light through the fiber cable at certain wavelengths, such as 1310 and 1550 nanometers (nm). The light energy transmitted down the fiber-optic cable contrasts greatly to the electron transmission methods of copper-based cables. Now that the basic fundamental explanation has been given (although it was an oversimplification), let's discuss a little of the actual configuration of a fiber-optic system.

A fiber-optic cable is made up of several layered components: an in-

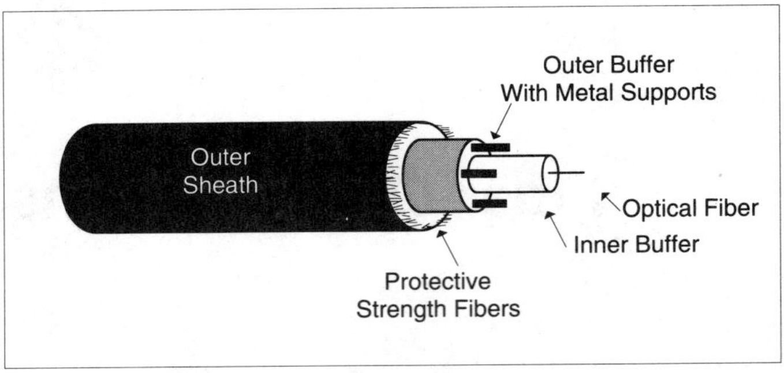

Figure 4.1 Fiber-optic cable.

ner core, a cladding, and an outer sheath. The inner core and cladding are made of optical glass or plastic. The outer sheath, which may consist of multiple layers, can be made of a combination of Kevlar, plastic, PVC, Teflon, and steel. See Fig. 4.1.

4.1.1 Fiber-Optic Cable Components

Each of the components in a fiber-optic cable must be made to exacting specifications to ensure the proper light-passing characteristics. The individual components, and their respective roles in fiber, are as follows:

Inner core: The inner core of a fiber-optic cable is the actual light path in which an optical signal is transmitted. This core must be incredibly pure, since any imperfection could distort the light signal, much the way the view through a drinking glass appears distorted. The diameter of the inner core may range from 5 to 100 microns (μ) for most glass-based fibers. Plastic fibers may go as high as 900 to 1000 μ.

Cladding: The cladding of a fiber-optic cable, like the inner core, is composed of glass or plastic. The main composition difference between the core and the cladding of a fiber-optic cable is the respective density. The higher density of the cladding serves as a boundary to keep light signals from escaping from the inner core (this will be explained in greater detail below). The diameter of the cladding (measured from the center of the inner core) can range from 100 μ to over 1000 μ depending upon whether the cable is glass- or plastic-based. See Fig. 4.2.

Outer sheath: The outer sheath of a fiber-optic cable may be composed of many different materials. These materials, or components, can include plastic sheaths, fibrous plastic strands, steel armor,

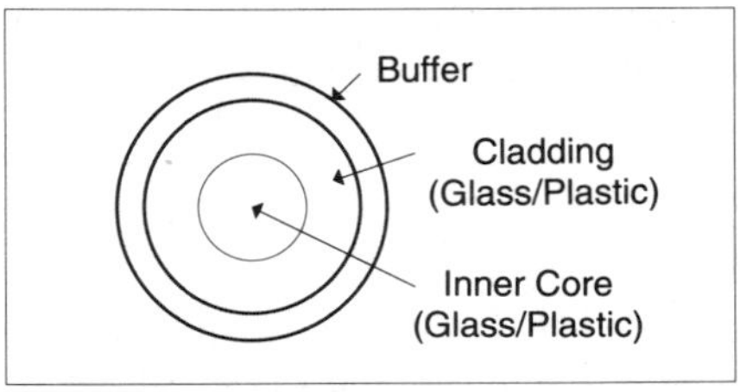

Figure 4.2 Fiber-optic core and cladding.

steel strands, and Teflon coating, all of which offer some value-added benefit to the cable, although they do not alter the inner core light-passing characteristics. For example, a steel strand wound around a fiber cable can provide extra support or strength to the flimsy fiber core. The outer casing of Teflon can provide plenum rating just as is provided for copper-based systems.

Connector: The fiber-optic connector (a plug and jack) is responsible for attaching a fiber-optic strand to a communications device, patch panel, cross-connect, or a new cable segment. The connector must be built to specifications that are even more stringent than coaxial or twisted-pair systems. In a fiber-optic system, two fibers must physically meet in perfect alignment in order to ensure a reliable connection without causing significant signal loss. Another example of the precision required for fiber-optic connectors involves the pressure placed on the ends of a fiber-optic cable. Too much pressure could cause the ends of the cable to bulge, changing the physical density and causing signal loss. Too little pressure and a gap between the two cables could occur, causing a complete loss of signal. For this reason, most fiber-optic connectors are spring-loaded to allow for a consistent fiber-to-fiber connection.

Light source: Every fiber-optic transmission system must have a light source connected to the cable in order to transmit information. Fiber-optic cables that have no light source attached are called *dark fiber* and are totally useless until a source is connected. The two types of signal generation systems commonly in use today are light-emitting diodes (LEDs) and light amplification by stimulated emission of radiation (lasers). An LED system provides a very broad light signal that consists of multiple light frequencies. Because of this broad spectrum of light that is generated by an LED, fiber-optic systems using LEDs cannot generally transmit a signal greater than 3000 meters (m). A laser, on the other hand, is capable of providing a very narrow frequency light wave. Laser-based fiber-optic systems can be capable of transmission up to 30,000 m. The costs of

LED systems are very low (a few hundred dollars per light source) as compared to laser systems, which can cost upward of $20,000 per light source.

4.1.2 Characteristics of Fiber-Optic Cable

Fiber-optic cable has several unique characteristics not available in twisted-pair and coaxial cable. These center around the physical characteristics (density, diameter, composition, manufacturing process) of fiber as well as the ability to pass large amounts of data.

The manufacturing process of fiber-optic glass involves heating and drawing refined glass bars, known as preform rods. Once a preform rod of purified glass has been heated, the glass is gently drawn into a thin flexible fiber. This process is very time-consuming and delicate since any imperfection in the diameter of the cable could result in a flawed product.

The majority of fiber-optic cables manufactured and installed today are made out of a single preform that is responsible for the formation of both the inner core and cladding. By manipulating the manufacturing process, changes to the diameter and density of the inner core and cladding can occur. These changes can have a profound impact on the light-wave transmission characteristics of a cable. As information signals are transmitted down an inner core, the light waves pass through the inner core and strike the inner wall of the cladding. See Fig. 4.3.

As light waves strike the cladding, the higher density of the cladding forces the light wave to reflect back in toward the center of the inner core. If the cladding were of the same or lesser density, the light wave would continue to pass through the cladding and leave the fiber. This would result in the overall signal degenerating because of high signal loss.

Some of the unique characteristics of fiber-optic cable include the diameter, flexibility, transmission speeds, and resistance to EMI.

Half-duplex transmission: Fiber-optic cables rely upon light-wave signals being sent down a single fiber-optic cable. If two signals

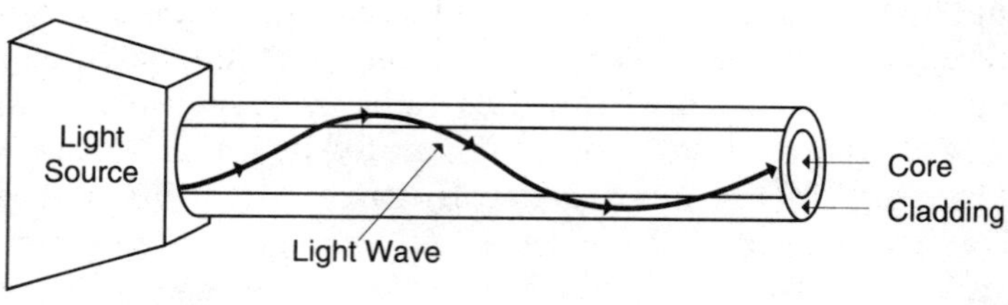

Figure 4.3 Fiber-optic refraction.

were sent from both ends of a fiber-optic cable, the two light waves would collide somewhere in the middle of the cable and thus cancel each other out. For this reason, fiber-optic transmission systems are usually (99 percent of the time) half-duplex; you can only transmit in one direction (unidirectional). This means that to communicate between two sites requires a minimum of two fiber-optic cables. There are some experimental fiber-based systems that are capable of simultaneously sending and receiving signals at each end. These systems generally use laser light sources, in combination with photoreceptors, to generate two distinct frequency signals, one for each direction. This *frequency-allocation* method is similar to the broadband LAN system in which individual frequencies (which do not collide) are assigned to multiple network systems. Unfortunately, these systems are very expensive when compared to most laser fiber-optic systems and as such are not used for communications systems.

Low cable diameter and high throughput: Compared to any other physical media, a fiber-optic cable (inner core and cladding) is one of the smallest systems available today. Even with an appropriate outer sheath, a typical pair of fiber-optic strands occupies about the same overall cable diameter as a four-pair twisted-pair cable or a two-pair shielded twisted-pair cable (≤0.25 in). This becomes very important when the data loading capabilities are presented. The twisted-pair approach (either shielded or unshielded) can reasonably offer up to 20-Mbps transmission speed. This is in great contrast to the throughput capacity of a typical fiber-optic cable which may be as high as 1 Gbps. This becomes mind-boggling when one considers that the actual throughput of a 1-Gbps fiber-optic network is roughly 62 times greater than a 16-Mbps twisted-pair network. This performance/size ratio is unique to the world of fiber-optic cable systems.

Bending radius and lack of flexibility: Compared to coaxial and some shielded twisted-pair systems, fiber-optic cables have far more flexibility in apparent physical configuration. Unfortunately, this is one of the least understood characteristics of fiber-optic cable. While it is true that a fiber can often be physically bent into a very small (few inches) diameter, this can severely damage the ability of the fiber to transmit a reliable communications signal. As a fiber is bent, the straight-line properties of a light wave make the wave attempt to go in a straight line. This can weaken the light-wave signal. Bends in a fiber-optic cable can also alter the density of a cable (something that the average copper-based system does not have to be concerned with). Additionally, since fiber-optic cables are made of "glass," they tend to be brittle and can break if stressed enough. As such, most fiber-optic cables are restricted to a 5- to 7-in-bend diameter. If a fiber-optic cable does develop stress fractures or cracks, the light wave can dissipate or reflect back into the cable in the original direction of the signal. Such a reflected signal could interfere with the original signal (remember that fiber is half-duplex).

Resistance to EMI: Fiber-optic cables, unlike copper-based systems, are not susceptible to EMI. This feature has made fiber optics a good choice for both commercial and military applications where EMI is a realistic impediment to standard copper-based cable systems. Examples of where fiber-optic cables might make EMI sense would include heavy machinery shops, large intrasite distribution systems (that may house thousands of power and telecommunications cables), and military applications in nonhardened facilities.

Security features: Fiber-optic cable provides some inherent security features not found in conventional copper systems. For example, fiber-optic cables are very difficult to tap into. Unlike copper-based communications systems where a communications signal can emanate from a cable (and thus be received by any listening entity), fiber-optic cable systems must be physically tapped into in order to extract data signals. There are two primary methods of accomplishing this. The first involves the insertion of a line-monitoring device directly into the path of a fiber-optic cable. Fortunately, this type of tap causes significant signal and line loss and can be detected with even basic fiber-optic test gear. The second method involves stripping away the outer sheath and part of the cladding to access a small portion of the cable. The cable is then bent with the hope that some of the signals passing in the relatively straight line of the inner core will continue off the fiber-optic cable. This type of tap is also fairly easily detected through line-monitoring for dB signal loss.

4.1.3 Types of Fiber-Optic Cable

As with most forms of telecommunications transmission media, there are numerous types and variations of fiber-optic cable available today. These different types, or options, involve all aspects of fiber-optic transmission, including inner core diameter, cladding diameter, material composition, and outer sheath design. These variations allow for a multitude of fiber-optic products tailored to meet specific applications. The following is a brief description of some of the different types of fiber-optic cables available today. Some of these characteristics are mutually exclusive (such as single and multimode) while others are not.

Multimode fiber: The two common classifications of optical fiber are single mode and multimode. These names refer to the type of light source used and the general diameter of the core and cladding. Fiber-optic cables that have a relatively large core diameter (i.e., ≥50 μ) are referred to as multimode fibers. Multimode fibers use an LED to generate a multiple-wave signal (hence the name). Since an LED generates a fairly wide-spectrum light signal, there is a tendency for the light wave to spread as the length of the fiber increases. As light waves are transmitted along an optical fiber, a certain amount of the light wave will strike the cladding material. Depending upon the angle at which the light wave strikes the cladding, the signal will be reflected back in toward the core at a related angle. Since the typical signal comprises many individual light waves, each striking the cladding at a potentially different point and angle, the light wave has a tendency to spread out over great distances (light waves that do not strike the cladding as often travel a shorter distance). See Fig. 4.4. This greatly limits the distance that a multimode fiber can support to under 3 kilometers (km). Despite this limitation, multimode fiber is very popular. This type of fiber is very well suited for building and campus installations and is not as expensive as single-mode fiber.

Single-mode fiber: Single-mode fiber is designed for long-haul transmission. It can easily support distances of 20 to 30 km. This type of fiber uses a small inner core diameter (~ 5 to 10 μ) and relies upon a high-powered laser for its light source. The laser provides a very pure single light wave (hence the name *single-mode*) that does not spread as rapidly as a multiple-mode light source. The process of manufacturing single-mode fibers is more refined than for multimode and as such is more costly. Additionally, the laser light

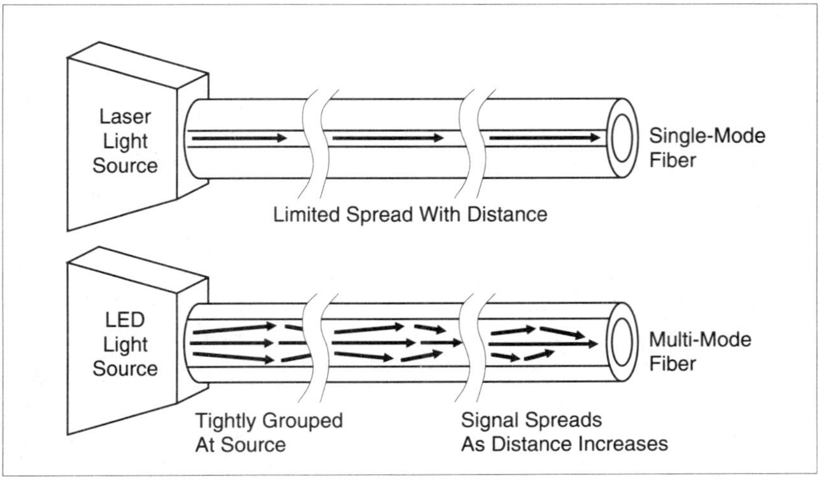

Figure 4.4 Fiber-optic signal spread.

source is much more expensive than the LED source used for multimode fiber. This type of fiber is used almost exclusively for long-haul transmission (i.e., MANs, multicity networks).

Step index fiber: A step index fiber is characterized by a large delta between the inner core and cladding densities. This drastic change in characteristic densities forces light waves to reflect from the cladding without passing into the cladding (known as a high refractive index). In such systems, the multiple paths of a light wave remain so configured. This can cause signal "spread" over great distances and effectively reduce the total bandwidth of the fiber. Step index fiber is used for both single-mode and multimode fiber systems.

Graded index fiber: In contrast to a step index fiber, a graded index fiber has a more gradual change in densities between the inner core and the cladding. Light waves that strike the cladding are gradually reflected back into the center core instead of being reflected sharply as in a step index fiber. This gradual change (in the refractive index) has the effect of evening out the light-wave signal and increasing the bandwidth and effective distance allowable by the cable.

In addition to these variations, which involve the inner core and cladding, there are also many configuration options that involve the outer sheath of the cable. For example, a fiber-optic cable might have an armored casing under the surface of the outer sheath. Such a casing is designed to protect the vulnerable inner core and cladding from outside damage. Another example might involve steel strands being placed within the cable to provide strength during cable installation. The following is a brief rundown of some of the more common fiber-optic cable installations.

Dual-strand fiber: Dual-strand fiber refers to a fiber consisting of two independent strands within the same outer sheath. This cable is typically used for horizontal fiber distribution and patch cables.

Multistrand fiber: This type of fiber consists of multiple fiber strands, each within an individual sheath, encased in a larger outer sheath. It is commonly used for riser (vertical) and campus distribution systems. The number of individual fiber strands supported in this type of cable usually range from four to several dozen.

Tight tube and loose tube cable: When a fiber-optic strand is placed within an outer sheath, it may either be tightly bound in place (tight tube) or loosely held (loose tube). While tight tube cables tend to hold the fiber in place better than loose tube cables, the loose tube cable allows for more flexibility during bends and installation. For example, as a multistrand fiber is bent, the fibers on the outside of the bend radius must travel farther than the fibers on the inside of the bend radius. The close-knit construction of a tight tube cable could cause stress on both inner and outer fibers. In comparison, a loose tube construction allows the individual fibers to flex or adjust to the difference in distance traveled between the inner and outer fibers.

Ribbon cable: Ribbon cables consist of a number of individual strands connected in a ribbon fashion. The ribbon cable is then encased in an outer sheath much the same as any other fiber-optic cable.

Slotted fiber: Slotted fiber is a multistrand cable that houses multiple strands in longitudinal slots, instead of individual fibers in a round group or cluster. The optical fibers in this configuration tend to have less individual sheath and outer casing than most other types of cable and can usually allow termination of individual fibers in a ribbonlike manner.

Multimedia cables: As a result of the increasing popularity in fiber installations to the desktop, many organizations are looking for ways to combine fiber and copper transmission media in a single cable. Although these types of cables are not very common today, they should become increasingly popular as more fiber is pulled to the desk in new installations.

Additional configuration options for fiber-optic cables may involve special weatherproofing outer sheaths (for outdoor installation) or special armoring for buried cable.

4.2 FIBER-OPTIC CONNECTORS

Fiber-optic cables rely upon advanced technology to provide ultrapure fibers and high-powered light sources. The tolerance of fiber-optic systems for variations in the performance metrics of any component is very small. This applies not only to the fiber and light source but also to the connectors used to terminate and splice the optical fibers.

Fiber-optic connectors (terminations) are designed to provide a fast, reliable mechanism that can allow a fiber to be connected or disconnected from another cable without additional work required. Splices, on the other hand, are designed to attach two fibers in a more permanent fashion. Where connectors are used in the connection of end-user equipment (LEDs, laser, and terminal), splices are used most often in the field to connect long segments of fiber (such as a break in a long-haul fiber) or at the building demarcation point.

4.2.1 Difficulties with Fiber-Optic Connections

Even slight deviations from the specifications of a connector or splice unit can cause a fiber to be improperly terminated, which could result in a poor and degraded signal. Some of the difficulties involved in terminating (or splicing) a fiber-optic cable are as follows:

Core size mismatch: In a conventional copper-based transmission media, all that is required for a connection between two cables is a good physical connection. The electrons can pass easily from one media to another as long as they have a physical match. Fiber optics, however, rely upon being able to continuously pass a light wave between the two cables. Any mismatch in the core diameters from one cable to the next would result in either a portion of the signal being reflected back into the originating cable or the signal spreading out as it encountered the larger-diameter core.

Improper core alignment: A perfect core alignment match between two connecting fibers is essential. If the cores do not perfectly match, there is the risk of light-wave reflection, which would disrupt the originating signal and weaken the remaining signal.

Ovality imperfections: Even with the best technology available, it can be very difficult to manufacture a fiber of glass that has the "perfect" characteristics as determined in the physical specifications; every fiber has some flaw or imperfection. Although these imperfections are usually very minor, remember that the inner core of a fiber can be as small as 5 μ. Minor imperfections, such as ovality (an oval shaped core or cladding) can result in a less than 100 percent connection between the fibers and can diminish the performance of the entire fiber-optic system. Although these imperfections are usually the result of the manufacturing process (and subsequently discovered during product quality testing), problems such as high ovality can occur after manufacturing if extreme pressure is placed on a cable.

Concentricity imperfections: Concentricity refers to an inner core that is not perfectly centered in the middle of the cladding and sheath. Connection of an overly concentric fiber to another imperfect fiber (or even a perfect one) can result in misaligned cores.

Lack of physical connection: Any gap between two fibers in a splice or termination connection would result in the light wave rapidly dissipating, resulting in a greatly weakened signal (if any) being passed from the originating to source fiber.

Excess pressure on connection: Much as a lack of physical connection between two fibers can disrupt a signal, too much physical connection can also have a negative impact on a light wave. Too much physical pressure on the ends of the fiber can cause minute chipping, fracturing, or bulging in the optical fiber. Any of these can disrupt a light wave.

Damaged optics: The ends of any two fibers being connected must be perfectly flat and polished before any connection can be made. Damage to the ends of a fiber can occur at several points during the connection. Improper cutting of a fiber can result in an uneven surface or chipped fiber. Improper polishing (sanding) of the end of the fiber can result in scratches to the surface. Any unintentional contact between the two fiber ends can cause scratching or chipping to occur. Minute particles of dust or dirt can also scratch the surface of a fiber.

Faulty connectors: Imperfections caused during the manufacture of a connector can result in many of the above problems. Connector materials such as cheap plastic and aluminum tend to have a higher occurrence of problems than materials such as ceramic or stainless steel. Additionally, many inexpensive connectors may not meet the original design specifications. The best approach is to select a connector from a reliable firm that is willing to guarantee the product for an extended period of time.

4.2.2 Types of Fiber-Optic Connectors

There are four types of popular fiber-optic connectors in use today. Each of these is presented below. To connect two connectors together requires the use of a special female-female connector.

Biconic connectors: Biconic connectors are tapered spring-loaded connecters that attach to a jack with a screw mechanism. The Biconic connector is very common in the IBM computer arena. See Fig. 4.5.

SMA connectors: The subminiature type A (SMA) connector uses an even or stepped ferrule with a screw mechanism for jack attachment. The two popular types of SMA connectors are the 905 and 906. The 905 uses an even ferrule connector typically made of steel while the 906 uses a stepped ferrule, also made of steel. The 906 SMA is the more popular of the two, although the SMA connector is not as popular as the other available connectors.

ST connectors: The straight tip (ST) connector is used predominately by AT&T in their PDS wiring system. The ST connector uses a spring-loaded bayonet outer shell similar to the BNC connector

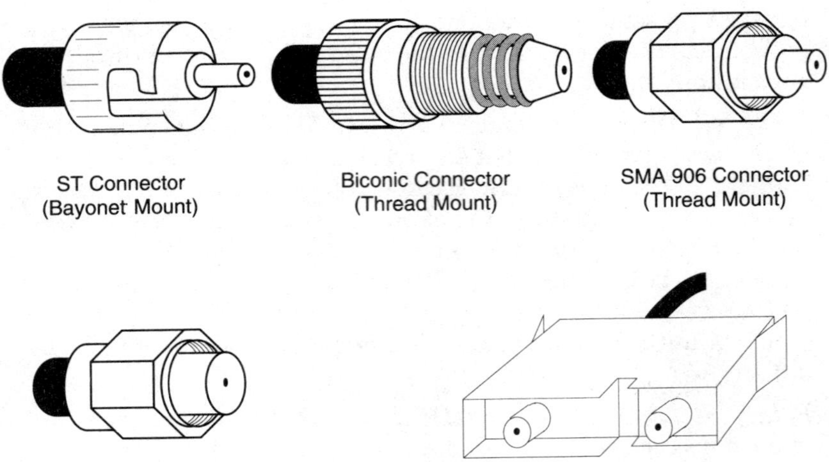

Figure 4.5 Fiber-optic connectors.

used for coaxial cable. The inner ferrule is even and may consist of a plastic compound, steel, or ceramic material. The ST connector is considered by many to be the optimal connector for use in building wiring systems. It is fairly small, offers high-performance characteristics, and is available from a large number of vendors.

FDDI connectors: The Fiber Distributed Data Interface (FDDI) defines a specific dual-fiber (duplex) connector for use only with FDDI systems (also known as a media interface connector, or MIC). The FDDI connector specification defines a rather large connector compared to the ST, SMA, and Biconic connectors. The connector is keyed on two sides to prevent the connector from being inserted improperly into a jack.

In addition to the above connectors, there are many other less common connectors available, many of which have been designed for specific applications and do not offer the universal appeal of the connectors listed above.

4.2.3 Fiber-Optic Splicing

There are many situations in which the use of established connectors to join two segments of fiber-optic cable may not be practical. The use of connectors may be too expensive, require too much room, or may result in an unacceptable loss of signal. Additionally, a permanent connection may be desired (the use of connectors establishes a temporary connection only). A fiber-optic splice is often used in these types of situations. A splice is the joining of two cable segments together in a permanent fashion. Splicing is most commonly used to repair cable that has been accidentally severed.

A splice typically involves a small physical component that is used to properly align and hold the two ends of the fiber-optic cable together. The actual method used to join the cables may involve pressure, chemical bonding (glue or epoxy), or physical bonding (laser or heating). All types of splices involve some level of insertion loss (degradation of the light signal).

Prior to splicing a cable, the ends of each segment are prepared in much the same fashion as for a connector termination (cutting, polishing, aligning, etc.). One of the strengths of a splice is that it usually provides a permanent means of adjoining two cable segments together.

For small splice installations there are self-enclosed splices available. The unit, perhaps as small as a cigarette and clear so that the splice can be inspected, completely encloses the actual splice providing physical protection after the splice has been completed. For larger splice installations, where dozens of fibers may be spliced together (i.e., the building network access point or distribution frame), special splice racks may be used. A splice rack contains multiple shelves which can swivel out of a large rack. See Fig. 4.6. Each shelf within

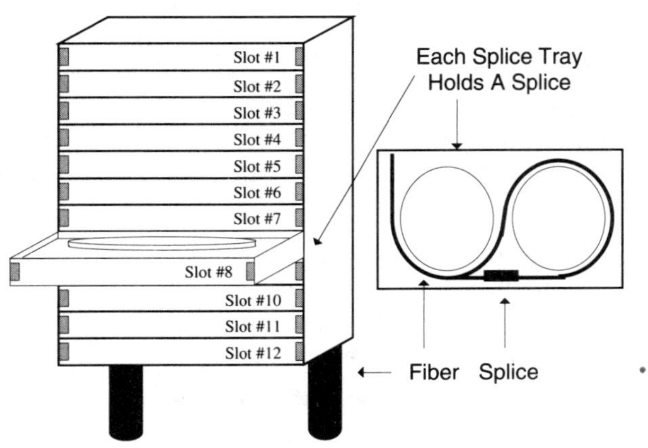

Figure 4.6 Fiber-optic splice rack.

the rack contains a looped path or groove in which the two ends of the fiber cables are placed. The splice itself is contained within a small splice block housed on the shelf. The rack provides physical protection for the splice.

4.3 COMMON FIBER-OPTIC APPLICATIONS

The development of fiber-optic cable has helped to improve the throughput, distance, and reliability lacking in many other telecommunications transmission systems. Many user organizations are planning extensive networking systems for the near future that will exceed the capabilities of most existing networking systems (i.e., twisted-pair and coaxial Ethernet/TRN). Although fiber is not yet ubiquitous in the inside building wiring systems, it most certainly will be within the next 5 to 10 years. This would not be possible without widespread support from a variety of networking vendors. This support comes from not only standards-based fiber applications (such as FDDI) but also from nonstandard systems such as Ethernet. Although fiber is not supported in the IEEE 802.3 specification, many vendors offer fiber transceivers and equipment to allow Ethernet to operate over fiber segments (such as in a backbone campus application). Some of the more common network applications follow:

Local area networks: Many popular LANs can be supported by fiber optics. For example, despite a fiber interface standard for Ethernet/IEEE 802.3, there are many vendors that offer fiber transceiver products. While the majority of local networks in use today rely upon copper transmission media, there is a growing number of installations that are relying upon fiber optics. These installations typically involve one or more of the following: extreme distances between nodes (in excess of 100 m), high EMI levels (such as found in factory environments), secure communications requirements (such as government installations), and/or near-term future requirements for high-bandwidth networking (such as a CAD requirement for FDDI within 2 to 3 years).

Campus networks: Campus networks typically require individual runs between buildings in excess of the "standard" copper cable lengths. Fiber-optic systems can usually increase the distance supported by an order of magnitude.

Metropolitan area networks: MANs must be able to serve a geographic region that is measured in miles and kilometers, not feet or meters. The distances required by MANs cannot be easily supported by twisted-pair or coaxial systems. Additionally, a single optical fiber can support a much higher traffic load than copper-based transmission systems. This allows for a network that can go farther on less material.

Wide area networks: Fiber-optic cable has become the media of choice for many long-haul vendors such as Sprint, MCI, and AT&T. In fact, most long distance telephone calls are transported over fiber either in the local loop (connecting local buildings to the main telephone company) or in the long-haul (intercity) system.

In addition to network applications, fiber optics is extremely well suited to supporting several types of end-user applications.

CAD/CAM: A common characteristic of many CAD/CAM systems is a large number of file transfers between individual computer systems for real-time use (file sharing) or for archival purposes. Such transfers require a high amount of bandwidth that cannot be supported by traditional networks (Ethernet, Token-Ring). Fiber networks, such as the 100-Mbps FDDI, can offer a legitimate solution to this problem.

T3 voice and data systems: The T-3 (DS3) standard specifies a physical coaxial interface using a BNC connector for all network connections. Coaxial cable was selected based upon its ability to support the 45-Mbps throughput of T3 services. However, most telcos are installing fiber to support T3 in place of coaxial cable. Fiber is smaller in size and can be later used to support higher-speed networks such as SONET and FDDI, provided the systems can use the same type of fiber (i.e., 5/100 μ fiber).

Secure government and military applications: Fiber optics is being installed in many military installations because of its resilience to EMI and its resistance to being illegally tapped.

Industrial systems: Fiber networks are ideally suited to industrial applications because of their high bandwidth (required for many robotics systems) and resilience to EMI caused by heavy machinery.

4.4 COST OF FIBER-OPTIC SYSTEMS

Fiber-optic systems are relatively expensive compared to other forms of telecommunications transmission media. There are three primary reasons for this: high manufacturing costs, high laser costs, and the immaturity of the market.

High manufacturing costs: The creation of fiber-optic cable, as described above, is a very delicate and precise process. The technology required to manufacture both the pure glass and the fiber strands themselves is very new and expensive. This cost should come down as the manufacturing process is refined and techniques improve (this is already happening).

High laser costs: Precision laser devices are required to generate a single-mode light wave. These devices are much more expensive than the electrical devices required to drive an electric-based signal (thousands versus hundreds of dollars). This only applies to single-mode systems; multimode systems use an inexpensive LED light-wave source.

The following table gives several examples of the cost of fiber-optic cable. When compared to the costs of twisted-pair and coaxial cable, they may seem very high. It is important to realize, however, that the cost of a building wiring system is only a very small component of the total building cost (including wiring support systems).

Product	PVC ($)	Plenum ($)
2-strand 62.5 μ/125 μ Indoor	0.70	0.90
4-strand 62.5 μ/125 μ Indoor	1.30	1.75
12-strand 62.5 μ/125 μ Indoor	3.50	5.00
12-strand 62.5 μ/125 μ Outdoor (armored)	7.00	

Fiber-optic connectors (i.e., SMA, ST, Biconic) range in price from $15 to $30 per unit depending upon the quality of the component. These prices reflect 1991 figures averaged from several sources and may vary depending upon source and quantity purchased (prices reflect minimum purchase quantity).

4.5 ADVANTAGES OF FIBER-OPTIC SYSTEMS

As described throughout this chapter, there are many advantages to the use of fiber-optic transmission systems over other available systems. These advantages range from the very important to the fairly insignificant. It is important to note, however, that every building environment (as well as every communications system) will not be the same as all others. What may seem to be a relatively minor advantage to one telecommunications manager may be a fairly significant advantage to another. It is important to keep this in mind when evaluating fiber and copper systems.

High bandwidth: There is no disagreement today that fiber optics can support superior bandwidth requirements over other transmission media. This factor alone has positioned fiber-optic systems as the media of choice for new and emerging networking systems.

Transmission distance: Fiber-optic cable is unparalleled in physical transmission distance without the use of repeaters. Individual fiber runs can be in excess of 30 km without the use of intervening signal regeneration devices. This is very important in the local networking arena where the ability to reach beyond a few thousand meters is key to installing campus and metropolitan networking systems.

Resilience to EMI: Since fiber-optic signals are based upon lightwave transmissions, and not electrical signals, they are extremely resilient to EMI sources. This includes not only electrical noise from manufacturing or power systems but also massive EMI sources such as nuclear explosions.

Security: The only practical way to tap into a fiber-optic system is to insert a device into the fiber or to bleed-off part of the signal (as described earlier in this section). Either of these two activities would result in either a complete or partial signal loss. Fiber-optic systems are based upon very fixed and limited light loss between source and destination devices. The signal loss between any two stations, once determined, can be used as a constant variable to determine if a tap has been placed into a fiber system.

4.6 DISADVANTAGES OF FIBER-OPTIC SYSTEMS

As with any telecommunications system, there are several disadvantages of fiber-optic networks that help to balance the positive points. It is important to note, however, that fiber-optic technology is still fairly new compared to copper-based technology. Most of the current disadvantages of fiber optics are expected to improve dramatically during the next 5 years as fiber technology and the marketplace mature considerably.

High cost: There is no question that fiber-optic systems are more expensive to both manufacture and install than traditional copper communications systems. The high manufacturing costs are caused by (relatively) low production of fiber and components and the sophistication required of the manufacturing process. The high installation costs are primarily caused by a shortage (albeit rapidly diminishing shortage) of qualified installation personnel. Installing fiber-optic cable (made of glass) can be a very delicate process because of the physical limitations as compared to copper (i.e., finite flexibility, limited tensile strength).

Splicing and termination problems: Although recent developments in this area have greatly reduced the amount of time and expertise required to splice or terminate a fiber-optic cable, the process is still much more delicate and time-consuming than terminating a twisted-pair or coaxial cable. While copper-based systems merely require a solid physical connection between the two components (cable-cable or cable-terminator), fiber optics requires either a fusion (heat) or epoxy (physical) connection. This increased difficulty in splicing fiber has also contributed to the high cost of fiber-optic installation.

Physical limitations: Fiber-optic cable (being made of either glass or plastic) is not as physically strong or durable as twisted-pair or coaxial cables. For example, twisted-pair cable can be easily flexed and bent without any significant change in the cable's ability to transmit an electrical signal. This differs considerably from fiber-optic cable, which can stress fracture or break if flexed (bent) too severely. Fiber-optic cable is also not as durable from a physical pressure perspective. While copper cables can withstand high levels of pressure on the outer sheath, fiber-optic cable can again be overstressed, leading to cable fracture or an alteration in the density or shape of the fiber (either of which would alter the light-wave transmission properties). This last disadvantage is important when fiber is being installed to individual desktop locations since many apparatus cables (connecting computers to wall jacks) are frequently stepped on, placed under furniture, or rolled across with chairs. Such activities could spell the end for many fiber-optic cables. Fortunately, there are many variations of fiber-optic cable that offer adequate outer-shell strength. However, these products can be expensive relative to other fiber, and copper, products.

Another physical limitation of fiber-optic cables is the tensile strength of the fiber itself. The tensile strength is the measure of the "pull" (pressure) that can be exerted on a cable (i.e., during installation) without damage to the fiber-optic material. Too much pressure can stretch or break the fiber. Stretching a fiber, even in minute levels, can significantly alter its transmission quality. This is important since many fiber-optic systems cannot tolerate even the smallest signal variance from the specified characteristics.

Lack of standardization: Although there are many "standardized" types of fiber and connectors, there is not yet a universal database of standards for the use of fiber optics in specific applications. For example, there are many different vendors who offer fiber-optic

Ethernet/802.3 networking systems. Many of these systems use different connectors (SMA, ST, Biconic) as well as different fiber, such as 100 μ/150 μ, 50 μ/100 μ, and 62.5 μ/125 μ (which is becoming a fast de facto standard for in-building wiring distribution). In addition, many of these product offerings use different light signaling systems and have different specifications for transmission distances. Building a universal fiber-optic Ethernet/802.3 network utilizing multiple vendors' equipment can be difficult, if not impossible. This lack of application-level standardization is an additional burden on a marketplace already trying to bring together different networking systems such as Ethernet, Token-Ring, Appletalk, and ARCnet.

4.7 SUMMARY

The following is a summary of the key points of this chapter:

- Fiber-optic cable is extremely well suited for high-speed voice, data, and video applications.
- Fiber-optic cable is not susceptible to EMI sources.
- Fiber-optic networking systems are very secure.
- Most data networking vendors are moving toward fiber support.
- Fiber-optic cable can be expensive and tricky to install and maintain.

Chapter

5

Secondary Transmission Systems

Definition: *Secondary transmission systems* The equipment and components used to augment and support the capabilities of telecommunications wiring.

In addition to the primary forms of telecommunications transmission systems, there are several other less common options available today. While these systems may differ considerably from traditional "wire-based" systems, they are viable options in many circumstances. This chapter presents a brief discussion of each of these "secondary" transmission systems, looking at what they are and how they can compliment twisted-pair, fiber, and coaxial wiring systems.

5.1 MICROWAVE/RADIO TRANSMISSION SYSTEMS

The concept of planning and installing a microwave, or wireless, transmission system as part of a building wiring system may appear a bit out of place. Is not microwave a nonterrestrial medium? Where are the wires? The cross-connects?

It is important to remember, however, that a building wiring system can often mean a campus wiring system. And in campus, or metropolitan, environments where cable facilities are lacking, a microwave/radio system can be an affordable alternative to installing new cable distribution facilities.

5.1.1 What Is Microwave?

A microwave/radio system has two basic parts: the processor unit and the transmission unit. The processor unit is a communications processor that is responsible for generating and interpreting communications signals. The transmission unit (such as a dish antenna ranging upward from 2 ft in diameter for microwave) is responsible for the actual transmission and receiving of a signal. While the processor unit is typically located indoors, the transmission unit is usually located on a tower. Microwave systems rely upon radio-frequency (RF) transmissions between two stationary endpoints (each endpoint consists of a processor unit and a transmission/receiving unit). See Fig. 5.1.

Microwave systems operate in much the same fashion as any other form of telecommunications media. The transmitting microwave system receives a data signal from a local communications device. The processor unit of the microwave system (typically located near the actual transmitter) then converts the data to an RF signal and transmits it to the receiver at the remote site. Each microwave system operates on a fixed frequency range (usually between 300 MHz and 100 GHz), which is controlled by the Federal Communications Commission (FCC). Just like a radio station, a microwave system must be licensed to operate at a specific frequency range. By licensing microwave systems, the FCC ensures that microwave transmission signals do not interfere with other free-space signals (such as aircraft communications) or other microwave systems.

The distances supported by a microwave system vary greatly with the transmission power and frequency of the signal. In addition, adverse weather conditions, such as heavy rain, snow, hail, or even fog, can sometimes reduce quality signal distance. The typical range of mi-

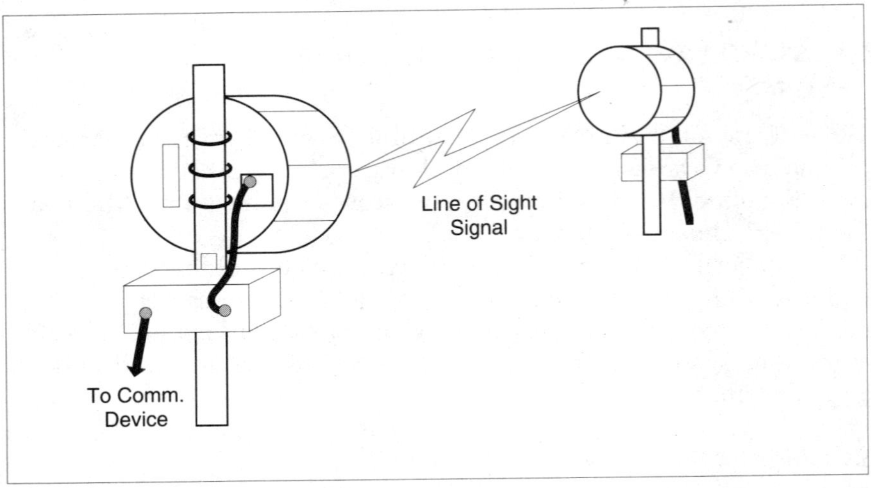

Figure 5.1 Microwave system.

crowave in normal applications peaks at approximately 15 to 20 miles (±30 km). Longer microwave transmissions require the use of intermediate microwave stations (commonly called *towers*). The endpoints of a microwave system must have a clear line of sight between them in order to properly communicate. There can be no interference, such as buildings, trees, or other microwave transmission systems.

Microwave systems are typically installed on the tops or sides of buildings. Such an installation usually places them over the top of most trees and smaller obstructions. While such a placement is beneficial for the operation of the microwave system, it does present some unique problems for the designer of a building or building wiring system. For example, adequate cable raceways must exist between the processor unit and the tower. This may not always be easy since the tower, located on top of a building, may not be near a standard cable distribution system. Additionally, microwave systems often require special support fixtures for stability in mounting (floor support as well as mounting hardware). Low-range systems, such as wireless LANs, usually do not support distances greater than a few hundred feet.

5.1.2 Microwave Connectors

While microwave systems rely upon RF signals between transmitting towers, a standard electrical signal is required between the processor and transmission units. This signal is usually carried over any number of types of coaxial cable. While the coaxial cable used for this function is often similar to the coaxial cable used for CATV and broadband networks, there is a very diverse range of cable offered by individual vendors. There is also an equally large number of connectors offered today. Unfortunately, there is very, very little standardization in the microwave cable and connector market today. Those standards that do exist, such as the United States Department of Defense MIL-specs, usually address cable and connectors that are application specific. Instead of specifying a single cable and connector, they specify a separate cable and connector for each type of application. Although standards bodies, such as the Electronic Industries Association (EIA), have defined physical and electrical standards for many different types of connectors, there is no real standardization as to which connectors a manufacturer can offer.

Some of the more common connectors in use today are variations on the N-series coaxial connector and the EIA 7/8-in connector [microwave system cables tend to be quite large and heavily shielded compared to most typical local area network (LAN) coaxial systems]. Fortunately, the development of universal standards is in process within the International Electrotechnical Commission (ANSI). This process is likely to produce some standardization within the industry by the mid-1990s.

As with any cable that is exposed to the outside elements, microwave cables connecting processor units to tower units are often armored and either gel filled or pressurized to prevent moisture from accessing the inner conductors.

5.1.3 Common Microwave Applications

Microwave is well suited to many different types of applications, ranging from military to commercial to private sector communications. The distances supported by microwave enable it to support both short- and long-haul networks. In the building wiring environment, microwave is typically used for one of several purposes, all involving campus or metropolitan area networks (MANs; microwave is never used within a building structure). These purposes include, but are not limited to, the following:

LAN interconnection: Microwave systems are often used to provide a point-to-point link between two LANs at different sites. The high bandwidth availability of microwave allows it to support the higher data rates required of LANs like Ethernet and Token-Ring. There is development work under way within the 800- to 1000-MHz range for wireless LANs. Wireless LANs generally use either point-to-point or point-to-multipoint systems to concentrate LAN stations into a single hub site. There are systems available today that use a range of frequencies, known as *spread spectrum,* in place of a single fixed frequency. Spread spectrum systems generally do not require FCC licensing.

Voice networking: Microwave systems are often called upon to link remote voice private branch exchange (PBX) systems together in metropolitan areas. Many metropolitan areas have a shortage of available cabling between buildings. Microwave systems can often provide a solution that does not require the acquisition of expensive cable right of ways.

Data networking: Microwave is very well suited to data networking and is often used to connect remote processing centers together.

Personal communications networks: Microwave systems, and variations on the cellular technology, are currently being investigated for use in providing personal communications capabilities (i.e., wireless voice/data devices that travel with a person, much like a cellular phone except they operate within a limited geographical area, such as a building or floor of a building). These systems, like cellular phones, are not restricted by line-of-sight requirements.

Backup communications: Many of today's large organizations place a very high value on the ability to transfer information rapidly between two points. This is especially critical of information technology companies, such as banks and financial institutions, that rely on the timely and accurate transfer of information for their daily revenues. With such importance being placed on the transfer of information, the importance of a backup communications system has risen greatly during the past few years. Microwave systems are often designed and installed to provide backup for more traditional cabling schemes (i.e., fiber or copper LANs and MANs).

5.1.4 Cost of Microwave Installations

The cost of a microwave system can vary greatly depending upon the size and function of the network. Typical market prices for a system (both ends of a point-to-point link) can range from $2000 to $30,000 for hardware alone. Installation, and of course maintenance (which is often not purchased for copper or fiber installations), can be significant recurring costs. Additionally, there is a one-time fee for the FCC microwave license. There are commercial consulting firms that can help a user organization with the FCC filing to obtain the license. This fee, including the firm's fee and license fee, can be as low as $500. (Systems that use spread spectrum generally do not require licensing since they spread the signal over a broad range of frequencies.)

There can be many hidden costs associated with a microwave system. In addition to installation and maintenance, the costs of training for personnel and the physical structure required to support a microwave system can often be prohibitive.

5.1.5 Advantages of Microwave

There are many situations for which microwave is an ideal transmission media for either voice or data applications. Factors to be considered when evaluating a microwave versus copper or fiber situation are the cost of the relative systems, the availability of cable and cable right of ways, and the suitability of microwave for the particular application.

Nonterrestrial design: Perhaps the most significant advantage that microwave can offer over a copper or fiber-optic transmission system is that it is not constrained by cable trays or right of ways. A microwave system can be used to connect any two points provided there is a direct line of sight between them (radio and wireless LANs do not require a line of sight). This makes microwave ideal for metropolitan applications (where cable right of ways may not exist) and campus applications (where cable distribution systems might not even exist).

Backup system suitability: When designing a diverse or redundant communications network, it is important to select alternate network paths that are both physically and electrically separate. Having a backup communications link that is located in the same cable or conduit does no good if the conduit is severed by a backhoe. The installation of a microwave system as a backup or redundant link can provide the necessary system separation required for a solid backup system. Such a system can also be used to provide temporary service for unplanned requirements.

System mobility: Requiring only a line of sight between transmitters, microwave systems can be installed very quickly (in a few hours) in mobile locations. This can allow emergency circuits to be created in the event of a primary system (copper or fiber) failure.

5.1.6 Disadvantages of Microwave

For every advantage that microwave has, there is an equal and opposite disadvantage. While the advantages of a microwave system can often make microwave ideally suited for a specific niche application, microwave is not the best alternative for most common networking applications. Some of the major disadvantages of microwave are as follows:

High cost: Although microwave can be cost effective in limited situations, it is relatively expensive to implement.

System size: The size of a microwave processor unit and transmitter is very large compared to the size of a similarly capable twisted-pair or fiber-optic transmission system. While this size difference may not be a problem for a single communications link, it can be a problem for larger installations. For example, given the point-to-point nature of microwave, a mesh network consisting of only five sites would be required between 10 and 20 transmit/receive units. Given that the smallest transmission units are approximately 2 ft in diameter, this could quickly become a very space-intensive operation.

Susceptibility to interference: Microwave systems are highly susceptible to external factors, such as other RF signals or extreme environmental conditions. This is not to say that microwave is unreliable but that it can be temporarily disrupted under many different circumstances.

Line-of-sight requirement: While the ability to operate under almost any condition provided line of sight is available can be considered an advantage, it can also be considered a strong disadvantage. If the two endpoints of a link are not visible (i.e., a building or natural obstruction is present), a microwave link is not possible.

When considering a microwave network, it is important to remember that in certain niches, microwave can be a viable alternative to physical media communications. However, for most day-to-day communications where physical media are readily available, microwave may not be the best alternative.

5.2 INFRARED TRANSMISSION SYSTEMS

Infrared communications systems have established a very small niche in the local networking arena. Several vendors now offer LAN support through infrared technology that only a few years ago was restricted solely to point-to-point applications. Although infrared systems are not considered to be part of a traditional building wiring system, they can be used as a wiring alternative and therefore can be considered part of a wiring system design.

5.2.1 What Is Infrared?

Infrared technology involves the encoding of information in pulses of infrared light that is transmitted point to point between two locations. This is very similar to microwave systems, with the exception that infrared uses an infrared frequency signal as opposed to a microwave RF signal. An infrared system is made up of a signal transmitter and a signal receiver, usually located in a single unit, and a processor unit. By directing the infrared light beam (representing digital data) at a remote receiver, information can be sent between communications devices. Infrared systems are often considered a distant cousin of fiber-optic transmission. The primary difference between the two systems is that infrared is not restricted to a fiber-optic light path. While this may seem to be an advantage, it does have its dark side (which is discussed further below). See Fig. 5.2.

5.2.2 Infrared Connectors

Infrared communications systems, like microwave systems, rely upon a nonterrestrial signaling scheme. Since signals are not transmitted

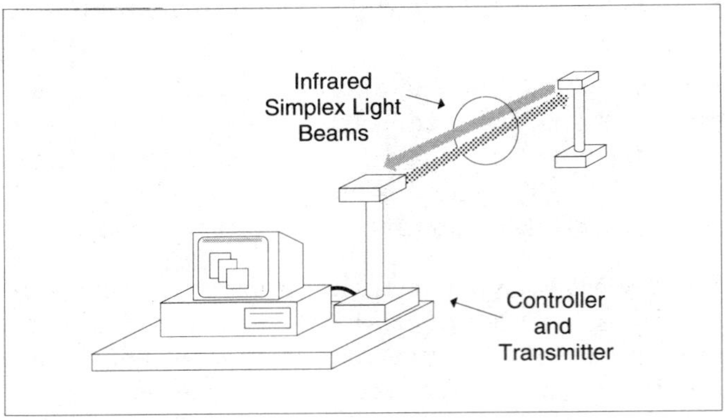

Figure 5.2 Infrared system.

via any form of cable or wire, there are no connectors. Connections between the processor unit and the transmit/receive unit are usually accomplished by coaxial cable. While the vast majority of infrared system vendors use standardized connectors and cables (such as BNC, TNC, RG-cables, etc.) for processor-to-transmission unit connection, there is no standardization regarding what type of cable or connector should be used between vendor organizations. Most cable and connector selection has been made independently by vendors to suit their particular application. Fortunately, most infrared processor and transmission systems are located within a few feet of each other, negating the requirement for cable distribution systems.

5.2.3 Common Infrared Applications

A very important characteristic of infrared technology is its susceptibility to signal interference. Infrared signals can be degraded by many sources, such as rain, fog, and snow. The sensitivity of infrared to such interference is markedly more extreme than that of microwave systems. Additionally, the distances supported by most vendor hardware today peak at around 300 m, far short of most outdoor microwave, copper, or fiber-optic transmissions. As such, infrared has been limited (at least in commercial applications) to indoor configurations.

The most common communications application for infrared technology is for connecting terminals or PCs in a LAN or multiuser computer configuration. These systems usually implement a centralized hub which contains multiple transmit/receive units (some more advanced units have a single receive unit which can differentiate multiple remote signals). Individual transmit/receive units, located alongside a PC or terminal, are directed toward the central hub, providing remote connectivity.

Since infrared technology requires a clear line of sight, this technology is well suited to large office rooms, warehouses, or industrial complexes that have a great deal of open space between communications systems. Individual transmit/receive units are typically affixed to the ceiling or to a power pole to provide a communications path unimpeded by office workers or factory equipment.

Examples of potential applications include factories (connecting machinery computers), department stores (connecting point-of-sale terminals), large offices (many terminals or PCs), and warehouses (connecting inventory systems).

5.2.4 Cost of Infrared Installations

Infrared costs can vary depending upon the size of the installation and the type of network being supported. Prices within the infrared industry for LAN-type applications range from a few hundred to a few thousand dollars per workstation. While these costs may seem high compared to a conventional transmission system (copper or fiber), such a system can be cost effective in environments where cable-based systems do not exist or where physical constraints prohibit the use of copper or fiber-optic systems.

5.2.5 Advantages of Infrared

There are several characteristics of infrared systems that can be considered strong advantages in many communications system installations. While some of these advantages may make the use of infrared appear extremely favorable, they must be tempered against the strong disadvantages discussed in the next section.

Mobile architecture: Infrared transmission units are fairly small (under a cubic foot). Additionally, infrared systems only require a direct line of sight for communications to take place. These characteristics allow infrared systems to be placed in almost any location within a large open facility. The small size, and the lack of any constraint typically associated with a fixed-cable system, make infrared a highly mobile technology. As communications equipment (i.e., computers) is moved around a floor, the infrared equipment can be easily repositioned to resume communications.

Nonterrestrial design: Infrared systems do not require any form of cable distribution system between the two communications points.

This allows infrared to be used in many situations where traditional cable facilities do not exist.

Immunity to EMI: Sources of EMI noise can seriously impair the ability of copper-based transmission systems to perform properly. Since infrared systems rely upon light transmission, and not electrical signals, they are not affected by EMI sources (radio towers, machinery, etc.). This ability, like that of fiber-optic systems, makes infrared transmission well suited to factory and other high-EMI environments.

5.2.6 Disadvantages of Infrared

There are several disadvantages of infrared transmission systems that seriously limit the applicability of infrared for most common networking systems. These disadvantages, including limited transmission distance, high cost, and lack of standardized support, are detailed below:

Limited transmission distance: The limited distance support of infrared systems does not allow infrared to be used for many campus transmission distribution systems.

Susceptibility to signal interference: The light-wave signal of infrared transmission systems can be easily disrupted by heavy rain, snow, fog, or other natural disturbances. For this reason, infrared is mostly limited to indoor applications or extremely short outdoor installations.

High cost: In situations where existing copper or fiber-optic transmission systems are available (as in most office environments), the relative cost of installing a new infrared transmission system can be prohibitive. Infrared is best suited to installations where cable transmission systems are either not currently installed or are operationally infeasible.

Lack of standardized support: There are currently no telecommunications industry standards to regulate or control the infrared market. This has resulted in the construction of proprietary systems that have little interoperability between different vendors. Additionally, infrared is not officially supported by most common networking systems (Ethernet, Token-Ring, ARCnet, etc.). This means that any vendor implementation is proprietary in nature and may not fully support the upper-layer network applications.

Line-of-sight requirements: Infrared systems require a direct line of sight between the two endpoints of the communications link. While this can be an advantage (infrared can operate in almost *any* environment that has line of sight), this can be a distance disadvantage in crowded buildings or campus facilities that have natural or artificial barriers. To get around this limitation, some vendors have been experimenting with passive reflective surfaces that allow signals to be bent (hopped) around minor obstacles. This technique has a lot of potential in the office (where the passive reflector is on the center of the ceiling of a large room) and in the campus (where the reflector allows signals to go around small buildings).

Equipment intrusion: A disadvantage of infrared systems is their physical presence. Each endpoint must have a processor and transmit/receive unit. This characteristic can be of little concern in open warehouses, but in a crowded office it can be too space consumptive.

Lack of vendor support: Infrared transmission has not established a major position in the networking distribution market. As such, sources of infrared products are somewhat limited. Some (certainly not all) vendors are startup organizations and may offer limited support for the long duration.

5.3 TRANSMISSION SUPPORT SYSTEMS

In addition to the basic building wiring components, such as cable and cross-connect systems, there are many other products that can be used to enhance the functionality of a wiring system. These products, known as support systems, include a wide variety of products and are available from a multitude of vendors. Some of the benefits that these products provide are the increased modularity of wiring systems, the reuse of older cabling technology, and the ability to support additional users with a system that has been used to capacity. Three of these products, patch panels, cross-connect adapters, and baluns, are discussed below. Each of these products is widely available and commonly used to enhance telecommunications wiring systems. As such, they should be considered part of a building wiring system during the initial design and installation process.

5.3.1 Patch Panels

A patch panel is a cable interconnection device that allows for two cables to be connected via a modular patch cable. Patch panels are very similar to cross-connect blocks in function, with the modular patch cable taking the place of the cross-connect cable. Any type of cable, including fiber-optic and coaxial, can be supported by a patch panel. For example, patch panels can be designed to support a wide variety of connectors. Fiber-optic connectors supported include Biconic, ST,

SMA, and FDDI. Coaxial connectors supported include BNC, TNC, N-series, and F-series. Twisted-pair connectors supported include RJ-45, RJ-11, and the IBM data connector. Some of the advantages of a patch panel include the following:

Increased system modularity: For example, if cable A is connected to patch panel A1 and cable B is connected to patch panel B1, a patch cord is used to connect the two cables together. In situations where many cables are connected to panels A1 and B1, a patch cable can easily be positioned to connect A(n) to B(n). In this manner, any cable on A1 can be connected to any cable on B1.

Interface to system equipment: Many computer systems use modular jacks (such as RJ-45) to provide connection to the wiring system. Patch panels can provide an interface to such equipment and allow for increased flexibility through system modularity.

Simplified network management: Proper cable management requires adequate labeling and ease of reconfiguration to accommodate moves and changes to the system. Traditional cross-connect blocks can often be difficult to both label and reconfigure. Patch panels can offer built-in labeling space as well as simplify the process of system reconfiguration.

One disadvantage to patch panels is the commitment required to install a patch-panel system. Most patch panels support only one type of connector (i.e., ST fiber-optic only). Once a specific connector has been installed in a patch-panel configuration, supporting a new connector can require the replacement of the entire panel or the addition of a new panel. Before installing a patch panel, the user must make sure that the connector to be used in the panel is the connector of choice for the wiring system. Maintaining continuity throughout the entire system is a requirement for proper patch-panel use. Installing patch panels with different connectors (such as RJ-11 and RJ-45) could lead to cable and connector mismatches and inhibit the flexibility offered by a patch-panel installation.

Costs for patch panels are usually fairly low, averaging about $3 to $5 per connector. Most patch panels can be either mounted on a wall or in a standard 19-in communications rack.

5.3.2 Cross-Connect Adapters

A cross-connect adapter is a modular device that attaches to an existing cross-connect block to provide increased modularity in either configuration or interface. Cross-connect adapters are usually only associated with twisted-pair installations. Adapters are readily available for 110-, 66-, and BIX-blocks, as well as some less common cross-connect block architectures. This type of device can often be used to upgrade older nonmodular cross-connect blocks to a newer modular plug architecture or to emulate a patch-panel configuration. For example, a cross-connect adapter may connect to a 66-block to provide an RJ-45 interface. Some of the advantages of cross-connect adapters include the following:

Simplified cable testing: Through a modular interface (i.e., RJ-11/45), cable pairs can be connected to test devices in a fast, efficient manner that is typically not associated with older cross-connect hardware.

Improved modularity: System reconfiguration with a cross-connect adapter can be as simple as changing the location of a patch cable. Punch-down cross-connect systems require individual cables to be repunched in order to reconfigure the system.

Simplified management: The use of modular adapters (with appropriate labeling) can make a wiring system simpler to reconfigure and troubleshoot. This simplification can reduce the chance of incidental errors resulting in system misconfiguration.

While modular cross-connect adapters can provide many benefits associated with patch panels, there are some disadvantages to such systems. They are:

Additional insertion points: The addition of a cross-connect adapter, or any type of device, into the path of a cable increases the risk of failure. The use of a cross-connect adapter adds at least one additional connection that would not exist in a traditional cross-connect configuration. This point can be an added potential failure point or can add to the overall dB signal loss of the system if the metallic material used in the system has a different electrical characteristic than the cable to which it has been connected.

Pair restrictions: In a standard cross-connect block, any twisted-pair can be cross-connected to any other twisted-pair. Since modular cross-connect adapters typically take two to four individual twisted-pairs and concentrate them into a single RJ plug, cross-connects can only be made in groups. This may be fine for systems that consistently require a fixed number of pairs. However, adding a connection for just a single pair can often be difficult.

5.3.3 Baluns

One of the most common additions to any established building wiring system is the balun, which is a device that converts a *bal*anced electrical signal to an *un*balanced signal and vice versa. In practical terms, this allows an unbalanced communications signal, such as that found in coaxial cable applications, to be transmitted over a balanced transmission media, such as twisted-pair. For this reason, baluns are often referred to as *coax-replacers*.

Baluns range in price from about $15 per pair to about $75 per pair depending upon the quality and the application for which they were designed. It is important to note that many baluns are specifically designed to support a single coaxial application, such as IBM 3270 terminals or Wang local networks. Baluns designed for one type of application are generally not able to perform properly with other systems.

Some of the advantages of baluns include the following:

Reduced cable size: The use of baluns to replace coaxial cable with twisted-pair cable can greatly reduce the amount of space required in conduit or ductwork distribution systems. This can be critical in many older buildings where bulky coaxial cable is already choking the cable distribution systems.

Media flexibility: Twisted-pair is a fairly easy transmission media to install. There are usually spare pairs available to most parts of a building. Coaxial cable, on the other hand, is fairly bulky and difficult to install. Since most coaxial cable systems can only support a single connection per cable, the addition of a second terminal to a given location often requires the installation of a new coaxial cable. With spare pairs typically available, a second terminal can usually be accommodated without the requirement for a new cable installation effort.

Network management: Most established coaxial cabling systems are not managed to the same degree as twisted-pair installations. This is primarily because coaxial installations were commonly ad hoc and installed on an as-needed basis in support of data applications. Twisted-pair, however, is usually installed as a complete system with adequate labeling and well-defined cable distribution paths (unlike the point-to-point nature of coaxial installations). Using baluns to replace segments of coaxial cable can allow coax-based applications to benefit from the network management systems currently in place to support twisted-pair.

In addition to the advantages of baluns, there are three major disadvantages that must always be considered:

Limited distance: Baluns usually cannot support twisted-pair segments in excess of 1300 ft (400 m). Additionally, this distance may be significantly lower (200 to 300 ft) depending upon the quality of the balun and whether or not it is being used with the type of twisted-pair it was designed for. This contrasts with the ability of coaxial cable with runs that often exceed 2000 ft (615 m).

Vendor quality: A balun is a fairly inexpensive and easy product to produce. Unfortunately, many baluns are of shoddy or inconsistent construction. Substandard baluns may not work properly, resulting in periodic failure or reduced distance support. For example, poorly designed baluns can significantly decrease the signal strength (dB) through impedance mismatches in hardware components.

It should be noted that not all applications can function with baluns. For example, baluns cannot be used with many LAN applications because of the specific electrical signaling requirements, such as propagation delay, that would be altered by the use of twisted-pair media. For this reason, baluns tend to be terminal and controller oriented where a single source and destination always control the complete coaxial or twisted-pair cable run.

5.4 TELECOMMUNICATIONS AND ELECTRICAL SYSTEMS

One of the most difficult tasks involved in the design of a telecommunications wiring system is the separation of telecommunications and electrical power systems. Unless the system is designed properly, electrical noise from power cables can wreak havoc on less-powerful communications signals. While the obvious answer would be to never place power and communications cables together, this is not a practical solution. Every communications device (terminal, computer, modem, etc.) requires ac current. Additionally, there are only limited areas in most building environments that can support a cable distribution system. This effectively forces the coexistence of telecommunications and power cables.

Given that the two types of cable systems must coexist, it is important to understand a bit about how the two systems may interfere with each other. This information can help not only during the design of new systems but also during the troubleshooting of existing systems. The following discussions present a brief overview of electrical noise and crosstalk problems.

5.4.1 Electrical Noise

Electrical noise is created when an electromagnetic field is generated around a copper or metallic cable. Any time an electrical current is passed through a metallic cable, an electromagnetic field is created.

The amount of power sent through the cable can be very small (such as a few volts typical of communications signals) or very large (such as a few kilovolts generated by some power lines). The amount of voltage does not matter; whenever a current is present, a magnetic field will be created.

The strength, or size, of the magnetic field is directly proportional to the amount of current being passed through the conductor (cable). A high current will produce a strong field while a low current will produce a weak field.

5.4.2 Crosstalk

Crosstalk occurs when an electrical signal in one cable interferes with the signal in an adjacent, or nearby, cable. Suppose two cables, A and B, carry 5 and 50 volts (V) ac, respectively. Crosstalk could occur as the result of the relatively strong electromagnetic field in cable B overpowering the weaker electromagnetic field in cable A. Since an electromagnetic field is directly related to the current (communications signal) being sent through a cable, any change in the outer magnetic field would cause an alteration or corruption of the electrical signal within the cable.

High levels of crosstalk can cause background noise (similar to the hum heard on some voice phone calls), which in turn can cause a decibel loss in the signal. The end result is that the signal on cable B cannot travel as far as it could in a no-noise environment. In the telecommunications wiring arena, cable modifications such as shielding have been developed to counter electromagnetic interference (EMI) and crosstalk problems.

As discussed earlier, fiber-optic cable and infrared systems rely upon a light wave, not an electrical signal. Thus, these transmission media are not susceptible to EMI or crosstalk.

5.4.3 Electrical Placement

In order to prevent crosstalk and other forms of EMI interference from occurring, there must be a minimum separation distance between most types of copper-based telecommunications equipment and power lines. This applies to both unshielded and shielded twisted-pair as well as all forms of coaxial cable. As a general rule, there should be at least 6 in between any telecommunications cable and power cable for each kilovolt-ampere (kVA) present in the power line, up to a maximum of 6½ ft for practical applications. A kilovolt-ampere is calculated by multiplying the voltage of a line by its current. For example, a 110-V line operating at 10 A would equal 1.1 kVA (110 × 10/1000).

This *rule of 6* for power and telecommunications separation is set forth merely as a practical guideline based upon existing installations and the distance required to ensure clean telecommunications transmission. In all cases, any local, state, or federal regulation (such as the NEC) which may supersede this guideline should always be adhered to without exception.

5.5 SUMMARY

The following is a summary of the key points of this chapter:

- In addition to fixed cable-based systems, secondary transmission systems, such as microwave (and radio derivatives) and infrared, can be used to support telecommunications within a building or campus.
- Secondary hardware, such as patch panels and baluns, can help extend the life of an existing system through modularity, increased management, and simplified reconfiguration capabilities.
- Telecommunications and electrical line integration is not possible in today's building environment.

Chapter

6

Vendor Wiring Plans

Definition: *Vendor wiring plan* A telecommunications cabling scheme developed and promoted by an equipment or wiring vendor. Usually designed to promote products sold by the vendor.

When it comes to devising a building wiring plan, there are many different items that must be considered, including the type of cable, the distances of each cable run, the type of station jack, and the type of cross-connect blocks. For the inexperienced telecom manager or other person placed into a "get to know wiring quickly" situation, these decisions can be overwhelming. Fortunately, many big-name organizations have developed specific wiring "plans" to help provide the novice cable manager with both guidelines and design tools.

This chapter is dedicated to a brief discussion of vendor wiring plans. When discussing vendor wiring plans, it is important to remember two key points: (1) vendor plans are designed for the vendor's benefit, and (2) vendor plans are typically in a steady state of change. For example, the Bell Atlantic Universal Information Transport Plan (UITP) has been in a state of steady change since its original inception in the early 1980s (this includes not only its components but also its status as a commercially available product). This chapter focuses on the intention and general characteristics of several of the more common wiring plans in use today. For readers who wish to delve further into a specific plan for information concerning design or implementation guidelines, it is recommended that you contact the vendor in question, who is usually more than willing to provide you with a complete set of documentation and your very own salesperson.

6.1 VENDOR PLAN CAPABILITIES

One of the most consistent and predictable concepts in the free-market business world is the idea of profit. Corporations and companies exist to make money for their owners or shareholders. Many people may argue that some corporations were created to "provide a better product to the consumer" or to "offer a service unavailable previously." However, their underlying motive is almost always profit (how many firms decide to operate with a negative cash flow? Even not-for-profit organizations generally must operate with a positive cash flow).

The concept of profit-making has not been lost on those vendors within the telecommunications industry, as evidenced by the engineers in the back rooms consistently developing new applications for sale to the general public. This concept is important since it reveals the underlying motive for the development of vendor wiring plans: to make money. Before I am accused of thrashing cabling vendors, let me say that the pursuit of profit is most admirable and has resulted in a technical revolution involving information technologies that has never before been seen in any market. What this pursuit does mean, however, is that the motives of the vendors may not always be in the best interests of the user.

What vendor wiring plans do offer varies considerably from vendor to vendor. There are three general "types" of wiring plans. These are the global, spec-based, and hardware-based types of plans. The general characteristics of each type are as follows:

Global plan: This plan offers guidance and products that support all features of the building wiring concept. This includes full life-cycle support including initial design considerations, installation guidelines, and hardware components. Management components may also be part of this type of plan. Two of the most popular global plans available today are the AT&T SYSTEMAX PDS plan and the DEC DECconnect plan. Global plans are usually offered by telco-related organizations (AT&T, RBOCs, etc.) and some of the larger cabling vendors (usually as an extension of a hardware-based plan) and systems integrators.

Spec-based plan: The spec-based plan offers information on design and cable specifications but does not actually specify individual hardware components (or if it does, it does not provide all of the components); the IBM Cabling Plan is an example of such a plan. Spec-based plans are designed to foster the sale of communications equipment, not cabling equipment. This concept will be discussed further in the section dealing with the IBM Cabling Plan.

Hardware-based plan: This type of plan is based upon a specific type of hardware and cable that can be adapted to fit any configuration or design. Hardware-based plans are commonly available from modular cable and connector vendors who have branched out from a hardware provider to a design and service provider. The concept of providing an "official wiring plan" has been highly profitable for such vendors who have used this type of service offering to get into the more lucrative component sales cycle.

Each of these types of plans can offer a specific type of functionality that can greatly reduce the amount of time and money that is spent both designing and implementing a building wiring system. While most wiring plans available today can be easily placed into one of the three categories listed above, there are several plans that either do not fit one of the general descriptions or are a combination of two or more type plans. For example, a hardware-based plan that provides some of the functions of a spec-based plan can be considered a type of global plan. However, simply because a hardware-based plan offers some type of off-the-shelf management system does not automatically qualify it as a global plan. Such hybrid plans are very common today as cabling vendors attempt to carve out a niche in the growing and increasingly competitive cabling and telecommunications market.

One of the more common emerging types of hybrid plan is one that combines traditional cabling-oriented functions (system design, installation, components) with actual networking systems. As an example, many private branch exchange (PBX) and local area network (LAN) system providers are also now providing complete building wiring system design services. This enables such vendors to offer a broader range of services when competing for communications system bids. One inherent disadvantage of such plans is that they tend to support only the products being offered with the specific bid or product line offered by the vendor. For example, a LAN vendor that specializes in Token-Ring networking may be inclined to suggest a shielded twisted-pair building wiring plan that, while suitable for a Token-Ring network application, might not offer the support for unshielded twisted-pair or fiber applications that may be required in the future.

6.2 VENDOR LIST

There are many different vendors that offer commercially available wiring plans. The services offered in these plans range from the basic to the complete, providing global, spec-based, hardware-based, and hybrid approaches. Established vendors currently offering off-the-shelf wiring plans (or equivalent combinations of services) include, but are certainly not limited to, AMP, AT&T, Bell Atlantic, British Telecom (BT), Digital Equipment Corporation (DEC), Hubble (formerly Brand-Rex), International Business Machines (IBM), Mod-Tap, Belden Cable, Thomas & Betts, and Siecor.

In addition to these vendors, many independent cable manufacturers and distributors have developed cable-plan products or offer services in design and management. These offerings, however, may be limited and do not always constitute a vendor wiring plan. To be considered a true vendor wiring plan, the plan must offer the following services:

Design guidelines: The vendor plan should offer specific guidelines for designing a building wiring system. These guidelines should include set parameters for sizing the number and type of cables installed in each location within a building as well as the length of individual cable runs. This should also include in-depth information on the cable distribution facilities (i.e., horizontal systems, risers, telephone closets, cross-connect blocks) as well as cable management considerations.

Multifunction support: Any good wiring plan will allow for multiple applications to be supported. For example, a vendor wiring plan that only supports voice or data systems is not a complete plan. The implementation of such a plan can prove costly if additional requirements must be added later.

Cable management: Regardless of the type of plan being selected (global, spec-based, hardware-based), a good wiring plan will always provide for some level of cable management. The cable management might be simple, such as color-coded components or a cable label system, or highly advanced, such as a computer-based tracking system. Any vendor offering that does not at least promote responsible cable management is not considered a true wiring plan.

Service features, such as installation and maintenance, while associated with many vendor offerings and certainly important when selecting an installation firm, are not considered part of a wiring plan. This is important since many cable vendors offer cabling plans that are little more than hardware and installation services (they do not provide written guidelines for configuration and design).

6.3 SELECTING A VENDOR PLAN

The selection of a vendor wiring plan for implementation within a given facility is a difficult and complicated process that must take into consideration many different variables. Considerable effort should be involved since the selection of a vendor wiring plan could mean the expenditure of hundreds of thousands of dollars. This investment can easily be doubled or tripled if future modifications are required because of deficiencies in the original plan.

Selecting a plan that is not responsive to your needs can mean not only recurring expenditures but also loss of productive time when normal business operations are halted for the installation of additional facilities. As an example, if a wiring plan is installed in a new building to support voice only, new cable may have to be installed as data applications arise, potentially disrupting normal business operations. If the original plan did not specify enough space in telephone distribution closets, additional rooms may require modification to accommodate the new data distribution systems. The same holds true for horizontal and vertical distribution systems. Inadequate planning for distribution hardware, such as insufficient conduit, can mean costly additions when new conduit or duct work must be installed.

Prior to the selection of any vendor plan, a requirements analysis of your immediate and long-term needs should be performed. Only after you understand your complete needs from a building wiring plan can you evaluate the offerings available from the different vendors. Key considerations when evaluating a vendor offering include the following:

Voice support: Does the system fully support your voice application needs? This should include all types of voice systems, including PBXs, key systems, and any other voice-related system (such as intercom or paging systems).

Data support: Can the system support all (not just some) of your data applications? This includes LANs and multiuser computer systems, such as IBM 3270 devices.

Video support: If you have video requirements, can the system adequately support them? If not, are there ways to modify the system to provide support?

Cable management: Any successful vendor offering must include a complete cable management scheme. This includes support for either color-coded or alphanumerically labeled cable and cross-connect block systems. Configuration worksheets, designed for installation but useful for system tracking, are also beneficial. If the system offers an automated system, it must be able to be implemented and easily maintained by your internal staff.

Plan modularity: Modular systems are usually easier to configure and implement than nonmodular systems. If the system is modular, it should allow for system implementation in stages, where each component can be selected independently of most other components.

Plan flexibility: Any acceptable plan must allow for a wide variety of physical system configurations. This includes support for most building and campus layouts. The plan should not restrict tall or unusually shaped facilities. The distribution hardware specified by the plan should support your building configuration without the need for major system add-ons.

Plan expandability: One of the most important characteristics of a building wiring plan is its ability to grow with you as your requirements grow. One of the basic rules of telecommunications is that communications systems never grow smaller. Any requirement, voice, data, or video, will increase with time. The actual hardware used may decrease, but the application requirement itself almost never decreases in size (at least not in a healthy corporation). Any plan selected must allow for new facilities to be installed without requiring reconfiguration of previously installed equipment.

Even with the guidelines listed above, the selection of a vendor wiring plan is not easy. Factors such as price, installation, support, and product availability must also be considered. It is very unlikely that a single plan will meet all of your requirements. But this is not a reason to be unwilling to implement part of a vendor plan or to implement a slightly modified version of an off-the-shelf vendor offering. As an example, how many people buy a new house without making some special modifications (such as a deck, skylights, landscaping, etc.)? The same holds true with vendor wiring plans. While the basic plan may suit your needs, vendor plans often require minor adjustments to fully meet your requirements.

When implementing a modified vendor plan, you must be careful that your modifications are both compatible with the plan and do not significantly change the plan so as to lose support from the offering vendor. What good is a vendor wiring plan if the vendor will not support it for you? Fortunately, most vendors, if approached up front, are willing to work with clients to ensure an end product that meets the user's requirements.

6.4 AT&T

AT&T offers one of the more popular wiring plans today, the AT&T SYSTEMAX Premises Distribution System (PDS). This plan, originally known as AT&T PDS, has been the model for many other vendor plans. The PDS plan has been developed gradually over many years. A sister plan to PDS is the AT&T Premises Lightguide System (PLS) which specifies the use of fiber-optic cable for backbone and riser applications. Prior to divestiture in 1984, AT&T had a virtual monopoly on the wiring business. Being the only provider of telephone service throughout the majority of the twentieth century certainly gave them an insight, if not an advantage, into how building wiring systems should be structured.

While inside building wiring was controlled by AT&T and local telcos during the predivestiture decades, control of inside building wiring now lies with the building owner. AT&T has realized that not every building owner understands inside wiring. The PDS system has been developed to not only maintain the AT&T name in the building wiring arena but to also capitalize on the large number of building owners who require assistance in developing, installing, and maintaining a building wiring system. As such, AT&T not only provides design guidelines but also provides equipment and services as a profit operation.

6.4.1 Components

The AT&T SYSTEMAX PDS plan is considered a global plan in that it offers complete design and management guidance as well as hardware and installation services. The plan itself comprises individual modules that address the end-user location, horizontal distribution, vertical distribution, campus distribution, and system administration (cable management). Areas supported by the PDS system include the following:

Configuration guidelines: PDS offers a complete set of guidelines for implementing both building and campuswide networks. Information on standardized equipment and distribution practices (such as distribution hardware and specific layout configurations) is also included. Given the experience of AT&T, it is not surprising that the PDS plan covers most aspects of any building environment.

Cable specifications: The AT&T plan is based upon the use of unshielded twisted-pair (four pairs of 24AWG wire) for both voice and data applications. In addition, the plan uses 62.5-μ/125-μ fiber-optic cable for riser and campus networking. The cable terminations use eight-pin RJ-45/48 and ST-connectors, respectively.

Cross-connect equipment: AT&T has standardized on the 110-block as the basic cross-connect component of the PDS plan.

Cable management: AT&T has included within the PDS plan a set of standardized guidelines for cable documentation and labeling. This system is an outgrowth of standard telephony practices which have been devised and implemented by AT&T for decades.

6.4.2 Applications

The AT&T PDS plan has been designed to support the majority of voice and data applications. However, by its reliance upon unshielded twisted-pair as the predominant transmission media, it has locked itself out of some applications that require shielded twisted-pair or coaxial cables. To compensate for this shortcoming, PDS allows for almost any media to be installed in addition to the unshielded twisted-pair (although support may be lacking for these non-PDS components). Specific applications that are well suited to PDS include the following:

- *Voice systems:* PDS can support virtually all voice systems in use today (including PBX and key systems).
- *Unshielded twisted-pair data systems:* This includes mainframe and terminal systems and LANs, such as Ethernet and Appletalk.
- *Fiber-based networks:* Systems such as Ethernet over fiber and Fiber Distributed Data Interface (FDDI) are well supported by the PDS plan.

By developing a plan that provides general guidelines to meet most building configurations, coupled with support for twisted-pair and fiber, AT&T has been able to provide a wiring plan that is used by many user organizations as the basis for their own internal wiring plans. The only real disadvantage of PDS is its lack of standardized support for custom applications or applications that rely upon shielded or coaxial cable. While PDS, like any other building wiring plan, can be modified to suit individual or special requirements, AT&T may not support the system if extreme modifications are made to the base plan.

6.5 DEC

DEC has long been a leading provider of computer equipment. DEC computer systems range from high-end mainframe and minicomputers (such as the VAX and MicroVAX) to the now defunct DEC Rainbow personal computer. Early in the 1980s, DEC realized the value of providing a complete system approach to computing devices. Under such an approach, not only would DEC sell computer systems, but it would also sell the communications systems required to network computers. DEC has relied upon the Ethernet LAN to provide this communications function for their computer systems. By providing a complete system, DEC is able to sell their computers into any environment regardless of the currently installed computer networking system.

The problem with this approach is that the building the computer system is being installed into must be able to support an Ethernet network installation. Unfortunately, Ethernet has not been the easiest of networks to install (pre-10BaseT unshielded twisted-pair). To help promote Ethernet, and the sale of DEC computer equipment, DEC developed the DECconnect wiring system. DECconnect was designed as a complete wiring plan that would provide for a multitude of communications requirements, such as voice, data, and video.

The DECconnect plan is a modified global plan. It provides a fairly complete system that deals with specific data needs (Ethernet and low-speed terminals) but does not address voice needs as completely as some other vendor plans. The plan itself is not designed to be a catch-all plan; it is clearly intended to service DEC equipment sites.

6.5.1 Components

The DECconnect wiring plan has four main components: the DECconnect faceplate (station outlet), the DECconnect wiring, the Satellite Equipment Room (SER), and the Cable/Facilities Manager (C/FM). Each of these components addresses a specific area, as discussed below:

Faceplate: While some cabling vendors strive for a single type of cable that can satisfy the majority of communications requirements, DEC has taken the approach that many individual requirements are unique enough to require a unique type of cable. To satisfy this requirement, the standard DECconnect faceplate consists of four individual connectors: voice, video, low-speed data, and high-speed data. See Fig. 6.1. The voice connector is a standard eight-pin connector designed to provide connection for most popular PBX vendors. The video connector is a standard F-series coaxial connector designed to terminate CATV coaxial cable. The low-speed data connector is a keyed eight-pin jack that provides unshielded twisted-pair connection. The high-speed data connector can be either a BNC

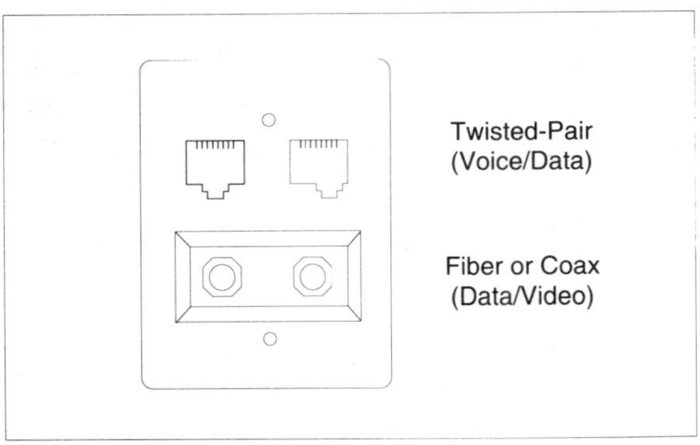

Figure 6.1 DECconnect faceplate.

coaxial unit designed to support 10Base2 Ethernet networks (thin-net/thin-wire), or a fiber-optic connector. One advantage of this type of modular faceplate is that individual connectors can be used depending upon the current networking requirements. In addition, the connectors in the faceplate are all modular and either snap in place or fit into a generic opening that can accommodate a coaxial or fiber connector. This allows special types of connectors to be used for specialized nonstandard applications.

Wiring: The standard DECconnect wiring plan defines four categories of wiring (voice, low-speed data, high-speed data, and video). These categories, which match the types of recommended connectors in the DECconnect faceplate, are very flexible in that there are many specific types of cable that can be applied to each individual category. For example, there are many different types of twisted-pair cable that can be used in the actual implementation.

SER: The SER is the rough equivalent of a telephone closet or intermediate distribution frame (IDF). The SER specifies the design and configuration of an equipment frame (rack-mount cabinet) which is intended to house DEC networking equipment (DEMPRs, DELNIs), voice system cross-connects, and patch panels. In addition to the standard star riser configuration for voice cable, multiple SERs within a single building are connected in a bus fashion with coaxial cable for Ethernet connectivity.

C/FM: The C/FM system is a VAX-based software program designed to manage cable distribution systems. Its primary function is to manage hub-based wiring systems (such as twisted-pair). Additionally, it has management functions which allow overall system capacity and planning activities.

The DECconnect system components are designed to bolster the DECnet networking system as a complete system. However, it does not contain any depth of information regarding overall system design (such as cable distribution methods, cross-connect block architecture, distribution frame specifications, etc.). This lack of completeness is a major disadvantage to the DECconnect system and does limit its applicability to many users.

6.5.2 Applications

The DEC-specific design of the DECconnect system enables it to support the majority of applications sold by DEC. Its support for unshielded twisted-pair enables it to support most voice systems (PBX, Key, Centrex). In addition, CATV and building security systems can be supported by the video cable and connectors. The data cables (twisted-pair, coaxial, fiber) can support most data applications, such as LANs and terminal connections. When it comes to design guidance, however, the majority of the DECconnect system addresses DEC networking products only. For example, shielded twisted-pair networks (such as Token-Ring) are not well provided for in the DECconnect architecture.

6.6 IBM

The IBM Cabling System is a detailed specification that outlines the majority of components required for a star-oriented cabling system. This includes station outlets, types of cable, and cross-connect and patch-panel architectures. It also includes detailed guidelines for cable run distances, overall system design parameters, and cable management systems. Areas such as specific distribution hardware are not addressed directly by the IBM Cabling System, although their use is included in the design guidelines. The basic cabling structure of the system is based upon the use of twisted-pair and fiber-optic cable. Although coaxial cable is required by many IBM computer systems, the use of twisted-pair and baluns is recommended throughout the system.

The purpose of the IBM wiring plan is to support the sales and installation of IBM computer and voice products. IBM does not push the cabling system as a stand-alone product in non-IBM environments. The cabling system is primarily supported by authorized IBM distributors who are certified to handle the design and installation of the complete system (usually in addition to IBM equipment). There are some components, however, that can only be ordered through IBM (such as some of the specific connection components). Likewise, there are some components, such as voice cabling components, that are better acquired through certified cable manufacturers (many of the components used in the IBM cabling system are manufactured by authorized manufacturers based upon IBM specifications).

The actual design guidelines for the IBM Cabling System are available in blueprintlike format. These guidelines are intended for use by building engineering and design personnel and cover all aspects of system design relative to IBM products. For example, detailed information on jack configurations and patch panels for most types of IBM products is available.

6.6.1 Components

There are four major components to the IBM Cabling System: station jack connectors, cable, distribution panels, and cable management systems. In addition, there is also a large amount of information available on system design (including planning and installation steps). Each of the four components is discussed below:

Station jack connectors: The IBM Cabling System relies upon several types of connectors to accommodate the many types of IBM voice and data systems. This includes support for Biconic fiber connectors, unshielded twisted-pair connectors, and IBM data connectors (shielded twisted-pair).

Cable systems: The IBM Cabling System defines a group of cable types that can be used to satisfy specific applications. Some of the more common cables, available from many different cabling vendors, are:

- *Type 1 cable:* Consists of two shielded twisted-pairs (22AWG) and is intended for use by data systems such as IBM Token-Ring. Spec number 4716748.
- *Type 2 cable:* Consists of two shielded twisted-pairs (22AWG) and two unshielded twisted-pairs (22AWG). The shielded and unshielded pairs are in the same overall sheath but independent from each other in terms of shielding and twist. This cable can accommodate both data and voice applications. Spec number 4716739.
- *Type 3 cable:* Consists of four unshielded twisted-pairs (22- or 24AWG). This cable is designed to support most common voice communications applications.
- *Type 5 cable:* Consists of two 100-μ/140-μ fiber-optic strands. This cable is intended to support high-speed data applications. Spec number 4716744.
- *Type 6 cable:* Consists of two stranded unshielded twisted-pairs (26AWG). This cable is only suitable for use as a patch-panel cable because of its limited distance ratings (which is due to its stranded conductor and thin gauge). Spec number 4716743.

Distribution panels: The IBM Cabling System explains in detail the configuration for the various distribution panels required to support the vast line of IBM computer and communications systems. This includes support for integrated baluns and coaxial, twinaxial, and fiber-optic systems.

Management: The IBM Cabling System relies upon a strict numbering system that clearly marks each component in a hierarchical fashion. Through the use of such a system, cable identification at any point in the cable system is possible.

While not directly specified as part of the IBM Cabling System, the 66-block is preferred for use in most installations.

6.6.2 Applications

The IBM Cabling System was designed not as a profit-oriented product but as an enhancement to IBM's existing line of communications and computer products. As such, it is designed to predominantly support IBM-related equipment (although it can function extremely well in many office environments as the complete voice/data wiring system). The IBM system is optimal for supporting Token-Ring networks. While Token-Ring can operate over unshielded twisted-pair, full utilization (maximum distance and number of users) can only be attained through the use of shielded twisted-pair. The IBM Cabling System is also ideal for users who must deal heavily with coaxial, twinax, or dual-coax networks in support of IBM computing systems. IBM offers a multitude of product solutions to network such systems using twisted-pair. Additionally, the IBM Cabling System, through its use of shielded twisted-pair and fiber-optic cable, is very well suited to installation in heavy EMI environments. This includes locations such as machine shops, factories, hospitals, and public utility environments.

6.7 MODULAR EQUIPMENT VENDORS

In addition to the mega-companies that offer building wiring plans, there are many smaller firms that specialize in building wiring systems. These firms, often referred to as modular equipment vendors, generally offer a wide range of products as well as many services, such as system design, installation, and management. To help differentiate themselves from other modular equipment vendors, many firms have developed standardized wiring plans.

The wiring plans offered by modular equipment vendors are usually quite different from the strict plans of IBM, DEC, and AT&T. Where the larger vendors offer a plan that must, for the most part, be taken as is with little customization, the smaller vendors offer plans that usually contain a wide variety of component and design options. For example, a vendor's wiring plan may contain four or five different types of station outlet or two or three types of cross-connect equipment. As such, the plans from modular equipment vendors tend to be less of a structured plan and more of a structured method for designing a wiring plan.

6.7.1 Components

Many modular equipment vendors, being independent from the large firms offering the major wiring plans, offer either all or part of several different major plans. For example, a wiring vendor may sell IBM Cabling System cables (type 1, 2, 3, etc.) as well as offer DECconnect modular components. Through this type of approach they are able to offer the best solution to meet a customer's requirement (the disadvantage here is that you often get a hybrid system that is not supported by the original plan's vendor).

Components, and services, that are typically offered by modular equipment vendors include the following:

- *Cable systems:* This includes everything from twisted-pair to fiber optics (not to mention many specialty cables, such as undercarpet cables).
- *Station outlets:* This includes RJ-type jacks, fiber-optic connectors, coaxial connectors, all in a multitude of configurations.
- *Distribution hardware:* Typical components include conduit, riser slots, ladders, rack equipment, ductwork, etc.
- *Cross-connect blocks:* This may include standard equipment, such as 110- or 66-blocks, as well as custom equipment utilizing printed circuit boards or other custom components (usually intended to provide a better electrical connection with restricted loss levels).
- *Design and implementation assistance:* Most modular equipment vendors will not only assist in the design and configuration of a wiring system, they will also install and maintain the system.

It should be noted that while most modular equipment vendors offer a wide range of equipment and services, some vendors specialize in individual wiring plans or in equipment specialty areas. For example, some vendors specialize in fiber-optic systems and have only limited offerings for twisted-pair systems. Other vendors may specialize in IBM Cabling System components and not offer any DEC or AT&T equipment. Additionally, some of the larger and more progressive modular equipment vendors have developed their own wiring plans or methodologies based upon the best characteristics of other commercial offerings (usually coupled with a few proprietary components of their own).

6.7.2 Applications

Because of the large range of equipment offered by modular equipment vendors, the number of applications to which their wiring systems can apply is also very large. Requirements which can typically be met by most modular equipment vendors include most LAN, computer, and PBX products. In addition, the flexibility of offering a wide range of product allows modular equipment vendors to support many nontypical requirements, such as extreme distance, heavy EMI sources, or custom data applications (for example, a special data connection that may require more than eight conductors, eight being the largest number of conductors usually supported by a single twisted-pair jack).

6.8 SUMMARY

Vendor wiring plans can offer a wide range of solutions to meet building wiring requirements. The key points to understanding vendor-offered wiring plans are as follows:

- There are many different vendors that can offer global, spec-based, or hardware-based wiring plans.
- Global plans offer a complete range of equipment and services that can support most wiring situations.
- Spec-based plans are primarily based upon wiring specifications or design guidelines that may encompass many different options for actual hardware implementation.
- Hardware-based plans are designed to sell hardware components (cable, cross-connects, etc.). Most plans of this type offer a wide range of hardware to fit almost any design criteria, and as such, become less of a standardized wiring plan and more of a catalog mix-and-match plan.
- Vendor plans often require slight modification or customization to fully meet user requirements.
- A good vendor wiring plan will provide support for both current and future applications (or at least provide a semipainless growth path).
- Support features, such as design, installation, maintenance, and management systems, can often be as important as the type of hardware chosen and should not be overlooked when selecting either a vendor wiring plan or an installation vendor.

Chapter

7

Standard Information Outlet

Definition: *Standard information outlet* An outlet consisting of a faceplate and termination jacks where the building wiring system terminates for connection to an end-user device. For example, the common telephone jack.

Perhaps the most important component of the building wiring system is the standard information outlet (SIO). Also referred to as a universal information outlet (UIO), the SIO is the point at which a user device, such as a phone or computer terminal, attaches to the building wiring system.

It is this point, the SIO, at which the user will form a basic association with the wiring system. If the SIO is difficult to access, or perhaps does not allow connection of the user device, the end user may form a less than satisfactory image of the building wiring. On the other hand, if the SIO is easy to access and can accommodate several different types of devices, the end user will probably form a positive opinion of the building wiring system.

The value of pleasing an end user is fairly evident; the better end users feel about the wiring system, the easier it will be to work with them during those difficult times that do arise. These difficult times, also known as adds, moves, changes, and problem resolution, can be very time consuming and destructive if you do not have the support of the individuals being added, moved, or changed.

7.1 SIO SELECTION

When designing a building wiring plan, it is necessary to address every detail in advance; once the process has started, changes in design can be costly. One of the first design activities that must take place is the selection of the SIO. The decisions made here, such as the number of jacks per SIO and the type of jacks required, will help to drive the design of other components, such as the type and amount of cable installed.

7.1.1 Type of Jacks

There are many factors which can influence the SIO selection process. These factors should each be considered very carefully; the replacement of jacks after installation is often costly and disruptive. It is most important for the SIO unit to provide for complete voice (private branch exchange, or PBX; intercom; key systems), data (local area network, or LAN; data switch; computers), and video (security, CATV) support. In addition, the selection of an SIO interface should not preclude the use of future technologies that can be planned for now. For example, the Integrated Services Digital Network (ISDN) specifies an eight-pin phone jack. Although you may not use ISDN today, chances are very good that ISDN will play some role in the future communications plans of your company. Specifying an eight-pin jack (which can accept six-pin plugs used by most phone systems) can enable you to take advantage of this technology as it is implemented over the next 3 to 5 years.

Before selecting any components, it is important to first list all telecommunications requirements and identify the jack options associated with each type of application. This process will help bring to the front any common jacks between similar applications (i.e., similar data jacks or similar voice jacks). See Fig. 7.1.

Common jacks should be selected wherever possible to reduce the number of jacks required to support your complete requirements. In addition to identifying common jack requirements, it is important during this process to identify unique requirements. Although there are many applications that share the use of a common jack, there are some applications that require interfaces not typically associated with telephone jacks (such as DB-25/RS-232). Although some RS-232 applications can be satisfied fairly easily with eight-pin phone jacks, many RS-232 applications simply cannot be forced to easily operate over other types of cable or interfaces. There are also applications, typically older communications systems, that use special connectors or cables that are proprietary. Identifying these applications up front can allow adequate planning for later system implementation.

Certain types of jacks have their own unique problems; there is not a common set of criteria that can be applied to twisted-pair, coaxial, and fiber. The following presents some of the unique criteria facing the more common types of cable and jacks.

Application	Possible Media	Possible Jacks
PBX	UTP	RJ-45
Ethernet LAN	UTP	RJ-45
	Coaxial	BNC
Token-Ring LAN	UTP	RJ-45
	STP	Data Connector
		RJ-45
	UTP	RJ-45
3270 Terminals	Coaxial	BNC
	UTP/Balun	RJ-45
FDDI LAN	Fiber	Duplex
		ST
	STP	RJ-45
Security/Video	Coaxial	BNC
		F-Series
	Fiber	ST

Figure 7.1 Common applications.

7.1.1.1 Twisted-pair jacks. Most twisted-pair jacks are RJ-11 or RJ-45 compatible. Very few twisted-pair systems, including data, use DB-type connectors (such as DB-25 or DB-9). When selecting an RJ-type jack, the following considerations must be applied:

Number of pins: This usually involves the decision to support either two, three, or four twisted-pairs per jack. While many voice and data applications require one, two, or three pairs, the selection of four-pair (eight-pin) jacks is generally recommended to allow for future expansion of applications. Very few twisted-pair applications developed today require in excess of four pairs. Additionally, smaller jacks (such as the RJ-11) can often fit into larger plugs (such as the RJ-45). It should be noted that this can result in loose connections if poor quality or old, worn jacks are used. Jacks used in this fashion can also provide poor connection if they are subject to outside stress, such as pulling on the external cable or repeated plug insertion. However, most smaller jacks from quality vendors can operate in larger plugs with no significant difficulties.

Pin configuration: The pin configuration is one of the most overlooked design parameters during the SIO design process. Many novice designers assume that all jacks are configured according to the "industry standard." Unfortunately, there are several industry standards, some of which even have different vendor interpretations. The most common configuration is the Universal Service Order

Code (USOC) standard. The USOC defines a pin configuration that pairs pins in a semisymmetrical design beginning in the center of the jack. The inner two conductors (4 and 5) are reversed in tip and ring polarity to prevent the inner two pairs from having crosstalk problems (a variation of this flips the sequence of tip and ring such that tip and ring alternate from left to right for the entire jack). See Fig. 7.2.

The other common pin configuration is the ISDN eight-pin configuration. This configuration specifies a configuration in which the pair sequence (i.e., the placement of pairs 1, 2, 3, 4) is less important than the basic conductor groupings. There are two major pair sequence options with this type of configuration. The first allocates pins 4 and 5 as pair 1, pins 3 and 6 as pair 2, pins 1 and 2 as pair 3,

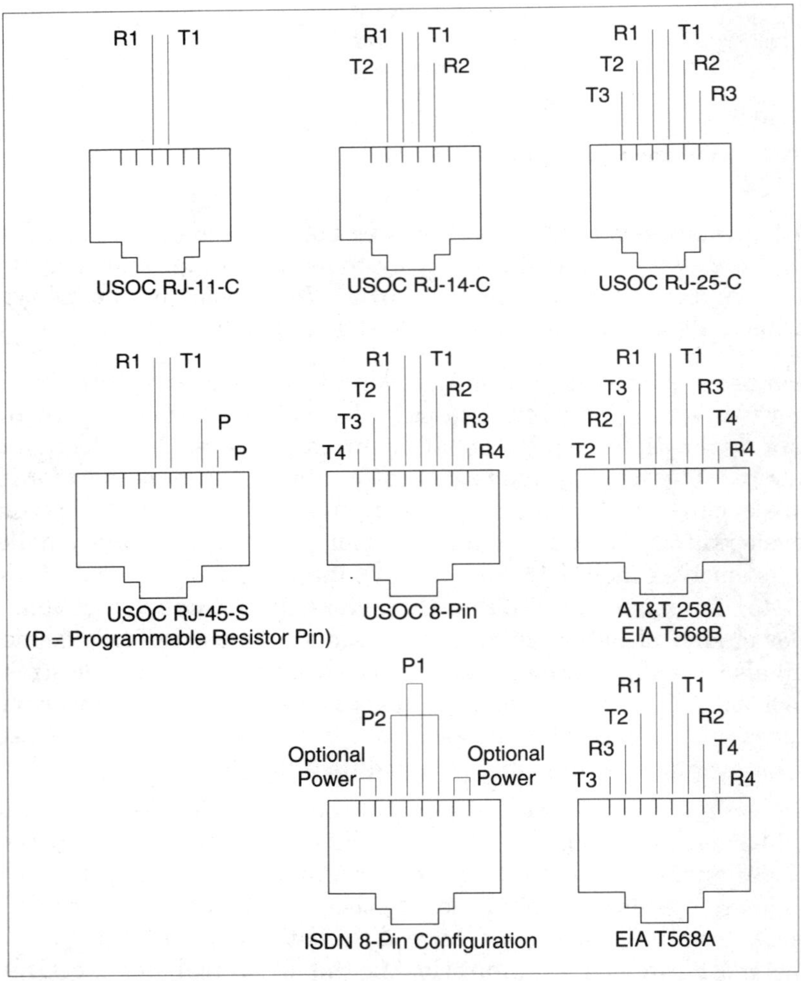

Figure 7.2 Twisted-pair pin configurations.

and pins 7 and 8 as pair 4. This configuration adopted by the EIA/TIA as the T568A standard, is the most common pin configuration in use today and is supported by many vendors for systems such as twisted-pair Ethernet (10BaseT). Additionally, this option is compatible with the inner two pairs of the USOC standard (allowing for backward compatibility for many older voice and data systems using an RJ-11 jack). The second option, supported predominantly in older systems and earlier versions of ISDN equipment, switches the sequence of pairs 2 and 3 (pins 1 and 2 are pair 2 and pins 3 and 6 are pair 3). This second version of the eight-pin jack, adopted by the EIA/TIA as the T568B standard, is not USOC compatible in any way and is generally not recommended for use in new installations.

Shift or keyed jacks: Many jacks can be shifted or keyed to allow only a special type of plug (matching the shift or key) to be inserted in the jack. Voice jacks are very rarely shift or key configured. When deciding to shift or key a jack, remember that whatever decision you make up front can greatly affect your operations later. For this reason, keyed jacks are preferred over shifted jacks. A keyed jack can accept a nonkeyed plug should you decide to migrate from a system that uses keyed plugs to using nonkeyed plugs. However, a shifted jack can only accept a plug that matches the shift exactly and will never support a standard nonshifted plug.

Jack connection terminals: Each twisted-pair jack in an SIO must be connected to the horizontal distribution system. This connection can be accomplished in a variety of ways, including screw-type and punch-down terminals. Screw-type terminals use a metal screw that uses pressure to hold a stripped conductor (twisted-pair lead) to a screw base or clip that is connected to the jack conductor. See Fig. 7.3. Screw-type terminals, common in most residential installations, are usually only associated with voice jacks because of the poor electrical characteristics they offer (too much metal increases line loss and increases the chance of crosstalk). A punch-down terminal is configured in much the same manner as a cross-connect punch-down block (i.e., 110- and 66-blocks and barrel-connector terminals). In addition, there are many other types of insulation-displacement terminals that are usually specific to an individual modular equipment vendor. It is generally recommended that the SIO jack connection terminal mimic the type of cross-connect block being used in the distribution frames. This will put forth a consistent punch-down architecture that will simplify management, installation, and system maintenance.

Component integration: Many vendors offer SIO units that allow support devices, such as limited-distance modems and modem-eliminators, to be installed as integral components. These support

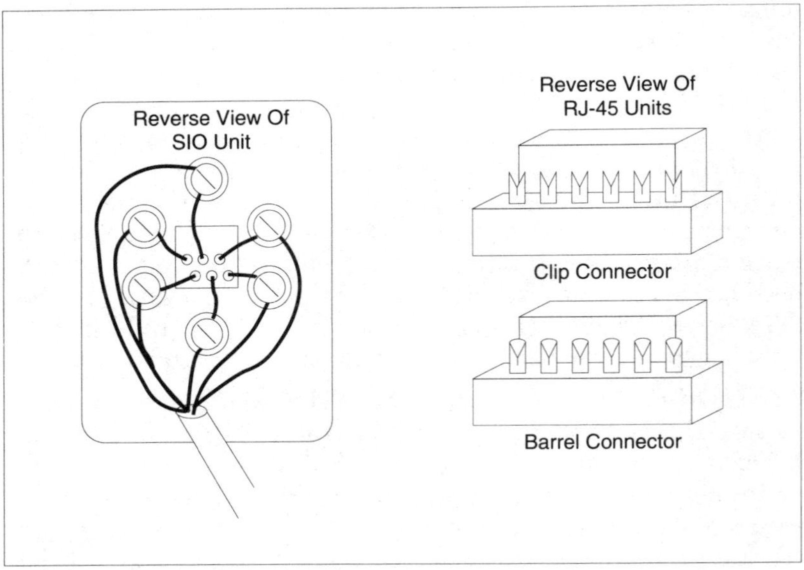

Figure 7.3 SIO jack configurations.

devices are installed on the backside of the jack and serve as the interface point to the horizontal distribution system. While offering an integral solution that simplifies user installation, these systems can also complicate system management. For example, fault identification of an integral component usually requires that the entire SIO be detached from the supporting structure (i.e., electrical box or wall-mount base). In addition, an integral component that has been fixed in place can preclude the use of the SIO jack for a noncomponent requirement (i.e., an application that does not require that specific integral component). The use of component integration techniques is generally not recommended unless the component is universal throughout an organization and will not become obsolete within 2 to 3 years.

Regardless of the component decisions made surrounding the SIO jack, twisted-pair jacks should always be installed with the conductors on the top or side of the jack. This will help to prevent dust or dirt from settling on the conductors, thus ensuring a solid connection between the jack and plug conductors.

7.1.1.2 Coaxial cable jacks. The selection of a coaxial cable jack and connector is almost entirely dependent upon the type of coaxial cable being used and the intended end-user application. Unlike twisted-pair and fiber-optic jacks, which can often be used to support a wide range of vendor products and applications (LANs, key systems, PBXs, modems, etc.), many coaxial jacks are associated with a specific vendor or application. For example, very few vendors use triaxial cable. Installing this cable and trying to implement a 10Base2 Ethernet (which uses RG-58 cable with a BNC connector) would simply not be possible. It is recommended that if a specific application is not known, the coaxial cable should be left unterminated behind the SIO (this may not be possible in all cases). If a coaxial termination is required, a common connector, such as the BNC, should be selected. The listing of coaxial applications (LANs, video, security, etc.) that you might use in the future may allow a common coaxial connector to be selected that would support most if not all of your future coaxial requirements.

Similar to twisted-pair jacks, many vendors offer SIO units that allow support devices (i.e., baluns) to be installed as integral components. This simplifies the connection process, allowing a coaxial segment to be connected directly to an SIO without the use of an adapter apparatus cable, which results in a coaxial–balun–twisted-pair connection. Since coaxial systems tend to be fairly universal throughout an organization (most firms do not have more than one type of coaxial system still in use), the use of integral baluns can be quite attractive. However, the use of such components requires that one of the SIO jacks be dedicated to a coaxial connector. For this reason, the use of integrated baluns is generally not recommended unless there is either a separate twisted-pair data jack or the balun-coaxial data device is universal throughout an organization and will not be replaced by a twisted-pair data device within 2 or 3 years.

7.1.1.3 Fiber-optic jacks. Fiber-optic jacks, while still not quite as universal as twisted-pair jacks, are more standardized than coaxial connectors. The two influencing factors in fiber-optic jack selection are the type of fiber being used and the intended hardware interface. The type of fiber being used will dictate certain parameters (such as the size of the fiber core) that must be considered in the selection of a fiber jack. However, the most important selection criterion is still the type of application being planned.

Many vendor products can only support a single type of fiber-optic interface (i.e., Biconic; subminiature type A, or SMA; straight tip, or ST; FDDI-duplex). When known, the application to be implemented should drive the connector selection. If the application is unknown, the fiber should be left unterminated behind the SIO. If a jack is required, perhaps for testing purposes or to take advantage of a limited-time fiscal budget, it is recommended that a common, high-performance jack be selected. The ST connector should provide a more-than-adequate connection for most fiber wiring systems. Remember that even if a non-ST interface is required in the future, the use of standardized fiber diameters (such as 62.5 μ/125 μ) and adapter cables can permit non-ST devices to interface with an ST-based SIO.

7.1.2 Number of Jacks Per SIO

Most SIO units contain between one and four jacks. The function of the individual jacks typically begins with voice support and expands to include data and video applications. The most common jack sequence is as follows (note that not all four jacks need to be used in every system; this is for structural planning only):

- *Jack 1:* Voice applications
- *Jack 2:* Data application (A)
- *Jack 3:* Data application (B) or video application
- *Jack 4:* Video applications

The number of jacks you select should be directly proportional to the number of simultaneous (or unique) applications you will have. For example, if you plan to support each user with a simultaneous phone, LAN, and printer connection, you might select one voice and two data jacks (two data jacks are required to support the concurrent use of the LAN and the printer). However, if each user workstation will have a phone and either a LAN or a printer, a single voice and single data jack would suffice.

Situations involving nonconcurrent but unique data applications can often require multiple data jacks. For example, suppose you plan to support either a coaxial LAN or an asynchronous modem connection (but never both concurrently) at a single user workstation. Two data jacks (such as an RJ-45 and a coaxial BNC) would be required since the two data applications do not use a common jack. It is important to note that even though the two applications are not planned for concurrent operation, all SIO units should be configured with both jacks to offer maximum flexibility. Failure to equip all SIO units with both jacks could severely limit the future movement or placing of data equipment (just think how many times people in your own organization move from office to office).

In addition to the above grouping, there are some SIOs designed to support more than one user. In these cases, the first two jacks are usually installed for voice applications (user A and user B). Additional jacks are usually designed to serve data or video applications. When selecting the number of jacks (based upon applications), it is advisable to overstate the number of jacks per SIO rather than underplan for future requirements. Each SIO should have a minimum of one voice

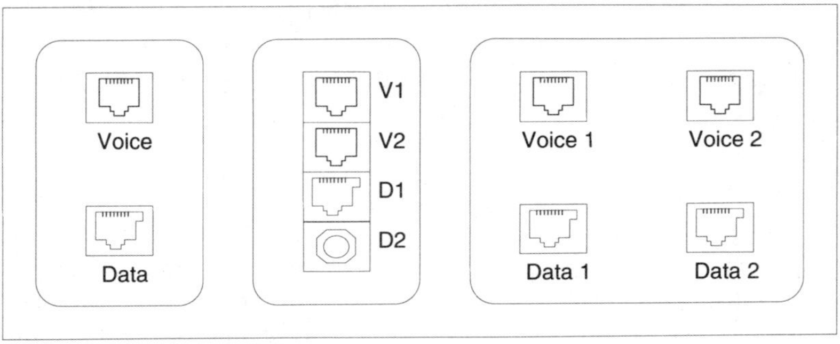

Figure 7.4 Multijack SIO units.

and one data jack. Additional jacks can be added based upon requirements.

If there is uncertainty as to the number of jacks that should be included, modular SIO faceplates can be used. Modular SIOs, in which the individual jacks snap into the faceplate, allow for jacks to be added or replaced at a later time without requiring complete replacement of the entire SIO unit.

When planning the layout of a multiple-jack SIO, it is important to follow standard conventions. Failure to do so can result in confusion when new users or maintenance personnel are required to interact with the system. In both single- and double-user SIOs, the jacks on the top are usually dedicated to voice applications. The lower jacks are usually dedicated to data applications. See Fig. 7.4.

The sequence of users in a multiuser jack is from left to right. It is important to note that ISDN applications involving combined voice and data support (integrated voice/data terminals) will typically be connected to the voice jack, allowing the data jack to support non-ISDN data applications (such as LANs).

Whenever a multijack SIO is used, the function of each jack should be clearly labeled (on the front of the SIO faceplate). For example, a three-jack SIO supporting one voice and two data jacks might have labels of "Voice," "Data 1," and "Data 2" placed next to the respective jack. Multiuser SIO units might be labeled "Voice A," "Voice B," "Data A," and "Data B."

7.2 OUTLET MOUNTING

There are three common methods of mounting, or installing, an SIO in an office environment, flush, surface, and pedestal; each uses a different approach to securing an SIO in place. Each of these methods is best suited to one or two specific types of building environments. Selecting the appropriate type of mount for an SIO can help to get the most out of a building wiring system. These three methods, and the type of requirements they are best suited for, are discussed in the following sections.

7.2.1 Flush-Mount Outlets

The most common type of SIO in use today involves flush-mount technology. The term *flush-mount* refers to the fact that the SIO is mounted flush within an existing structure. The only part of the SIO that is visible from the exterior is the faceplate and the individual jacks. See Fig. 7.5.

The flush-mount SIO is attached, through a wall or other solid fixture, to an electrical box or other similar mounting equipment. Although an electrical box offers maximum protection for the telecommunications cables inside, such a box may not always be a practical solution (especially in renovations). In such cases, a variety of special mounting brackets or clamps can be used to flush-mount an SIO into a wall or similar unit.

The flush-mount configuration is well suited to office environments in which SIO units can be placed within a wall (such as in a dry-wall office environment) or within power or communications poles. Occasionally, flush-mount units are placed within the floor of a building, as an alternate solution to pedestal-mounting (this is usually discouraged because of the increased risk of physical abuse placed on these

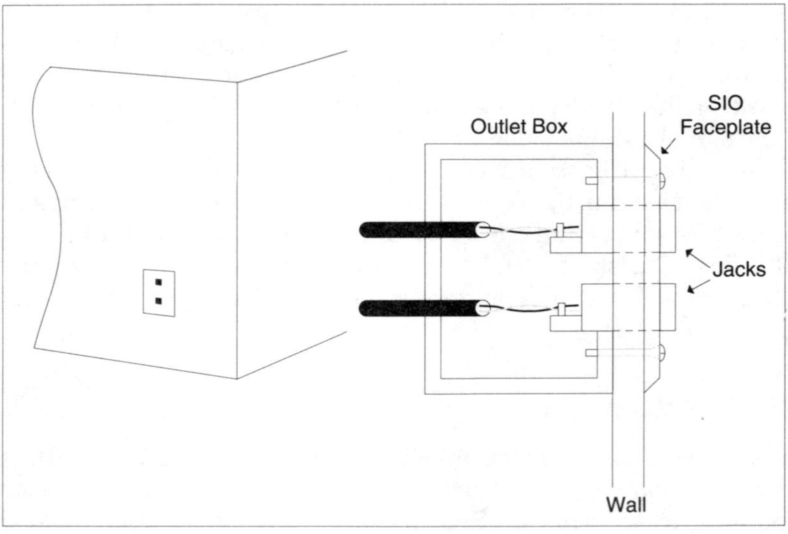

Figure 7.5 Flush-mount SIO unit.

units). Flush-mount units that are placed within the floor of an office should have spring-loaded covers that seal off the jack and connectors when not in use. Such a device prevents dust, dirt, and small objects from damaging the jack and/or conductors.

In general, flush-mounting has several advantages. SIO units that are flush-mounted offer a great deal of protection to the connection between the SIO jack and the distribution cable since the connection takes place within a wall or other enclosed structure. Such a connection is not subject to physical tension or movement. The flush-mounted SIO also does not significantly protrude from the wall. This allows furniture to be placed reasonably near the wall (within an inch or so) without damaging or stressing the connection. This nonobtrusive aspect also makes the flush-mount SIO the most appealing type of mount from an appearance perspective.

The flush-mount SIO should be used whenever possible. The only strong disadvantage of the flush-mount method is that the SIO is fixed in location and cannot be easily moved or expanded. Nonetheless, this approach should always be used unless building design constraints, such as office layout, horizontal distribution, or building materials, do not permit. For example, if an office building has solid concrete or wooden walls, the installation of a flush-mount unit may not be feasible or desired (especially in historical or restoration structures). Likewise, if your building uses underfloor distribution to a large open office, and power poles are not permitted, pedestal outlets may be the only feasible solution (this is discussed in the following sections).

An interesting variation on the flush-mount SIO is the inset-mount unit. This outlet consists of an SIO housed within a box or unit located flush within a raised floor. Inset-mount outlets differ from flush-mount units in that they are placed exclusively within a floor and commonly feature flip-up designs. See Fig. 7.6.

It is important to note that an SIO should never be mounted on an electrical box that currently contains electrical wiring in any form.

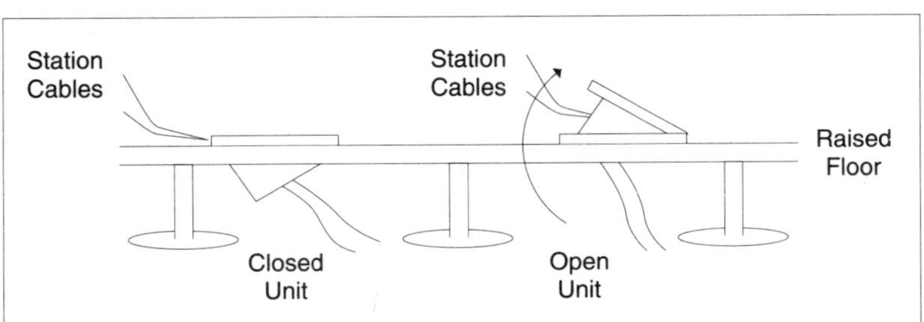

Figure 7.6 Floor inset-mount SIO unit.

Electrical and telecommunications wiring systems are not to be intermixed for distribution or outlet devices (the National Electrical Code has specific regulations against this).

7.2.2 Surface-Mount Outlets

A surface-mount outlet consists of an SIO that is part of a stand-alone unit or box. This unit is attached to the desired mounting surface directly; no portion of the surface-mount SIO is located within the wall. This contrasts sharply with the flush- and pedestal-mount configurations. The jack components of the SIO may be located on either the side, bottom, or front of the unit. Jacks are very rarely located on the top of a surface-mount unit to prevent dust or other particles from entering the jack.

Surface-mount outlets can be attached to any solid surface using screws (primary) or double-sided adhesive pads (secondary). This type of unit offers several key advantages over other types of units.

The surface-mount unit offers a great deal of location flexibility since the unit can be placed on any solid surface. This includes walls, floors, desks, modular furniture, and ceilings. This flexibility in installation is very well suited for installation in (1) environments in which jacks relocate frequently (such as mechanical shops or buildings under renovation), (2) buildings that use solid wall construction, and (3) modular furniture configurations that do not provide sufficient cable troughs or outlet boxes for telecommunications wiring.

There are many situations, especially involving historical buildings, in which the telecommunications systems must have a minimal lasting impact upon the structure. Surface-mount units, with the use of double-sided adhesive pads, can be easily installed (and later removed) without drilling into wall or surface structures.

Many building wiring systems are called upon to support greater numbers of users than originally intended. In cases where fixed flush-mount units cannot be installed (such as a power pole with a limited number of preinstalled outlets, all of which are occupied), a surface-mount unit can provide a short-term solution to SIO installation. This technique should always be thought of as a temporary measure until more adequate systems can be installed.

While there are several advantages to the surface-mount unit, there are some disadvantages as well. For example, the surface-mount unit is highly prone to physical abuse. These units often protrude more than an inch from the surface to which they are attached, which can subject them to impact from furniture being moved or even accidental physical contact from persons nearby.

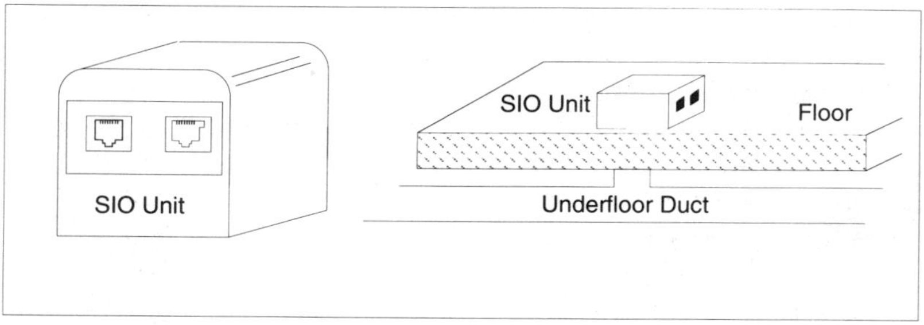

Figure 7.7 Pedestal-mount SIO unit.

7.2.3 Pedestal-Mount Outlets

The pedestal-mount outlet is similar to the surface-mount outlet in that it consists of an SIO housed within a box or unit that is attached to the surface of a structure (usually a floor). However, where the surface-mount outlet is flexible in terms of placement, the pedestal-mount outlet must be placed on the floor directly over an underfloor or undercarpet cable duct. In this manner, the pedestal-mount serves as the user access point for floor-based horizontal distribution systems. See Fig. 7.7.

Pedestal-mount outlets usually support both telecommunications and power system access. Care must be taken during the pedestal selection process to ensure that adequate separation (i.e., metal barrier) is present between the two different types of cable. This type of unit is almost always manufactured from metal to provide structural protection to the unit (being on the floor of an office tends to invite physical contact with chairs, furniture, and feet).

7.3 STATION OUTLET CABLES

The station outlet cable provides the physical connection, or link, between the end-user device and the SIO. Often referred to as an *apparatus*, *phone*, or *patch* cord, the outlet cable must be capable of terminating a wide range of equipment. Outlet cables can be constructed of twisted-pair, coaxial, or fiber-optic cable. With each type of media, there are many different possible configurations.

7.3.1 Twisted-Pair Outlet Cables

The most common outlet cable in use today is the *silver/satin* cable used to connect most telephone, data, and LAN equipment to the SIO. It typically consists of four to eight solid copper conductors terminated in two RJ-type jacks. There are several options available when designing or selecting a twisted-pair outlet cable.

An important point to note here is that the use of any keyed jacks may affect the type of modular units used in conjunction with outlet cables. When keyed or shifted jacks are being used for an outlet, the user-device end of the outlet cable should never be keyed or shifted unless absolutely necessary since this restricts the diversity of devices that can be used in conjunction with the outlet cable. Keying or shifting the SIO end of an outlet cable is sufficient to differentiate data and voice outlet cables.

The individual copper conductors can be either solid or stranded, ranging from 26AWG to 22AWG in diameter. The conductors can be grouped into separate pairs and twisted to match the characteristics of twisted-pair cable. Untwisted outlet cable pairs can impair the signal of data systems that require twisted-pair for electrical reasons (if untwisted outlet cables are used with twisted-pair systems, the length of the cable should be restricted to about 10 ft or under 3 m). Most outlet cables in use today do not use twisted-pairs.

Two major variants on outlet cables are the straight-through and crossover cables. A straight-through cable is one where the pins (i.e., 1 through 8) of a jack on one end connect up to the same pin number on the opposite end. For example, pin 1 is connected to pin 1, pin 2 to pin 2. This is the preferred cable configuration since it is transparent to the application and serves as a pass-through device for end-to-end communications.

A crossover cable connects opposite pins together in a cable. For example, pin 1 on one end of the outlet cable is connected to pin 8 on the other end. Pin 2 is connected to pin 7. These two types of configurations are detailed in Fig. 7.8.

Figure 7.8 Outlet cable configurations.

7.3.2 Coaxial Outlet Cables

The use of outlet cables for coaxial connections is typically restricted to the specific application and type of coaxial cable being used for horizontal distribution. Modular connectors are almost never used with coaxial outlet cables. Additionally, the use of modular baluns located within SIO units is usually not recommended unless the application is clearly defined and universal throughout an organization. To restrict the incidence of damage to coaxial outlet cables, cable length should not be more than 16 ft (less than 5 m).

7.3.3 Fiber-Optic Outlet Cables

Fiber-optic outlet cables, much like coaxial outlet cables, must often be tailored to meet the specific application. This requires that special cables be used for jack conversion. For example, an SIO with an ST fiber connection that must be connected to a Fiber Distributed Data Interface (FDDI) device will require a specific ST-to-FDDI-duplex outlet cable. The use of modular adaption units, such as those used with twisted-pair outlets, is not recommended because of limited insertion loss budgets of most fiber systems (every cross-connect or splice reduces the strength of the light signal). To restrict the incidence of damage to coaxial outlet cables, cable length should be less than 10 ft (approximately 3 m).

7.4 DESIGN PARAMETERS

In addition to the application-specific decisions that must be made concerning an SIO unit, there are several basic decisions that must be made concerning the overall design of a building wiring system. These decisions, such as the number of stations to be installed or the location of the installed SIOs, can have a major impact on the ability of a wiring system to actively support the entire user base. Design parameter decisions are fundamental in nature and must be present in any system regardless of the type of jacks, cables, or applications being used.

7.4.1 Station Sizing

Whether planning a new building or a renovation, the number of stations to be supported by the entire building wiring plan must take into account a number of factors, including the function of the building, the type of office layouts, the distribution system to be used, and the number of occupants that can be supported by the building. While none of these factors alone can determine the number of SIOs that should be installed, together they can be used as a very effective guide.

7.4.1.1 Building function. The type of activity intended for the building can be used to gauge the number of workstations and SIO outlets that must be present. For example, a building designed to serve as a general office would probably require a high density of SIOs, roughly one per worker. In contrast, a building designed as a maintenance facility would probably have a limited requirement for SIO units since most of the space will be dedicated to mechanical functions.

Some building requirements can be initially deceiving. For example, a medical facility may not have a large staff compared to the size of the facility. However, because of the information-intensive nature of medical care (medical monitoring systems, intercoms, telephones, computers, etc.), a large number of SIO units may be required. Other building functions that typically require excessive SIOs include research and development facilities, government administrative offices, and financial institutions.

7.4.1.2 Office layout. The type of office layout can help to determine the density of users in a building. Most office layouts tend to have specific user densities associated with them. For example, open offices that use modular furniture, where a single user has an office area of less than 80 ft^2, tend to have a high density rating. Similarly, offices that have fixed-wall offices generally allocate 80 to 150 ft^2 per office. An office with a lot of open areas that will not be used for offices (such as demonstration areas, auditoriums, meeting rooms, or large entranceways) requires fewer overall SIO units. The important point is to gain insight concerning the number of users that will be placed in each office and the number (density) of offices that can be expected.

7.4.1.3 Distribution system. The type of distribution system used will typically affect the number of SIO units required for renovation installations only. New building installations can usually be designed to accommodate any reasonable number of users. Renovations, however, are often limited by the capacity of the existing distribution system.

When dealing with renovations, it is important to determine the cable capacity of the distribution system. This involves determining the number of cables that can actually be placed within the system (the designer must first determine the type of cable required for distribution). Additionally, the designer must check to determine if the distribution system has access to all locations where user offices or workstations are planned. In cases where the anticipated number of end users exceeds the requirements of a building distribution system, it may be necessary to use multiple distribution systems.

7.4.1.4 Number of occupants. Perhaps the most important factor in determining the number of users that must be supported by a building wiring system is the occupancy rating established for the building. This rating, determined by government organizations (e.g., fire marshall), is based upon a number of factors that include the size and purpose of the building and the structural support. This is a number that cannot ever be exceeded. Building to this number will always ensure that you have sufficient SIO units (or at least sufficient capacity in distribution systems).

Whenever a number less than the full occupancy rating is used for planning the number of SIO units, additional spare capacity for growth must be built into the system. Most experienced designers typically add anywhere from 30 to 50 percent spare capacity. For example, a building that is intended to support 1000 office workers should have a building wiring system capable of supporting 1300 to 1500 users. This is very important since the number of SIO units installed will greatly affect the sizing of the entire building wiring system (horizontal distribution, riser systems, phone closets, etc.). This spare capacity should be distributed evenly throughout a facility.

7.4.2 Station Placement

It is essential that any building wiring plan stipulate that the placement of SIO units occurs in a consistent manner that provides the best possible user access and unit protection. User access refers to the ability of an end user to quickly locate and interface with the SIO. Unit protection ensures that the SIO is not placed in a location where, even though it is user-convenient, it will be subject to undue stress or physical damage.

The placement of an SIO is highly dependent upon the type of office environment in use. There are, however, some general guidelines that pertain to any station placement.

The SIO should never be placed within an outlet containing non-SIO equipment (specifically, any item pertaining to electrical power distribution or other system that may either cause signal disruption or cause accidental injury to a person during SIO installation or maintenance). As a general rule, an SIO should be placed within 2 ft, but not less than 1 ft, of any electrical outlet.

The SIO, where possible, should be placed in such a location as to be easily accessible to an end user. A user should not have to move or crawl under furniture to connect an outlet cable. This will also make troubleshooting and maintenance much easier.

The SIO should never be placed in any location where it will be subject to physical abuse. This includes locations directly at floor level or where items, such as furniture, are likely to be moved around with any frequency. In store- or stockrooms, for example, it may be necessary to locate an SIO near an electrical switch to prevent equipment from being stacked against it.

7.4.2.1 Standard office placement. A standard office is defined as any office with four fixed walls. The walls may be constructed of a variety of materials, including plasterboard, drywall, concrete, and wood. A standard office should have a minimum of one SIO per 60 ft^2 of open office space. For example, a 6- × 10-ft office requires a single SIO (yes, there are offices that small). Similarly, an office between 60 and 120 ft^2 should have either two or three SIO units.

SIO units should be located at least 1 ft, but not more than 1.5 ft, from the base of the floor. Units should be placed in the middle of open walls, preferably not a wall with a door or window (i.e., placement should be on the sides of the room). Multiple SIO units should be installed opposite each other such that user workstations or desks can be evenly located throughout the room. See Fig. 7.9.

SIO placement should be within 1 to 2 ft of a power outlet. This will provide adequate power supply for information terminals associated with the SIO. Note that SIO units should never be placed within 1 ft of

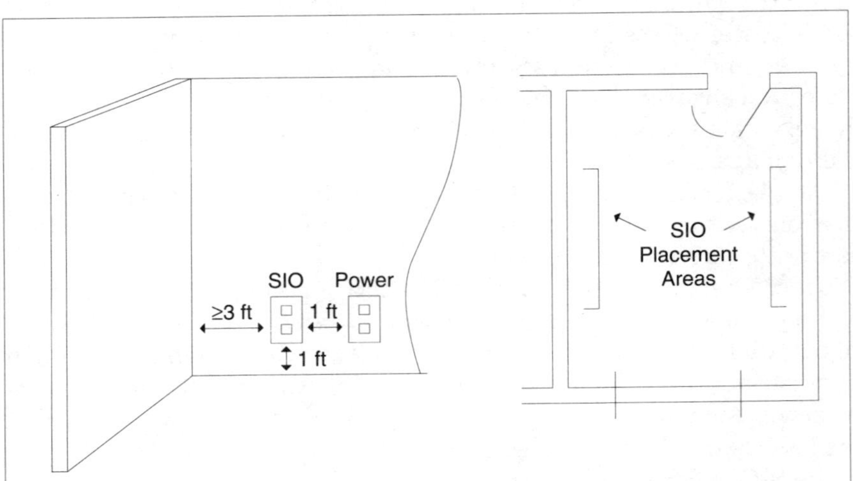

Figure 7.9 SIO placement.

a power outlet. The placement should be on a level line with power outlets whenever possible. For example, if power outlets are located 16 in from the floor, SIO units should also be placed 16 in from the floor.

The location of an SIO unit should be carefully planned and coordinated with the space or office planners associated with the building design. Such planning can sometimes lead to an enhanced work environment. For example, planning the location of SIO units that may vary from office to office can add aesthetic value to the office layout by not restricting all desks or workstations in each office to the same location.

7.4.2.2 Modular office placement. Modular offices are defined as offices composed of movable partition material or modular components (including walls, cabinets, workstations, etc.). Modular office furniture is becoming increasingly popular as organizations seek to minimize office construction expenditures. Modular offices are usually designed to support only one user at a time. For this reason, multiple SIO units are almost never required.

Since modular units typically have only three walls and a limited amount of available space, placement of the SIO unit is extremely important. There are two common locations for SIO placement. The first is under the desktop where a computer terminal or phone will be located. The height of the SIO is usually restricted to the placement of the premade SIO housing box, commonly in the metal cable conduit directly at floor level. When placed under a desk, it is important that the unit be placed slightly to the right or left on the wall facing the user. Such placement will ensure close access for information devices but also prevent the SIO and associated cables from being struck by users' feet or chairs. See Fig. 7.10.

The second location is just above the desktop or work surface. Such a placement will improve user access, which is sometimes difficult in crowded modular offices when SIO units are at floor level. The SIO unit should be placed within 6 in of desktop level in the nearest corner (right or left). It should not be placed near the end of a wall or in a location, such as near the edge of the desktop, where it may come into frequent contact with either the user or other objects, such as books, calculators, coffee cups, or other office equipment.

As with standard office outlets, any outlet located with a modular furniture unit should be within 1 to 2 ft of a power outlet. This will provide adequate power supply for any information terminals associated with the SIO.

Many new modular furniture units have the capability to support an in-bound SIO, which provides connection from the horizontal distribution system to the modular unit. The in-bound SIO is then con-

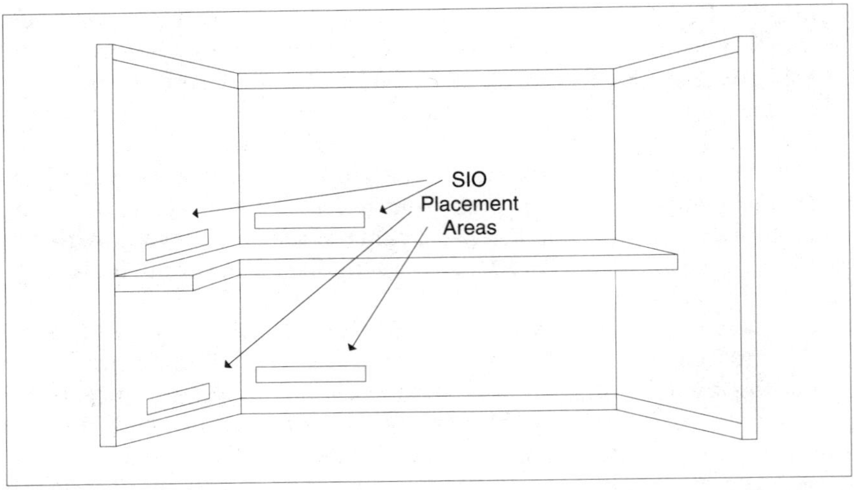

Figure 7.10 Modular SIO placement.

nected via internal cables in the modular furniture to the actual workstation or end-user SIO. In such situations, the in-bound SIO should be located in such a manner as to be easily accessible to the outside cable source, such as a power pole outlet. The in-bound SIO, however, should never be placed in such a location as to be openly susceptible to physical or environmental damage (i.e., on an open wall facing a hallway).

The use of modular furniture can create several problems with SIO placement and performance. Some modular furniture equipment does not provide utility boxes or outlet locations for flush-mount SIO units, thus requiring the use of surface mounts.

Many modular furniture units do not provide adequate space for cable installation. Additionally, many have sharp 90° bends that can damage cable during installation. Although prewired furniture may seem to eliminate this problem from an end-purchaser perspective, it is important to carefully check such products for the quality of both installation and cable characteristics.

Older modular furniture systems may not provide adequate separation (i.e., metal shield) between communications and power cable conduits. Units that route electrical cable through the bottom of the furniture and communications cable through the top are becoming increasingly popular as a solution to this problem. Additionally, many older furniture units can be easily retrofitted with top-mount conduit systems.

7.4.2.3 Power pole placement. A power, or utility, pole is defined as any unit that extends from the ceiling or floor to provide a pathway and outlet for electrical and/or communications cable. Power poles are extremely common in open offices and offices that use modular furniture. The following guidelines exist for the placement of SIO units:

- SIO units should be located a minimum of 1 ft and a maximum of 4 ft from the base of the power pole.
- SIO units should never share outlet boxes with electrical power receptacles.
- SIO units, where possible, should be located a minimum of 1 ft from any power receptacle.
- Power poles should be evenly spaced throughout an open office and never placed in such a manner as to impede access to doorways, stairways, windows, or emergency equipment such as telephones, fire alarms, security devices, etc.

A preferable configuration for a power pole is to have electrical outlets (when required) located within the bottom 1 to 2 ft of the pole while SIO units are located within the 3- to 4-ft range. The clustering of multiple SIO units in a small area is acceptable with power poles. However, care should be taken to evenly space the units or to place them on multiple sides of the power pole where appropriate (i.e., a single power pole centered between two to four desks).

7.4.2.4 Underfloor or undercarpet placement. Underfloor or undercarpet systems are defined as any cabling system which requires that communications cable be distributed beneath the carpet or within the flooring of a building. Such systems are most commonly associated with open offices; however, they can be found in many older facilities that have individual fixed-wall offices. As with power pole placement (an item often used with such systems in place of pedestals), the outlets should be evenly spaced throughout the office environment. Care should be taken to ensure that the SIO unit is not in a location that is subject to office traffic or obscured by office furniture. The general power guidelines described for power poles apply in this situation.

7.4.3 Special Cases

Any building wiring design is going to have special cases involving SIO units. These cases involve any situation where typical "office" environments do not apply. Some of the more common special cases, and recommendations as to how they may be handled, are presented in the following text. Note that multiple recommendations may apply to a single case (i.e., a public pay phone that is located outdoors will have two sets of guidelines listed below). Also note that the following guidelines are recommendations only and are not intended to fully address every possible configuration.

7.4.3.1 Reception areas. Most offices, whether commercial, health, or industrial in nature, have a separate location within the office dedicated to reception purposes. A switchboard console may or may not be located in this area. It is important that any reception area have a minimum of one SIO unit dedicated to providing telephone access for guests. This outlet should be totally separate from the SIO provided for the receptionist. The most common reception configuration offers one guest SIO near or at the reception desk while a second guest SIO is placed adjacent to a chair or couch.

In addition to telephone access, it may be necessary to provide special SIO facilities for security or control devices. Such special cases need to be clearly identified in advance of system installation to ensure that the special requirements of these systems can be met.

7.4.3.2 Switchboard facilities. Many offices have a room or area (possibly a reception area) dedicated to telephone switchboard or operator access. It is not uncommon for switchboard consoles to require multiple connections to the in-house PBX or key system. In such an event, multiple SIO units may be required. Multiple SIO units installed for switchboard access can be located in a single cluster (unlike multiple SIO units located within a typical office).

7.4.3.3 Public pay phones. Every office building should have some number of public access pay phones for use by both employees and visitors. The following guidelines should be considered when planning for pay phone service:

Public pay phones should be located in or near the lobby of any building. The phone should be wall-mounted if possible and located in such a manner as to not impede the flow of movement through the area.

All public pay phones should be readily accessible to all persons. This includes handicapped workers. This requires that some public pay phones be installed at such a height (approximately 3 to 4 ft) to be accessible from wheelchairs or by young persons.

Public pay phones should never be placed in such a manner as to prohibit access to emergency exits, such as doorways, stairways, and windows, or emergency equipment, such as fire extinguishers, fire alarms, and security equipment. Additionally, no public pay phone should be placed in such a manner as to put a user at potential risk (i.e., the placement of a public pay phone on the wall of an underground garage ramp).

7.4.3.4 Mechanical shop phones. SIO units located within any mechanical facility should be located a minimum of 10 ft from any industrial equipment or machinery. No SIO should ever be placed in such a location as to either interfere with, or be susceptible to, the normal working operations of the facility. Care should be taken to locate SIO units where they will not be exposed to adverse conditions, such as EMI, loud noise, or physical abuse. An SIO should be located in any "office" workspace that may be located within the facility.

For security and emergency purposes, there should be a minimum of one common access phone (or intercom) located within any closed-shop environment. Such an SIO and phone should not be a public pay phone unless there is no accessible organization phone system available.

7.4.3.5 Common access phones. In addition to SIO outlets located within individual offices, many organizations are locating SIO units in hallways or other common access areas such as cafeterias. Every public room (such as recreation room or conference facility) should have at least one SIO present. The location of these SIO units should generally follow the guidelines for standard office placement where applicable.

7.4.3.6 Outdoor phones. The placement of SIO units, usually for public pay phones or security phones, is becoming increasingly common within buildings or campuses. There are several important guidelines that should be considered when planning an outdoor SIO and phone installation.

Any telephone located outdoors on building grounds should be easily accessible from a common walkway or entrance. Ideal locations include open areas within 10 to 25 ft of building entranceways. Outdoor pay phones should never be placed closer than 10 ft from an entranceway to avoid congestion and ensure phone access during busy hours. Phones should not be located near any bushes or shrubs that could pose a security risk to phone users. SIO units should always be installed in well-lighted areas. Covered facilities are recommended when possible.

Any telephone placed outdoors should have adequate lightning protection. This includes surge and lightning protection on the cable serving the unit and protection from lightning striking the unit itself. This requires that any outside SIO unit not be placed within 25 to 30 ft of any large body of water (fountains, wading pools, swimming pools, etc.). Additionally, an SIO unit and associated telephone should never be placed in the middle of an open area; outdoor phones should be placed close to, or attached to, a building. Phones should never be located in the middle of an open area or attached to tall or thin objects that are likely to facilitate a lightning strike.

Any SIO unit located outdoors must be protected from the physical elements. This may require covering the unit or providing a waterproof enclosure. Additionally, an outdoor SIO unit should never be located or positioned in such a manner as to be susceptible to accidental physical abuse. SIO units mounted outdoors should generally be located a minimum of 4 ft from the ground to prevent dirt, water, or ice and snow from collecting within or around the unit.

7.4.3.7 Garage phones. Telephones, and thus SIO units, are commonly installed in garage facilities for a number of reasons, including staff access and security. There are three important locations for SIO units: garage staff locations (gate booths, repair facilities), exterior entranceways and exits, and interior entranceways and exits (stairways, elevator shafts, etc.). All covered or underground SIO locations should be brightly lighted and have lighted signs directing users to the SIO unit location. SIO units should never be located in driveways or areas where automobile traffic would pose a risk to users.

7.4.3.8 Recreational area phones. A recreational area in a building is any location where workers may gather for either recreational activities or a break. This includes areas such as cafeterias, smoking lounges, snack areas, gymnasiums, and exercise rooms. The placement of SIO units in such rooms can be extremely beneficial. For example, they can allow employees easy access to a phone system in response to a page, improving employee accessibility. Additionally, SIO units in these areas can provide emergency access if required.

7.4.3.9 Security access phones. Many organizations employ security access systems for buildings or individual offices. These systems may require SIO facilities for both electronic devices as well as phone access. In addition, the placing of an SIO unit directly outside the main doors of an office can provide after-hours visitors access to the organization's telephone system (most PBX systems can restrict specific phone extensions to internal calls only).

7.4.4 Surge Protection

It may be necessary to locate a surge protector in tandem with the SIO unit (separate from any power line protection). This is advisable for any SIO units that are served by horizontal distribution cables that run outside the building or for SIO units that service sensitive electronic equipment (i.e., computers or network communications equipment). Such protection will help to guard against powerful electrical currents passing through the communications cable into an end-user device.

Power surges on a communications line can be caused by lightning strikes on external (or improperly grounded) cables, accidental contact with power lines, or through EMI noise generated by power cables placed too close to communications cables. There are two types of physical units for this service: internal and external.

Internal units involve a surge protection unit built directly into the SIO jack. There are a number of vendors that provide this type of product. This type of unit is recommended for mass installations since the integral unit can be more cost effective to install than stand-alone units.

External surge protectors may be located either within the SIO box (or immediate area) or attached to the station or apparatus cable. This type of installation is most common where only a few devices within a facility require protection. Protectors located within the SIO unit (but not as an integral component) can often be difficult to install and do not offer easy access should the protector be triggered by a power spike or surge.

Station protectors can guard against rapid spikes in current as well as gradual increases. Regardless of the type of installation used, the protector should be properly grounded.

7.5 INSTALLATION GUIDELINES

The most careful planning and design can easily be rendered pointless by an improper SIO installation. For example, installing an SIO unit prior to the installation of the horizontal wiring can hamper the installation of cabling and can also risk damage to the SIO from repeated removal (during cable installation and testing).

7.5.1 Installation Scheduling

An SIO should only be installed after all major interior building construction has been completed and there is no chance of physical damage from either physical contact or environmental access (such as water or dirt). Additionally, the SIO should never be installed until the complete horizontal distribution system (including cable) has been installed. If an SIO is being installed into a new wiring installation, it is important that the unit not be installed until the horizontal distribution cable has been adequately tested for continuity, appropriate cable pairing (if twisted-pair), and electrical and light-wave characteristics. It is much better to find out that an SIO cable requires replacement before you have installed the SIO unit.

7.5.2 Configuration Testing

Prior to installation, all details concerning the pin or cable configurations should be established and documented. Failure to do so can result in SIO units being improperly configured. After installation and horizontal wiring connection, each unit should be physically inspected for proper pin and cable alignment. Additionally, a random sample of units should be tested for proper cable continuity.

It is important to also verify the physical integrity of the SIO unit. This includes an evaluation of the stability and alignment of the SIO in the outlet box (if appropriate) and an inspection of the modular components (i.e., a modular jack that snaps into place within an SIO faceplate).

7.6 SUMMARY

The following are the top key points of this chapter:

- The SIO is the main user interface into a building wiring system.
- SIO units can be configured to accommodate voice, data, and video in a single unit.
- Modular and multifunction SIO units can increase flexibility and options.
- The selection of an SIO must take into account all current and potential future requirements.
- The number of SIO units installed must take into account the maximum number of users permissible during the life of the wiring system.
- Proper system testing of SIO units should always take place prior to system cutover.

Chapter

8

Horizontal Distribution System

Definition: *Horizontal distribution* The distribution of telecommunications wiring along the horizontal axis from a centralized distribution point to individual SIO units.

The horizontal distribution provides connectivity, on a single floor, between the distribution frame or telephone closet and the associated SIO units. This typically consists of individual cables run from the central distribution frame to the SIO units. The types of cable used for horizontal distribution may include twisted-pair, coaxial, and fiber-optic cable. There are many other components in addition to cable, such as the distribution hardware, fire-stopping systems, and design layout, that make up the horizontal distribution system. Only by working together can these individual components form a solid horizontal distribution system.

8.1 CABLE SELECTION

One of the fundamental decisions that must be made in the design of the horizontal distribution system is the type of cable to be installed. This decision, in tandem with the type of jacks used in the SIO, is usually fairly independent of most other components in the building wiring system. The strong exception to this is the type of cross-connect equipment used in the distribution frame architecture.

8.1.1 Type of Media

The selection of a distribution media is based on the present and future requirements of the user's voice, data, and video systems. When evaluating a set of requirements, voice, data, video, and utility applications must all be taken into account.

The appropriate cable must be determined for any voice system in use. This is usually a very easy component to design since the majority of voice systems are based exclusively on twisted-pair cable.

The key to evaluating data requirements is to find a common denominator between the different types of data applications. The question that must be asked is, "Is there a common media that all my data applications can utilize, and if so, is this media acceptable for component installation?"

Video systems often use coaxial or fiber cable as a distribution means (the exception here is video signals that have been prepared for remote transmission over twisted-pair telco circuits). Finding a common media may be difficult since media standardization between video vendors is still taking place. Although many utility systems (security, HVAC, etc.) are not compatible with most telecommunications facilities, it is important to look for those exceptions that exist. One of the most common areas of interoperability is in the use of computer control and monitoring of utility systems.

Once the requirements of your information systems have been established, the selection of a media type is fairly straightforward: twisted-pair, coaxial, and/or fiber. The quantity of cable pulled to each SIO is based on requirements for individual applications.

Twisted-pair: Most standardized wiring plans in use today recommend the installation of between three and four twisted-pairs for each SIO for both voice and data.

Coaxial cable: Given practical constraints on the amount of space available within a distribution system (and in the SIO unit faceplate), only one coaxial cable is usually pulled to a single SIO.

Fiber-optic cable: Fiber-optic systems (the majority of which are half-duplex) usually require a minimum of two strands pulled to each location. Some devices, such as Fiber Distributed Data Interface (FDDI) dual-attached stations (connected directly to the FDDI ring) will require four strands.

The quality of the cable is very important. While many vendors can provide low-speed cable, far fewer vendors can provide high-quality cable that can support Ethernet, 16-mbps, 16-mbps Token-Ring, and 100-mbps FDDI (over twisted-pair). High performance cables, such as the Belden DataTwist Cable (1456–1458A) and the AT&T and Northern Telecom data-grade cables, are highly recommended.

8.1.2 Shielding Requirements

Many building environments dictate the use of shielding for twisted-pair installations [shielding is not required for coaxial cable given its unbalanced signal and immunity to most electromagnetic and radio frequency interference (EMI and RFI) sources—fiber-optic systems, being based on optical signals, do not require shielding]. There are many possible sources of EMI and RFI that can disrupt the signal on twisted-pair (see Chap. 2), including radio towers, power lines, and mechanical and medical equipment.

Unshielded twisted-pair cable installed in high EMI and RFI environments can result in poor or limited performance for data equipment [local area networks (LANs), data switches, digital private branch exchange (PBX) systems, etc.]. If you are designing a wiring system that involves close proximity to any of the above items, shielding may be required. Testing for high EMI or RFI levels can be performed by a specialty signal detection firm that will locate sources of signal leakage or EMI radiation.

In addition to large sources of EMI or RFI, placing an unshielded twisted-pair cable alongside a regular power outlet cable can also result in EMI interference. It is important to consider this when planning the distribution route for telecommunications cable. This is often a problem with modular furniture that does not provide adequate separation (steel barrier) between cable conduits for telecommunications and electrical wiring systems.

There are a number of systems that require shielded twisted-pair for normal operation. These usually require shielding to restrict the amount of EMI being radiated from the cable itself and not to protect the cable from outside EMI sources. Unless the manufacturer allows the use of unshielded cable (which usually requires additional signal filtering hardware), shielded cable should always be used.

There are a few physical characteristics of shielded twisted-pair that should be considered prior to the selection and installation of this type of cable: increased cost (because of an increase in cable materials), increased size of cable (an issue in small conduits), and difficulty of installation (because of rigidity of the shielding).

Additional information about the characteristics of shielded cable can be found in Chap. 2.

8.1.3 Plenum Requirements

As discussed previously, a plenum is an area or duct through which air is passed for circulation within a building. The installation of non-plenum-rated cable in a plenum environment is not only hazardous but against the law as well (see the National Electrical Code for more information). There are two methods for distributing a cable through an air plenum: the use of conduit or ductwork and the use of a plenum-rated coating on the cable. The use of conduit or ductwork can involve the installation of conduit from the distribution frame to the SIO unit, some additional design constraints as a result of the fixed nature of conduit, and additional cost expenditures of conduit installation. In contrast, the use of a plenum-rated cable can involve the use of an existing cable distribution system, a marginal increase in overall cable diameter (which decreases the cable flexibility), and an increase in the cost of cable (approximately $1/ft increase).

Of the two methods, the use of a plenum-rated cable is preferred over the use of conduit unless conduit is required for other reasons. The costs associated with plenum-rated cable, while higher than nonplenum cable, are typically not as high as the additional costs and loss of design and installation flexibility associated with conduit systems. The plenum ratings CL2P and CMP are certified for horizontal air plenum, CL2R and CMR are rated for vertical distribution, and CL2 and CM are rated for general open air distribution.

8.2 DISTRIBUTION SYSTEMS

The term *distribution system* refers to the type of physical hardware used to distribute the horizontal distribution cable from the distribution frame to the SIO unit. There are five major types of distribution systems: underfloor, ceiling, molding, undercarpet, and raised floors. The following section presents a brief discussion on the various types of distribution systems and their relative applicability in the typical building environment. Detailed installation and design criteria for such components may vary by manufacturer. The information and guidelines presented here are intended for generic use only and do not replace or overrule any manufacturer's documented instructions for proper system usage. Specifically, systems such as underfloor distribution, poke throughs, and any other system which physically alters any building component should not be installed unless structural building engineers have been consulted during the design process.

8.2.1 Underfloor Distribution

One of the most common distribution methods throughout the early and mid-twentieth century has been the underfloor distribution system. Although not common in new installations, this type of system is found frequently in building renovations and offers some distinct advantages for the right building environment. Under such a system (no pun intended), cable paths such as ducts, conduits, troughs, or raceways are placed within the solid body of a floor for cable distribution when the concrete is poured. Cable is then installed in the duct after the floor and building interior has been completed.

There are several advantages to the use of underfloor distribution hardware in a building wiring system. First, the underfloor system is hidden from sight, unlike other systems such as molding or power poles. Second, underfloor distribution systems are difficult to tap into since access to the floor is generally both difficult and obvious. Third, cable installed in underfloor distribution systems is not likely to be disrupted by construction or maintenance work such as the case with ceiling distribution systems. And fourth, underfloor distribution systems generally offer protection against many EMI sources as a result of the cable being encased in a metal "shield."

There are several disadvantages to the use of underfloor distribution hardware in a building wiring system. First, underfloor systems can often be difficult to access, which in turn can restrict the placement of SIO units. Second, since underfloor systems are generally installed during the pouring of the floor concrete, it is very difficult to expand the size or quantity of these systems. Third, any flooding of a floor (i.e., broken plumbing fixture) can result in a flooding of the underfloor distribution system (yes, this can ruin the cabling). And fourth, many existing underfloor duct and trough systems do not offer separate raceways, or paths, for electrical and telecommunications cables. The following is a discussion of characteristics typical of underfloor distribution systems.

8.2.1.1 Underfloor distribution hardware. There are several different types of hardware that can be used for underfloor distribution. The most common are ducts, conduit, and troughs. Each of these components can be used with any of the other systems to provide a complete distribution system. This is very important in large installations which may require multiple distribution components to satisfy diverse requirements. Each of the components is described as follows:

Underfloor ducts: An underfloor duct system consists of a rectangular metallic duct placed within a concrete floor structure. Duct hardware typically ranges in size from 1 × 3 in to 3 × 6 in. Within a duct, partitions may be installed to separate electrical and telecommunications cables. See Fig. 8.1. Although larger systems are available, they are usually not used because of the structural problems involved with placing such an object into a floor. There are two classifications of distribution ducts: main ducts and feeder ducts. Main ducts are large ducts that are used for the central distribution of cable from a distribution frame to central locations throughout a floor. A feeder duct is the duct used to distribute cable from a main duct to an individual SIO unit location. Main and feeder duct systems are connected through duct junction boxes. A junction box is a

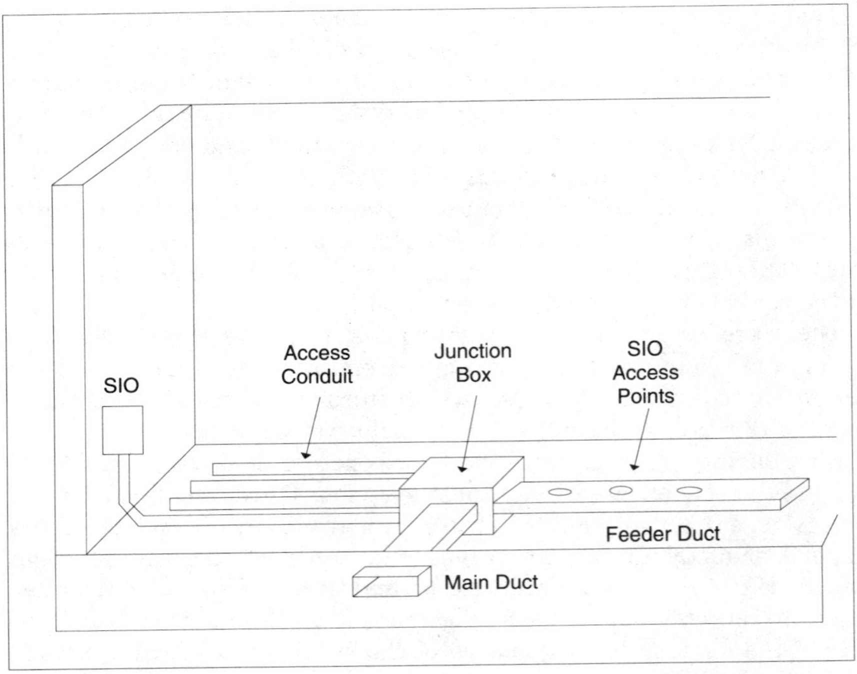

Figure 8.1 Underfloor hardware.

large box (a diameter twice the width of the duct) that provides interconnection between the two types of duct. Junction boxes are also used for connecting segments of similar duct and for access to the inside of the ductwork during cable installation. Preset openings called *inserts* (a prepunched disk averaging 1 in in diameter that can be knocked or twisted out) are generally placed in the top of the duct approximately every 12 to 36 in.

Underfloor conduit: Underfloor conduit consists of segments of circular tubing placed within the concrete of a floor during floor installation. This is very similar to underfloor duct systems. Conduit is generally ≥1 in in diameter and metallic in construction. Underfloor duct systems often use conduit to connect main or feeder ducts to individual SIO units (commonly via junction boxes).

Underfloor troughs: Unlike conduit and ducts, underfloor trough systems consist of cable troughs, or raceways, inserted into the top of a concrete floor. Trough systems are not totally enclosed by concrete; the top of a trough consists of a steel plate that acts as the top of the floor upon which carpeting or other flooring material is placed. Trough systems may be either rectangular or V-shaped in design.

With the exception of trough systems, underfloor distribution hardware can usually be installed in either one or two levels. A single-level system has duct or conduit on one level (two dimensions, X and Y). A two-level system has duct or conduit on two levels (three dimensions, X, Y, and Z where Z = depth). Two-level systems are usually connected via junction boxes or inserts aligned between the two systems. In most two-level systems, the upper level will follow one axis (i.e., X) while the lower level will follow the opposite axis. It is not uncommon for one level to consist of main distribution ducts and the other level to consist of feeder ducts or conduit.

8.2.1.2 Underfloor distribution layout. Given that underfloor distribution systems are usually installed during the construction process and cannot be easily altered, they are usually very orderly in their design and layout. The design of such systems is usually done on a zone basis where sections of the building are labeled by zone and fed by large main ducts or conduit. Within each zone, individual feeder systems reach out to individual SIO units.

Underfloor systems, often constrained by the floor structure they are placed in, are designed to provide grid coverage of all open floor space. The quantity and exact location of underfloor distribution systems depends upon the number and location of SIO units. A typical office layout is detailed in Fig. 8.2.

Underfloor ductwork should be designed to allow access within 1 ft of each planned workstation. In large open areas where worker floorplans are not available, the underfloor duct system can be de-

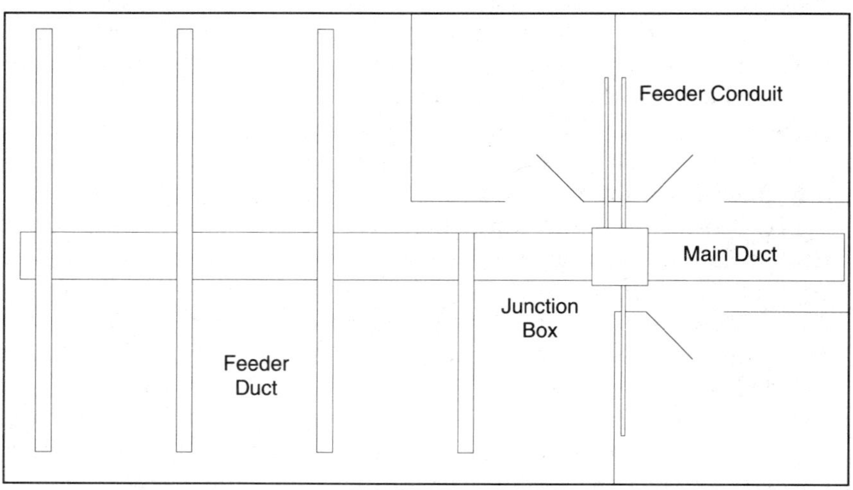

Figure 8.2 Typical underfloor distribution.

signed to effectively saturate the open area so that desk placement is not contingent upon distribution system. Caution should be used to not place feeder duct too close to any exterior wall for structural reasons (always consult a building contractor before designing or installing any ductwork around the perimeter of a building).

8.2.1.3 Underfloor distribution access. Underfloor duct systems can be accessed by three common methods: inserts, junction boxes, and termination boxes. Each of these systems has a particular function that is best suited to certain applications. These methods generally do not apply to trough systems, which usually offer open access to any portion of the trough (a big advantage during cable installation and maintenance). However, remember that trough systems are generally used as main distribution channels feeding many smaller feeder duct systems and thus will interact with the following access methods.

Inserts: An insert is a precut hole (disc) in the top of a duct that can be easily removed by either twisting or punching the disc. Access to the disc requires a portion of the concrete floor to be removed (chipped away) around the immediate access point. Inserts are usually used to provide access for SIO unit pedestals or power or utility poles. Inserts are generally not large enough to be used for cable installation or pulling purposes.

Junction boxes: Junction boxes are square or circular boxes located within a floor and used to connect segments of underfloor duct together. They are generally deeper than the duct they serve and are not covered by poured concrete. The top of a junction box usually consists of a steel plate or disc that is attached to the box with an inset screw or bolt. Junction boxes are usually a minimum of 6 in and a maximum of 18 in in diameter. Their function is to allow access to the inner duct system during cable installation (to facilitate cable pulling), to prevent sharp 90° bends in cable runs, and to facilitate cable fault detection and correction. Junction boxes are usually placed a minimum of 10 ft apart (but never more than 30 ft apart). Since the top of a junction box is flush with the top of the floor, it is important that any floor covering (i.e., carpet or tile) adequately cover the steel plate as well. Flooring can either be cut to fit the steel cover (common for tile or carpet) or can be placed over the entire surrounding region. The latter method is only recommended for use with carpeting. Regardless of whether a modular carpet or a large section of carpet is used with a flap opening, the edges of the carpet should have some form of resealing adhesive to ensure that the carpet does not pull up or curl at the edges. A junction box may also be used to connect two different distribution systems together. This is common in buildings on the West Coast where earthquake protection is built into many building wiring systems. Many large quake-resistant buildings are built in sections or modules that are on different slabs. This allows large flat buildings to shift up or down during a quake while minimizing structural damage. In such buildings, the border between building sections may contain a large junction box (i.e., several feet in diameter) where excess cable can be stored. The excess cable allows the building modules to rise and fall without breaking the cable, thus ensuring that interior wiring systems do not fail (although the exterior lines may not be so lucky).

Termination box: A termination box is a square unit used to terminate, or cap off, a duct. Such boxes can be considered a form of junction box since they allow a fixed, secure place for cable to be pulled from. In addition to providing an installation function, these boxes can also be used to house SIO units.

It is important to provide adequate access mechanisms to your underfloor duct system. Insufficient access points can severely increase the level of difficulty of cable installation or maintenance. Additionally, access points should never be placed in locations that could prove hazardous to users or maintenance personnel, such as in front of a stairway.

8.2.2 Ceiling Distribution

One of the most common distribution methods in use today is the ceiling distribution system, which relies upon cable paths such as ducts, conduits, or raceways located above a suspended (drop-tile) ceiling. Access from the ceiling distribution system to SIO units is accomplished through the use of power or utility poles, in-wall conduits, or other local distribution method.

There are several advantages to the use of ceiling distribution hardware in a building wiring system. First, the ceiling system is hidden from sight with the exception of some local distribution methods, such as power poles. Second, cable installed in ceiling distribution systems is generally safe from most office activities such as office furniture installation or regular office activities. Third, ceiling distribution systems, in conjunction with the use of suspended ceiling tiles, are generally easy to access for installation or maintenance purposes. And fourth, since the space above a suspended ceiling is generally quite open, distribution hardware can be easily installed or reconfigured. This flexibility allows ceiling distribution systems to adapt to new or evolving user requirements.

There are some disadvantages to the use of ceiling distribution hardware in a building wiring system. First, cable installed above suspended ceilings can be easily accessed and tapped unless the cable is completely enclosed in a conduit or duct. Second, while ceiling distribution protects cables from office activities, it does not afford the same level of protection for activities that occur above the suspended ceiling. Third, suspended ceilings often result in telecommunications, security, control, and electrical wiring systems being installed in a noncohesive manner where individual systems may encounter other systems in a random fashion, creating EMI. And fourth, much of the open spaces above suspended ceilings are plenum-rated due to their use for air circulation purposes.

8.2.2.1 Ceiling distribution hardware. Ceiling distribution offers perhaps the most flexibility of any hardware distribution system. This is a result of the characteristics of most drop-ceiling installations, which have a large quantity of open space with a fairly unobstructed path between most distribution points and end-user SIO units. The most common types of hardware used are ducts, conduit, ladders, and raceways. Several of these hardware systems are very similar to underfloor distribution and can be treated in a very similar fashion. Additionally, these components are very often used with other systems to provide a complete distribution system. This is very important in large installations which may require multiple distribution components to satisfy diverse requirements. Each of the components is described as follows:

Duct systems: Ceiling duct systems consist of rectangular metallic ducts suspended from the main structural ceiling above a drop-tile, or suspended, ceiling. Duct hardware, usually larger than underfloor duct, may be as large as 6 × 12 in depending upon the function of the duct (i.e., main distribution or feeder). The use of large ducts is limited only by the requirement for a stable support system for the ducts; each duct and related hardware component must be anchored to the structural ceiling with a metal support rod or metal strap. A single ceiling duct will usually not support both electrical and telecommunications cables. See Fig. 8.3.

There are two classifications of distribution ducts in use today: main ducts and feeder ducts. Main ducts are large ducts that are used for the central distribution of cable from a distribution frame to central locations throughout a floor. A feeder duct is the duct used to distribute cable from a main duct to an individual SIO unit location. Main and feeder duct systems are connected through duct junction boxes located throughout the duct system. A junction box is

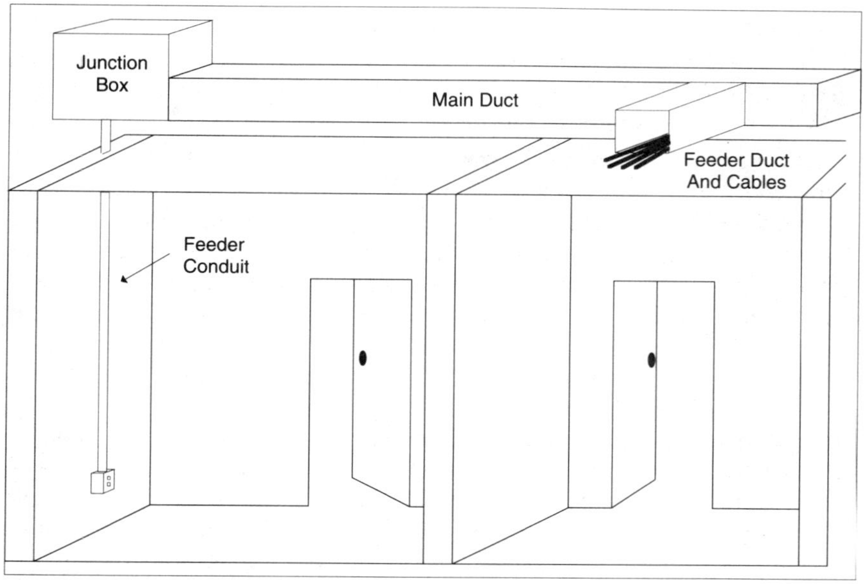

Figure 8.3 Ceiling ductwork.

a large box (a diameter up to twice the width of the duct) that provides interconnection between the two types of duct. Junction boxes are also used for connecting segments of similar duct and for access to the inside of the ductwork during cable installation. In addition to junction boxes, cut-out squares with removable covers are often used to provide access to cable pathways. Pull and splice boxes should be located approximately every 3 to 6 m along the cable path.

Ceiling conduit: Ceiling conduit consists of segments of metallic (or special plenum-rated hardened plastic) tubing suspended from the structural ceiling above the suspended ceiling. Conduit is generally between 1 and 5 in in diameter. Conduit is generally supported by support rods or straps attached to the structural ceiling above. Extremely large-diameter conduits are usually not used for main distribution in favor of duct systems. Small-diameter conduit is commonly used as a feeder distribution channel with duct used as the main distribution hardware. Access to cable pathways within a conduit are usually limited to junction boxes and conduit sleeves. A conduit sleeve is a movable sleeve that covers an opening or incomplete segment of conduit. The sleeve is moved to one side of the opening for cable access. Conduit systems are often used to connect main or feeder ducts to individual SIO units (commonly via junction boxes). Conduit systems should always have adequate pull-box locations used for cable installation and cable splices. Whenever possible, a sharp bend in a conduit (i.e., 90°) should be accomplished with large junction boxes and curved conduit. Pull and splice boxes should be located approximately every 3 to 6 m along the cable path.

Ceiling ladder: Ceiling ladders offer an easily accessible distribution method for main distribution cables. A ceiling ladder consists of a metallic ladder suspended horizontally from the structural ceiling above the drop-tile or suspended ceiling. The ladder is constructed of two supporting bars connected by a series of metal bars, resembling a climbing ladder. Ceiling ladders offer a large number of cable-tie points where cable can be tied or lashed to the bars or rails. Additionally, cable can be woven between alternating bars to provide additional pull reinforcement. Ladders generally range from ≥1 to ≤2 ft in width and are supported by support rods or straps attached to the structural ceiling above. Ladder distribution hardware is usually not used for feeder distribution because of its large size and support capabilities (it is primarily designed for large cable loads and is considered too costly and structural support overkill for small installations). As a result, small-diameter conduit, duct, or raceways can be used as a feeder distribution channel with ladders used as the main distribution hardware. Given the open-air nature of ladder systems, the use of plenum-rated cable is required whenever ladders are installed in plenum-rated environments. Ladders are most prevalent in large cable distribution rooms (such as computer or distribution frame rooms) where large amounts of cable must be routed from an originating device to a distribution system or between communications systems.

Ceiling raceway: Ceiling raceways, like ladder systems, offer an easily accessible distribution method for horizontal distribution cables. A ceiling raceway consists of a metallic tray (open on the top) suspended horizontally from the structural ceiling above the suspended ceiling. The raceway, consisting of a bottom and two sides, offers an open tray in which cable can be placed with a minimum of difficulty. Raceways are very similar to ducts and range in width from 6 to 12 in and are supported by support rods or straps attached to the structural ceiling above. Raceway distribution hardware, although lightweight and available in small sizes, is usually not used for feeder distribution. Feeder distribution in raceway systems is usually accomplished with small-diameter conduit or free distribution. Since raceway systems do not offer any cable enclosure, the use of plenum-rated cable is required whenever raceways are installed in plenum-rated environments.

Miscellaneous support hardware: In addition to the commonly used components listed above, a great deal of cable is installed using a wide variety of cable distribution components. These components usually involve a simple piece of hardware that is easy to install, cost effective, and easy to work with. Examples of this type of component include the numerous variations on the cross-bar support, which is a fixture that consists of a single horizontal bar (or hook) that is suspended from the structural ceiling via one or two support rods (or straps). Although this type of hardware does not offer any physical or plenum support, it is fairly common in small installations or buildings where cable distribution hardware must be added after building completion.

Three problems encountered with ceiling distribution hardware involve inadequate support, improper labeling, and interaction with power systems. The structural equipment used to support any ceiling distribution hardware must be more than adequate to support a minimum of 3 times the weight of any anticipated installed cable. Cable weight must take into account the possible use of plenum-rated or shielded cable, which may greatly increase the weight/foot ratio of the cable.

Any ceiling distribution hardware must be properly labeled as to the function of the hardware, including the type of cable installed and the areas serviced by the hardware, and to prevent the hardware from being used as a structural support for either persons or any other nontelecommunications function. Improper labeling can often result in personnel or hardware damage.

Equipment used for the distribution of telecommunications cable should never come in physical contact (within 6 in) with any type of power (electrical) cable. Electrical systems (including power cable or lights) should never rest on or be supported in any way by telecommunications cable distribution equipment. More information on this can be found in Chap. 8 of the National Electrical Code (NEC).

8.2.2.2 Ceiling distribution layout. Suspended ceilings offer an almost ideal environment for cable distribution. There is usually ample area with mostly unobstructed paths between any two points. Of course, there will always be ceiling obstacles such as fire walls, ventilation ducts, lighting systems, etc. However, these can usually be worked around in a very easy and straightforward manner. The key to successful ceiling distribution is a proper design layout.

There are two main types of ceiling distribution layout: freestyle and zone. Freestyle involves a direct point-to-point linking of distribution panels and SIO units. Zone distribution requires the division areas of space within a building into zones. Each zone is served by a bulk distribution channel. There are two subsets of zone distribution: slotted and block. Both are detailed in greater detail below.

Home-run connections. Independent of the type of distribution layout being used, there is one overriding design criterion for all distribution cable. Simply stated, all cables between a distribution closet and an SIO unit should be home-run cables, which are single cables, with no breaks or cross-connect points, between two locations. A distribution closet is defined, for this purpose, as any major distribution point, such as an intermediate distribution frame (IDF) or satellite distribution frame (SDF).

The intent of this "home-run" rule is to ensure that all cables in the horizontal distribution are clear and without any potential failure points (such as cable opens, shorts, line loss, etc.). This rule prohibits the placement of any type of splice or cross-connect between the distribution frame (DF) and the SIO.

As with any good rule, there is always an exception or two that seem to defy the logic of the original rule. And the home-run cable rule has one major exception. There are many older buildings that, because of previous design constraints, cannot support single cable runs for all SIO units. This will typically be a result of limited cable space in either conduit or duct systems (a single 100-pair twisted-pair cable takes up less room than 25 four-pair cables). Only when cable distribution systems do not allow for individual SIO cables should minor distribution points be used. While it is not possible in many older renovations or in historic building sites, every effort should be made to design a horizontal distribution system that can support single SIO cables.

When minor distribution points are necessary, there are some general guidelines that should always be followed. The minor distribution point should use a consistent type of cross-connect block, be located in a consistent and readily accessible location, be housed in a locking cabinet to prevent unauthorized access, and be clearly labeled as to its function and contents to prevent accidental mishaps. A minor distribution point should never be placed in such a location where accidental contact or exposure to environmental sources may damage the unit and should never be used for any other purpose other than the distribution of telecommunications cable.

Freestyle distribution. Freestyle distribution of horizontal cabling involves the installation of cable in a point-to-point configuration, typically a star, between a major distribution point and an SIO unit. Freestyle, most suitable for small installations, is so named since there are no set patterns or zone restrictions that must be followed for distribution layout. Cable is installed between two points in the path that offers the shortest distance or least resistance. This requires a complete, and unique, distribution path (i.e., hardware) for each SIO location. See Fig. 8.4.

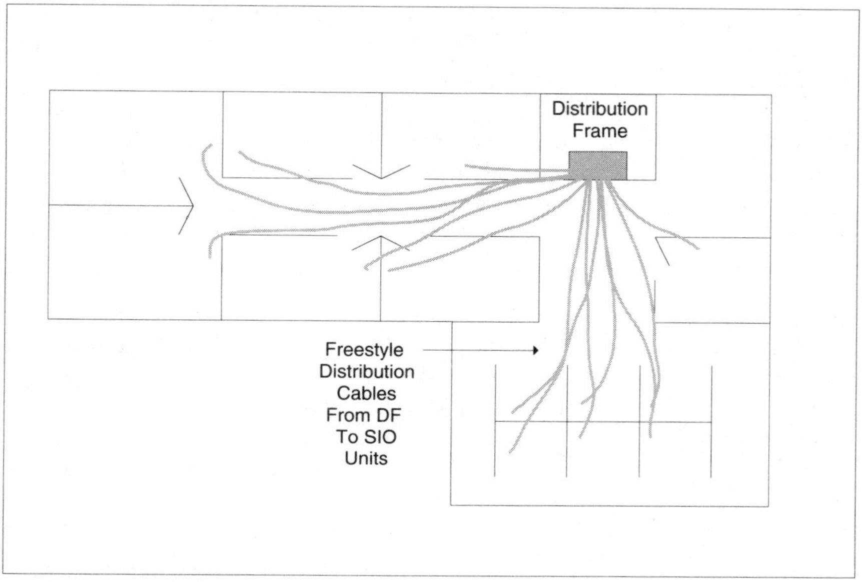

Figure 8.4 Freestyle distribution.

As shown in the figure, a freestyle installation is not constrained by any zone or fixed distribution layout. It is common, however, for individual distribution components to follow a similar path for ease of installation purposes.

Freestyle distribution layout is most commonly associated with either the use of conduit or the lack of any fixed distribution hardware. The association with conduit is due to the characteristics of conduit, which include a somewhat flexible media ideally suited to supporting a relatively minor number of small-diameter cables that would be associated with a single SIO unit. The lack of any fixed distribution hardware requires that individual cables be either supported by or fixed to miscellaneous support systems, such as steel rods used to suspend ceiling tile grids.

The use of a freestyle approach to cable distribution layout is generally discouraged except in those few instances where the building environment is very small, the number of SIO units to be supported is relatively few, and the ceiling distribution area is free of any major obstacles or points of electrical interference. Problems typically encountered with freestyle distribution include difficulty in fault location, EMI sources, and accidental damage. For example, cable that is installed indiscriminantly throughout a building can be difficult to locate. Also, when multiple cables are installed in a somewhat random fashion, the chances of a single cable encountering an unplanned-for

EMI source, such as fluorescent lights or high-voltage power cables, are greatly increased. Additionally, since it is very difficult to map out the installation of a freestyle distribution, it is common for maintenance repair personnel to accidentally damage telecommunications cable during the repair or installation of other ceiling-based equipment (i.e., A/C, power, or structural systems). The implementation of a freestyle distribution layout should only be accepted if the layout meets the size criteria established above, is cost efficient (including maintenance costs), and is clearly marked and documented.

Block zone distribution. The most common form of distribution layout in use today is block zone distribution. Under this type of layout, each floor, or DF serving area, is broken up into several blocks. Each block represents a single geographic region that often coincides with a particular workgroup. The size of a single block usually averages around 500 ft^2 but may be as large as 1000 ft^2 (only in very large buildings).

Each block zone is served by a single distribution path, such as a raceway system, that is responsible for delivering all horizontal cabling to a central point within the block zone. Under this form of distribution layout, individual SIO cables are used instead of bulk cables. See Fig. 8.5.

Once a cable has been delivered to the block zone, a secondary form of cable distribution is used to provide a path from the central point to the SIO unit. This secondary form may involve a combination of sev-

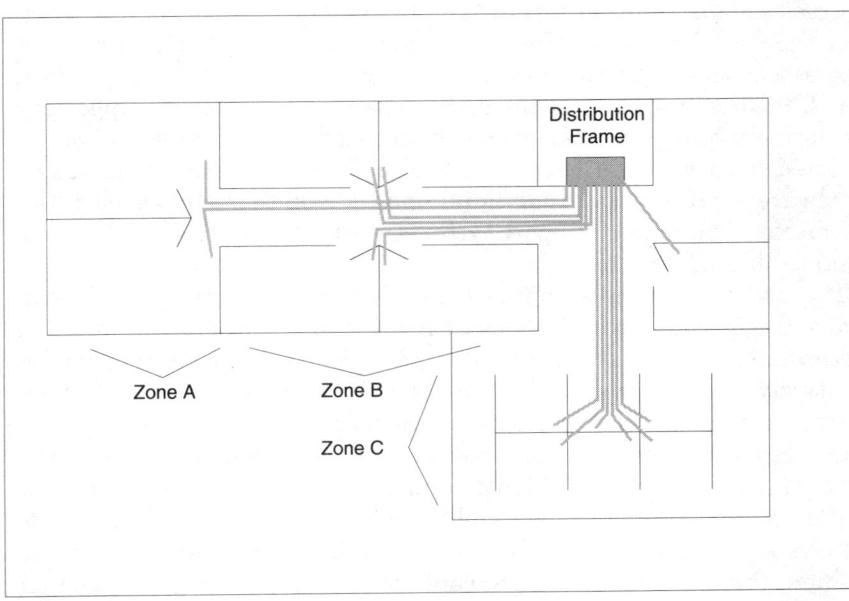

Figure 8.5 Zone distribution.

eral distribution methods, including conduit, raceways, ducts, and freestyle distribution. Given the difficulty of installing the secondary distribution hardware, it is advisable to keep the size of the serving area associated with any given block zone to a relatively small region (≤ 500 ft^2).

The block zone approach provides for a cost-efficient distribution that offers numerous advantages. For example, it is much easier to install a single larger distribution component or group of cables than it is to install many smaller individual components. Since the installation of additional cables after the fact can often be difficult, the initial installation of multiple cables in one single effort can help reduce installation time and costs. Additionally, a large group of cables is much more readily marked and identified than a large group of single cables, each following a unique path. This also helps to facilitate fault isolation and repair since fewer cable paths can result in improved documentation and control.

As with most forms of cable distribution, the recommended path for distribution hardware follows the layout of hallway systems. Such an approach usually results in easy access to the distribution hardware and cable (since individual office workers do not need to be disrupted to trace a cable through a building) and provides the best method for reaching the greatest number of SIO units in the simplest configuration (since that is the function of most hallways).

Slotted zone distribution. A less common form of zone distribution is slotted zone distribution, which is similar to block zone distribution in that specific geographical regions of a building are served by a single distribution path. However, where block zone distribution is designed to provide a single path to the center of an isolated distribution zone, slotted zone distribution is designed to serve a single slot, or spoke, region that originates from a main distribution point in an outward fashion to the perimeter of the building. See Fig. 8.5.

Slotted zone distribution uses a drop-off method of cable distribution where individual cables are "dropped off" from the main distribution path as they encounter their destination (SIO). This method works well in large square office buildings. Hardware distribution systems that are flexible in terms of access, such as raceways or ladders, are most commonly associated with slotted zone distribution.

Because of some severe design and access constraints, slotted zone distribution is most commonly used in conjunction with block zone distribution and is only used for locations relatively close to the originating DF.

8.2.2.3 Ceiling distribution access. One of the great advantages of ceiling distribution systems is the easy access afforded by the suspended ceiling. This not only increases the ease of system installation but also helps to promote rapid fault isolation. To further take advantage of the access offered by ceiling distribution, it is recommended that the design layouts used for ceiling distribution focus on common access areas, such as hallways or other open public areas.

Access to individual cables within a distribution hardware component depends totally upon the type of hardware being used (more information on individual component access can be found in the preceding sections). It is recommended that there be some method of cable access at least every few meters along the entire cable path. This applies almost solely to conduit and duct and is intended to facilitate the resolution of problems such as jammed or knotted cables. In addition, an ample supply of cable access points can be most beneficial if cable must be installed after a cable path is already in use (most cables are installed at the same time in bulk distribution hardware such as ducts or large conduit).

8.2.3 Molding Distribution

Many buildings are not constructed in such a manner as to allow conventional cabling distribution (i.e., underfloor or in ceiling). This is very common in older buildings, renovations, historic buildings, and special-use facilities, such as schools and maintenance shops, where walls may be either wood or concrete and ceilings are typically hard plasterboard. The use of molding systems for cable distribution can offer a fairly flexible solution to these types of building situations. Molding systems, which are attached to the inner surface of a wall or ceiling, can offer many advantages for the right situation.

Since it attaches directly to the face of a wall, it is fairly easy to install and no preparation of the ceiling or floor is required. Molding systems can be installed anywhere there is a fixed wall or ceiling surface. Compared to most other systems, molding can be very inexpensive. Many molding systems use a latch-cover approach where access to any portion of the cable within the molding is achieved by simply removing or lifting the outer cover of the molding. Additionally, molding is available in many different types of material and in many different configurations. This is ideal for older or historic buildings. For example, a woodgrain molding (emulating a crown molding) may be used to blend in with a natural wood wall surface.

While there are several advantages to the use of molding, there are also several important disadvantages. Since molding systems intrude upon the common open-user space within a building, there is a limited number of moldings that can be put in any one place. Where a large number of cables can be easily hidden within a ceiling, allocating the same amount of space on the wall of a hallway is often impractical.

Unless the molding is of extremely high quality and designed to blend into the environment, it can be very obtrusive and unsightly. Additionally, any cable distribution system that is installed on a wall (either in an office or a hallway) is at greater risk of accidental physical damage than a comparable system housed within a floor or ceiling structure.

Molding is very often used as a supplemental form of cable distribution to serve areas within a building that are not accessible via the main distribution hardware (i.e., conduit or duct within a ceiling).

8.2.3.1 Molding distribution hardware. Molding distribution hardware typically consists of ducts or conduit (squared) affixed to the interior (user) side of a wall or ceiling. Junction boxes are used to join sections of molding together and special corner boxes are used to prevent sharp 90° bends in the molding. There are two main types of molding in use today: tube and raceway. There are also many other specialty types of molding designed to support custom applications.

Tube molding. Tube molding consists of a series of small-diameter tubes connected together with junction boxes to form a distribution path. The tubes, typically 1 in in design (i.e., not round), may be manufactured from metal or hardened plastic. See Fig. 8.6. Given the relatively small size of the tubes in use today, a single tube usually cannot support more than two or three SIO units. Thus, the support of several dozen SIO units can require a dozen or more independent tubes. This can take up a lot of space and usually relegates the use of tube molding to either very small installations or as a secondary distribution mechanism for either raceway molding or other forms of distribution hardware.

Tube molding is perhaps the most flexible type of molding in use today because of its small size and ease of assembly and configuration. Many of the plastic molding systems available today can be easily cut to any length and combined with a vast number of attachment units (i.e., bends, splitters, junction boxes) by even the most novice cable installer.

Raceway molding. Tube molding consists of a series of raceways encased in a protective shell to form a distribution path. The raceways may be manufactured from metal or hardened plastic. Within the main unit, there may be several raceways, each designed to support a different type of cable or application. This enables a single molding system to support multiple SIO units, and it also segregates voice and data cabling and telecommunications cable from other types of cable (power, security, control, etc.). The dividers between different cable raceways may consist of solid wall partitions or may be nothing more

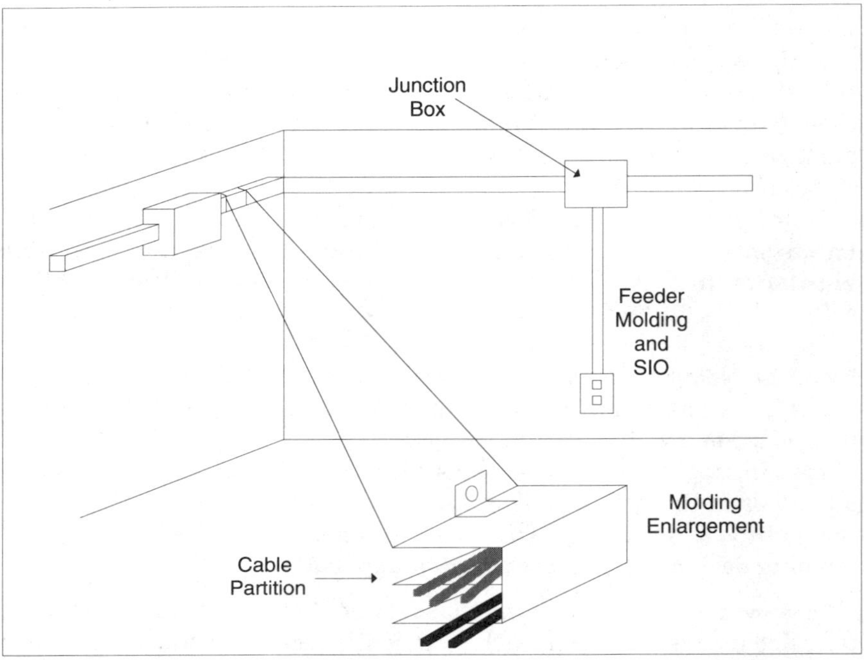

Figure 8.6 Tube molding.

than physical separation as a result of raceway placement (a raceway may consist of a series of bars or hooks on which cables are placed).

Raceway molding can be substantially larger than tube molding but is usually not more than 1 ft in width or more than 6 in in depth. Raceway molding systems are affixed to the wall with either bracket attachments or internal mounting screws.

Given the larger size of raceway molding, it is commonly used as a main or primary form of cable distribution and is often supplemented with tube molding for actual SIO integration. Like tube molding, individual segments of raceway molding are connected together with junction boxes. Unlike tube molding, however, raceway molding usually offers cable access along the entire molding path. This is usually provided by lifting up the outer shell or cover of the molding to expose the interior raceway systems.

8.2.3.2 Molding distribution layout. Molding distribution systems are, by their very nature, confined to a distribution layout that follows the configuration of hallways and offices within a building. The major layout considerations of molding distribution involve the use of raceway (primary) and/or tube (secondary) systems and the placement of the molding system (baseboard, crown, or mid-level).

As with any form of distribution system, it is advisable to construct the system so that main distribution channels are used to feed individual zones, and feeder distribution channels are used to connect to individual SIO units. Additional information on zone distribution can be found in Sec. 8.2.2.2.

Any type of molding layout in use should follow several basic guidelines that will help to ensure the viability of the distribution system. Each type of molding used must be securely fastened to the wall in accordance with the guidelines established by the molding manufacturer for the type of molding and the type of wall.

Molding should never be used for other, unintended, purposes or used to support other objects (pictures, other molding, etc.). There should be regular access to the cable at preset intervals (typically not more than 3 m). No main access point, such as a junction box, should be placed in such a manner as to be located in a heavy traffic area (i.e., top or bottom of a stairwell entrance) where maintenance personnel run the risk of either personal injury or injury to others in an emergency situation. Additionally, all connections from feeder or main molding systems to SIO units should be vertically aligned (diagonal molding across a wall is never recommended).

Baseboard molding layout. Baseboard molding layout places the molding system (either raceway or tube) along the baseboard of the wall, making the molding system relatively easy to access and install. A major disadvantage of this type of configuration is the relatively high incidence of physical contact associated with anything at or near foot level. Such installations are likely to encounter physical contact from users and furniture, not to mention physical elements such as water (flooding) or slush and salt (tracked in from users' boots during winter months, if you happen to live in a snowbound region).

Crown molding layout. Crown molding layout, the antithesis of baseboard molding, places the molding along the top of the wall. This type of molding is typically more difficult to install and maintain given its placement at the top of a wall. This can be acutely apparent in buildings with 10- or 15-ft ceilings. Access in these types of buildings requires ladders and can result in increased cost and time required to maintain the system. However, crown molding is much more secure from unauthorized access and physical damage than baseboard molding layout.

Mid-level molding layout. Mid-level molding layout permits the placement of cable at an appropriate level (not top or bottom) depending upon the given circumstances. This usually results in molding systems being installed between 3 and 6 ft from the floor. Please note that this applies only to horizontal molding designed for distribution.

Vertical molding, serving an SIO unit, must be installed in such a manner as to permit the location of the SIO unit in an ergonomic fashion. Mid-level molding is very easy to install and maintain given its placement at a comfortable work level. However, this type of molding is much more apt to incur physical damage than other forms of molding layout. Additionally, unless it is carefully controlled in regard to the size, color, and placement, it can often be aesthetically unpleasing. Mid-level molding is generally not used in most installations involving office environments.

8.2.4 Undercarpet Distribution

Undercarpet distribution is one of the less common, and more specialized, forms of cable distribution. This type of distribution involves the placement of special flat cables beneath the carpeting in an office environment. The cable associated with undercarpet distribution must be specially designed and manufactured to be as thin as possible so as not to be either disruptive to the carpet or subject to undue stresses from traffic on the carpet.

The placement of cables underneath a carpet requires a modular carpet that is installed in squares or strips that can be easily removed for cable access. For this reason, undercarpet distribution is generally used only for large open office environments. In addition, undercarpet distribution, given the flat requirements of the cable, is unsuitable for bulk cable distribution. This usually relegates undercarpet distribution to a secondary, or feeder, role to support a main or primary distribution method (typically underfloor distribution).

8.2.4.1 Undercarpet distribution hardware. As mentioned above, undercarpet distribution requires the use of specialized cable that has been designed as unobtrusive as possible. The cable is also designed to withstand the shock of daily foot, chair, and furniture movement associated with most offices. This specialized cable is generally available in short (≤ 10 ft) segments to promote modularity. The cable strips are relatively narrow (≤ 6 in) and designed to attach to other segments with simple snap-on components. As such, there is no major distribution hardware associated with undercarpet distribution systems.

There are several types of protective hardware associated with undercarpet distribution. These components fall into one of two categories: cable protection and connector protection. Cable protectors consist of flat segments of either plastic or metal that sit under and over the cable. These protectors are designed for installation in high-traffic areas, such as a main hallway. The protectors are designed to fit as an integral part of most undercarpet cable systems.

Most undercarpet cable systems have a segment connector approximately every 3 m. These connectors are constructed of metal and plastic and serve to connect multiple segments of undercarpet cable strips. These connectors are especially vulnerable to physical contact since they are made of breakable plastic snaps and plugs. Connector protectors consist of metal or hard plastic squares (or discs) that fit over the top of the connector. Like cable protectors, connector protectors are usually designed to fit as an integral part of the connector.

8.2.4.2 Undercarpet distribution layout. The physical characteristics and lack of a buildingwide hardware mechanism can restrict both the scope and layout of an undercarpet installation. Almost without exception, undercarpet cable is restricted to open hallway channels and offices. Such a configuration is required for installation and trouble isolation and resolution since cable is installed in short segments, all of which must be accessible.

The typical undercarpet layout centers around a zone approach with the main distribution mechanism involving either ceiling or underfloor distribution. Once primary distribution has taken place to a centralized zone location, a small cross-connect is used to connect the main distribution cable to the feeder (undercarpet) distribution cable. While secondary cross-connect points (secondary distribution frame) are generally not acceptable, they are required in this instance since the undercarpet cable is a specialty cable separate from the main distribution cable.

One of the strengths of undercarpet cable is its flexibility in small-area configurations. This includes small turns and bends. For larger, multipair cable, special modular units can provide an adequate turn radius. For smaller cables, 90° bends can be accomplished through a flip-twist of the cable. (Please note that the flip-twist can only be done with certain cables—only the manufacturer can tell for sure. Additionally, if you flip-twist once, you will probably have to flip-twist a second time to maintain the pin order of the cable.)

One of the difficult goals of an undercarpet layout is to avoid configurations in which wall partitions or furniture may obstruct access to any segment of the cable. This requires a detailed, orderly layout, even though the first design impulse may be to configure an open room in a star configuration similar to slotted zone ceiling distribution.

8.2.5 Raised Floor Distribution

Raised floor systems have become very common in most large office buildings. A raised floor consists of a platform floor (individual tiles) supported by a grid of structural stantions. It is commonly used in computer and PBX rooms where large cable distribution or special air conditioning requirements exist. Each floor tile is approximately 2 × 2 ft. Raised floor systems commonly have air intake vents cut into selected tiles through which cold air is drawn.

8.2.5.1 Raised floor distribution hardware. Most raised floor systems are placed approximately 1 to 2 ft above the base floor. This allows ample room for bulk cable distribution. Given the open space beneath such a floor, a wide variety of distribution hardware is available. This hardware virtually mirrors the hardware used for suspended ceiling distribution. There are two important points to note with raised floor distribution. First, many raised floors are considered plenum areas and require the use of plenum-rated cable or an enclosed distribution method (i.e., conduit or duct). Secondly, most raised floor installations are the result of a requirement for extreme flexibility in cable distribution. This results in free-cable distribution (without the use of any distribution hardware) in most existing raised floor installations. This is generally acceptable given the enclosed and limited nature of most raised floor rooms.

8.2.5.2 Raised floor distribution layout. The lack of any solid distribution hardware in most raised floor rooms necessitates a structured cable distribution layout. Individual sections of the room should be defined as zones wherever possible with main distribution channels established for common distribution. This is very similar to ceiling distribution and will help greatly in cable identification and fault resolution.

The most practical approach to establishing zones within a raised floor is to first segment computer and SIO systems by communications requirements and second by geographic areas. Since many computers in a raised floor room only communicate with one or two other systems (i.e., mainframes all communicate to a group of front-end processors), you may want to have separate distribution channels for point-to-point communications. It is important, however, to consider geographic location since you do not want a large number of point-to-point connections.

8.2.6 Power Pole Distribution

Power poles are extremely popular in cable distribution systems. A power pole is a unit that extends between the ceiling and floor to provide a pathway and outlet for electrical and/or communications cable. Power poles are extremely common in open offices and offices utilizing modular furniture and can be used in conjunction with virtually any type of main distribution hardware. This includes underfloor and ceiling distribution as well as undercarpet and raised floors to a lesser extent.

The function of the power pole is very straightforward. However, the selection of an appropriate power pole can be complicated by the extreme number of variations in pole design and configuration. The following are guidelines for the general selection and design of power poles for horizontal cable distribution:

Prewired units with preconfigured snap interfaces to the wiring system should be avoided whenever possible. The use of a prewired unit adds a cross-connect point, which can add to line loss. Additionally, prewired units can be difficult to work with for fault isolation and cable replacement. The only real exception to this guideline is when power poles are being used with a modular distribution media such as undercarpet cable, which typically cannot interface with a power pole without an adapter connection.

Power poles should always have separate distribution raceways for telecommunications and electrical cable. This includes separate outlet boxes for the two types of cable whenever possible.

8.2.7 Poke-Through Distribution

It is important to discuss poke-through distribution, which involves the boring or cutting of a hole in a wall, floor, or ceiling in order to poke a cable through. This method of cable distribution is often used with other acceptable distribution methods. Unfortunately, poking holes in building structures is not a generally acceptable thing to do. This can result in structural damage to a building and can create fire hazards; it is not recommended except in the rarest of cases when it is the only acceptable method for distributing a cable to a given point and that point cannot be moved to a different location. Additionally, making holes for cable distribution implies an unstructured approach to building wiring that can lead to installation and management difficulties.

If a poke-through approach is to be used, there are two activities that must take place. First, a qualified building engineer must be consulted to determine if there is any possible risk of weakening the structure of the building. Second, the hole itself must be sealed (and fireproofed if applicable) to prevent any possible water or environmental damage. Failure to perform these two activities can result in an unsafe and unmanageable wiring system.

8.3 DESIGN CRITERIA

There are certain design criteria that apply to the majority of horizontal wiring installations. These criteria apply regardless of the type of distribution method being used or the size of the installation. While

OPTIONS	BUILDING TYPE			
	Open Ceiling		Fixed Ceiling	
	Non-Plenum	Plenum	Open Floor	Solid Floor
Ceiling Duct	Y	Y	N	N
Ceiling Raceway	Y	Y*	N	N
Ceiling Ladder	Y	Y*	N	N
Ceiling Conduit	Y	Y	N	N
Undercarpet	Y	Y	Y	Y
Underfloor	N	N	Y	N
Molding	Y	Y	Y	Y
* Requires Plenum Cable				

Figure 8.7 Distribution options.

there may be other important points to ponder during the design of a distribution system, the following most certainly cannot be forgotten. More detailed information on most of the following topics can be found in the preceding sections since they relate to specific types of distribution hardware and techniques.

8.3.1 Distribution Hardware

Figure 8.7 shows a sample hardware configuration based upon building type and layout.

8.3.2 Cable Selection

It is important for cable selection to be based on all current and future applications. This includes voice, data, video, and, where applicable, security and control systems as well. Plenum-rated cable (or encasing distribution hardware) should be used for cable placed in any plenum environment. Shielding should also be used for any building environment where EMI is a major concern (power facilities, machine shops, factories, etc.). Failure to select the proper cable can result in an inadequate system that almost certainly will be costly to upgrade in the future.

8.3.3 Power Integration

8.3.3.1 Underfloor distribution electrical integration. The distribution of cable in underfloor conduit or ductways may provide some amount of shielding against external EMI sources. However, telecommunications cables are still very susceptible to EMI sources located with the duct or conduit. Any underfloor system used should provide separate ducts or partitioned areas for electrical and telecommunications cabling. The requirement for this is clearly identified in Chap. 8 of the

NEC. This requirement may force modifications to underfloor systems where possible (such as troughs that offer complete access).

8.3.3.2 Ceiling distribution electrical integration. As with any type of communications cable, it is imperative to keep both communications and electrical power cables in their proper places. Although there are modular equipment vendors who offer integrated solutions, they should be avoided for two reasons. First, you do not want to have an electrician who knows nothing about telecommunications cable being forced to interact with your communications wiring because it happens to be in an integrated distribution platform. Secondly, you never want to have a telecommunications staffer involved in any cable system carrying live voltage. Minimum separation levels should be established and adhered to on an organizationwide basis. This will not only improve the integrity of your telecommunications wiring system but also will make it easier to install and maintain.

This requires (in accordance with the NEC) that power-bearing cables be physically separated from telecommunications cables. Some of the more common electrical integration problems associated with cable distribution include improper placement of cable alongside power or lighting equipment and the placement of the two different types of cable in a single distribution component. Basic guidelines for integrating telecommunications and electrical wiring in a ceiling area include:

Cable integration: Never house telecommunications and power cables in the same distribution hardware unless the hardware is specifically designed for cable integration and offers adequate physical separation (i.e., metal barrier).

Proximity to lighting: Telecommunications cable should never be placed directly above or rest on any type of light fixture. This is especially true for fluorescent lights, which are extremely disruptive to telecommunications signals (beware of older "ballast" type lights).

Cable proximity: Communications and power cable should never be tied together or use the other cable for support. This not only makes sense from an EMI perspective but will also help to prevent potentially life-threatening accidents should a nonelectrical maintenance person come in contact with a live power system. The best way to avoid accidental contact is to clearly plan and label distribution channels for both types of cable. The rule of 6, which gives inches per kilovolt-ampere, should be used as a general guideline when possible.

8.3.3.3 Undercarpet distribution electrical integration. The building constraints that force the use of undercarpet cable may also force the use of underfloor power cable as well. Fortunately, many undercarpet systems are designed to support both types of cable. When selecting an integrated cabling approach, the use of independent undercarpet strip systems for each type of cable is recommended.

Most cases, however, will not involve any large-scale integration of the two cable types. For example, many undercarpet installations are a result of insufficient underfloor duct systems; the ducts are usually still used for power distribution. Additionally, the vast difference between the majority of power installations and the configuration of undercarpet cable virtually rules out any significant interference between the two cabling systems. Nonetheless, the guidelines set forth for minimum cable spacing (see Chap. 5 for the rule of 6) should always be followed, as well as any applicable codes or regulations.

Please note that the above guidelines are not intended in any way to supersede or replace any local, state, or federal codes or guidelines regarding the integration of power and telecommunications cables.

8.3.4 Sizing

One of the most consistent trends in telecommunications is that applications always get larger. Very rarely do we see an application, such as LAN access or the number of workstations, decrease. Additionally, more and more offices are down-sizing the amount of space allocated to individual workers. Many firms looking to conserve capital or add staff without increasing office space are using modular offices in an open configuration to replace the traditional single office approach.

When you are planning a distribution system, it is important to design the system with sufficient capacity to not only handle the current applications and office load but to also be usable when you either add staff or add cabling requirements for the new LAN or data system.

The first goal of horizontal distribution design is to determine the number of cable pairs that must be installed. This number should be based directly on the number of SIO units planned for current or future installation. This number should be checked against the total building occupancy limits to make sure that the number of SIO units is not being underestimated. It is usually much more reliable to make a decision on a fixed number that cannot be exceeded (the occupancy limit) than on a number that is most always exceeded (the office manager's floor plan).

Spare cable should be installed whenever possible and practical. For example, spare cable can be installed with ceiling distribution ducts or raceways without significant overhead. Conversely, installing additional cable in a slotted zone conduit system may prove more difficult (a strong argument for not using home-run conduit over more open distribution channels such as duct or raceways). When spare cable is installed, it should be pulled to a centralized location with its given distribution zone and left in a coiled spool. The length of the cable should be sufficient to reach the farthest point within the zone. This may not always be possible with many distribution forms, such as underfloor ducts, but should be followed whenever there is adequate hardware.

The second goal of the horizontal hardware design is to determine the quantity of distribution hardware that is required. This usually requires that the type of hardware be predetermined. However, given the cost differences between the various forms of distribution hardware, the type of hardware selected can often be influenced by the quantity of cable being installed. For example, installing an enclosed duct system in a ceiling can be considered cost inefficient when the pair count installed is relatively small.

When designing the layout of the distribution hardware, it is important to have sufficient hardware so that any single component does not exceed 60 to 70 percent of capacity (50 percent is optimal). Leaving excess room in the hardware will make additional cable installations or troubleshooting much easier than if the system is filled to capacity. This spare capacity should be evenly distributed throughout the building and not concentrated in any single component or location.

In order to properly design a distribution system with spare capacity, it is important to know the type of cable being installed. This includes not only the base media (i.e., coaxial, fiber, twisted-pair) but also the shielding, plenum rating, jacket type, and pair count since these factors can greatly change the diameter of a given cable, which directly affects the number of cables that can be placed within a distribution component. For example, a 1-in conduit can usually handle five or six typical four-pair cables. *Typical* refers to 24AWG unshielded twisted-pair found in the majority of installations. If you add shielding to the cable, however, the same conduit may be able to only support four or five cables. Changing the cable from 24AWG to the thicker 22AWG can also reduce the number of cables. Additionally, bundled cables that include multiple types of cable (such as IBM Type 2 cable, which includes two shielded 22AWG pairs and four unshielded 24AWG pairs) can be difficult to work with because of inflexibility and increased outer sheath thickness.

As a general rule, larger-pair cables can provide significant pair advantages over smaller cables. While a 3-in conduit can support 25 to 30 unshielded 24AWG pairs in individual sheaths, the same size conduit can support over 300 similar pairs encased in a single sheath. Please note that this pair increase should not be used to support mixed-pair distribution between a DF and an SIO (i.e., using a 100-

pair cable for main distribution connected to 24 individual four-pair cables for feeder connection to the SIO). It should also be noted that squared duct, raceway, and ladder systems can generally support larger quantities of cable than rounded conduit systems, especially when small-pair (four or fewer) cables are being used.

The only way to accurately determine the quantity of hardware required is to first determine the quantity and type of cable to be installed and then compare this with the various systems, and their relative cost, offered by established hardware vendors (i.e., compare various duct offerings). This decision is usually independent of the type of hardware being used, which is based upon the building characteristics. However, it can be an influencing factor and should be considered when possible.

If no cable type has been determined, the hardware distribution system should be designed to provide support for a minimum of six cables, four pairs of unshielded twisted-pair each, to every potential SIO location. Providing for such a cable installation will help to ensure that the distribution hardware can meet at least the minimum requirements likely to occur with most cabling systems today.

8.3.5 Building Segmentation

Many buildings are constructed so that the total building is made up of individual segments, or modules. This segmentation is usually transparent to the common office worker. Constructing a building in segments can be the result of structural add-on architecture, such as occurs during renovation and restoration processes. As an old building is updated and modernized, new segments are often added to increase the overall floorspace. Many new buildings are intentionally designed in this fashion to compartmentalize the building utilities, such as power and water service. This is popular in large facilities that house multiple long-term tenants. In such arrangements, each module within the building is constructed so that it could stand alone without the other modules. This typically includes a solid wall partition, except for doorways, between different modules.

Another reason for building large facilities with multiple modules is to enhance the survivability of the structure in case of natural disaster, such as an earthquake. A good example of this would include some of the newer office buildings in such earthquake-prone areas as Anchorage and San Francisco. It is not uncommon to find buildings designed in these areas with a modular approach to structural support. The benefit of modular segmentation is that any one segment can rise, fall, or move laterally (to a certain extent) without the collapse of the entire facility.

The modular design of a building poses some interesting architectural and wiring situations. The most pressing issue with building wiring design in such a facility involves the module-module interface. If the buildings are designed to allow some level of flexibility between modules, the interesting question is "What happens to the distribution hardware or cabling that passes between or connects the two facilities during a live test of this modular survivability theory?"

Given that communications can be of vital assistance in any crisis situation, it is up to the building wiring designer to ensure that the wiring structure can function properly during times of stress. This requires a flexible interface between building modules. This flexibility requires modification of both hardware and cabling systems.

The hardware system, duct, conduit, or other method, must provide a movable interface. This means that any openings between the facilities used to house conduit, for example, must be large enough that minor relative movement of the two structures will not damage or crush the conduit. With underfloor distribution systems, this usually requires an underfloor junction box that spans the two modules. Again, these junction boxes should be both wide and deep enough to allow minimal movement without cutting any cable installed between the two modules.

Where cabling is concerned, the only reliable way to ensure that cable is not stressed or severed during up and down or lateral movement is to provide a supply of excess cable length at each module interface. This typically consists of a few feet of coiled cable within each junction box. Again, it is preferable to have a single junction box that crosses both modules (split in the middle) than to have two completely enclosed boxes on either side of the module interface.

As with any building constructed of multiple segments, there may be fire-code regulations pertaining to the use of any common interface wall as a firewall. Local building codes should always be consulted before installing distribution hardware between building modules.

8.4 INSTALLATION GUIDELINES

Once a horizontal cabling system has been designed, the installation process must begin. It is important to install both the hardware and the cable at the proper time to prevent unnecessary work or system damage. Additionally, there are certain activities, such as fire-stopping, checking for asbestos in renovations, and system testing, that should be performed for many installations.

The following sections set forth some basic guidelines that should be followed when planning an installation for horizontal distribution.

8.4.1 Installation Scheduling

A common cabling fault involves the timing of when a cabling system is actually installed. The installation of a system too early in the building construction cycle can result in damage to the system as construction continues within the building. Likewise, if a cabling system is installed too late in the cycle, it can be overly costly and time consuming (imagine trying to install a ceiling distribution system after the ceiling is in place).

The following guidelines should be considered to ensure proper distribution hardware installation:

Hardware installation: Hardware should be installed only after the building exterior has been completed and the risk of damage to the system by adverse weather conditions has been eliminated. This may not seem a major problem; however, excess water in a conduit or duct system can alter the electrical characteristics of cable placed within the hardware. In any event, only rust-resistant metal should be used.

Ceiling installation: Ceiling distribution systems should be installed prior to ceiling tiles being installed. It is preferable for the system to be installed prior to the ceiling tile grid installation but not mandatory.

Conduit and duct installation: Any conduit or duct system installed within a wall (i.e., feeder conduit between a main duct and an SIO unit) should be installed prior to the drywall being installed.

Pull cables: An appropriate pull cable (12AWG or larger) should be installed in any enclosed distribution hardware (duct, conduit, or molding).

Undercarpet installation: Undercarpet cabling should only be installed after the building interior is fully completed. This type of system is easily installed after carpeting has been put in place (assuming that removable carpet tiles have been installed rather than single-piece or glued carpet).

Underfloor installation: All underfloor ducts should be installed during the floor pouring process. This may seem a bit obvious, but it is surprising the number of telecommunications personnel who inquire after a building has been occupied if they can bore holes into concrete floors for conduit or duct. This also applies to trenches that are inset into concrete floors. Although it is possible to cut trenches into established concrete floors, it is often expensive or prohibited because of structural support requirements.

Cable access: Any distribution hardware should provide sufficient access (≥ 2 ft) into any DF location to provide adequate access to cable installers. The exception to this is conduit which should be terminated within 1 ft of the entrance unless the conduit must reach a specific room or frame location prior to termination.

Hardware finishing: After the installation of any hardware system, all rough edges, such as a connector or pullbox joints, should be filed, deburred, or covered with plastic coating (plenum-rated if required). Smooth grommets should be installed in any location where cable must pass through rough-cut metal, such as the entrance into a duct system or where cable may pass through metal studs or partitions. Failure to perform this activity can result in cuts into cables, which can sever a conductor and/or diminish performance characteristics.

Labeling: All telecommunications-related hardware should be clearly labeled as such and free from any unauthorized cable or equipment (it is surprising how many A/C and power systems are tie-wrapped to telecommunications ladders or conduit for support).

Stress testing: Any hardware installed in a ceiling should be stress-tested to ensure that the hardware is capable of supporting the maximum cable weight as specified by the hardware manufacturer (it is a good idea to add 25 or 50 percent to this weight given system overloading and those surprise cases where the system must perform an unexpected function).

Equipment grounding: Any metal distribution equipment should be properly grounded if it terminates/originates in any common distribution room (and if required by local, state, or federal codes).

The following guidelines should be considered to ensure proper cabling installation:

Installation timing: Cable should be installed only after the building exterior has been completed, the distribution hardware has been completed and stress-tested, and the interior of the building completed to such a point where the risk of damage to the system due to adverse weather conditions has been eliminated.

Cable protection: Any cable installed while building construction is still in process should be adequately protected. This includes cables left exposed during the installation process.

Simultaneous installation: All cable to be placed within a single distribution component should be installed at the same time when possible. This may not always be possible for mixed-media installations, such as twisted-pair and fiber. However, it should be noted that it can be very difficult to install cable into a system already housing cable because of decreased pull-room, the risk of cable knots (where multiple cables get woven together), and the increase in cable friction between established and new cables (relative to cable being installed in a properly designed and lubricated distribution hardware component).

Cable testing: Proper cable testing, as described in the following sections, should always be performed immediately prior to cable installation.

Grounding: Any grounding requirements of a cable should be satisfied as soon after cable installation as possible. This will ensure that the procedure takes place in an orderly fashion and will help facilitate proper testing procedures.

Cable stress: Care should be taken to not overstress any cable during installation. Stress can occur when a cable is pulled with too much force or when a cable is pulled around a sharp bend (where pull boxes should be used). Special tension meters (similar to a spring scale) are available to measure the amount of tension being placed on a cable during installation.

Equipment access: Sufficient excess cable should be available at each end of the horizontal distribution. Approximately 2 to 3 ft should be available at the SIO unit and approximately 15 to 20 ft should be available at the DF location. This spare cable will not only help to make the installation process easier but will also allow system reconfiguration in the future. Each time a reconfiguration takes place, old cross-connects must be trimmed back, shortening the excess cable. Additionally, spare cable will allow cross-connect blocks to be placed in the most work-efficient location, which is typically not next to the cable access point.

The above guidelines are intended to represent a list of items that should be considered during installation. They do not represent a complete set of all activities nor are they intended to supersede any codes or regulations pertaining to building wiring design or installation activities.

8.4.2 Fire-stopping

One of the most important, and simplest, activities that must be performed after a horizontal building wiring installation is the fire-stopping, which is intended to prohibit the spread of a fire from one location to another. As such, fire-stopping involves the filling of any holes in building materials that occurred as a result of hardware or cable installation. An example of areas that should be fire-stopped includes any point where conduit or duct passes through a wall or floor. Any hole left open could allow flames or smoke to pass through.

It is important to note that fire-stopping a hole in a wall or floor is not the same as fireproofing or fire-rating (where a type of material is certified to withstand flames for a certain period of time). While some may argue that only holes in fire-rated or fireproof structures should be fire-stopped, there are benefits to fire-stopping any hole. These advantages include prohibiting air or moisture flow, soundproofing, and (the most overlooked) keeping small rodents from using conduit or duct access points as miniature superhighways.

Fire-stopping applies to any break in a wall or floor. This is extremely important for complete breaches that pass through an entire structure as well as partial breaches that only pass through one side of a structure.

8.4.2.1 Fire-stopping materials. There are two common types of fire-stopping systems available today: preconfigured and flexible. Both of these product classifications are monitored and evaluated by many organizations, such as the American Society for Testing and Materials (ASTM), the Underwriters Laboratories (UL), the Canadian Standards Association (CSA), and the National Fire Protection Association (NFPA—through the NEC).

Preconfigured fire-stopping systems involve mechanical equipment designed to provide a fire-resistant seal around cable and conduit. This type of system is usually only used for situations involving a complete breach since a mechanical device is required on either side of the wall or floor (as an option, the preconfigured unit can be inset into the structure, thus requiring only one unit). In addition to providing fire-stopping protection for a breach in a structural component, preconfigured systems can also be used to house a complete structure, such as a splice or junction box.

Flexible fire-stopping systems involve several types of materials which can be fitted either into or around a cable passageway. There are many different types of recommended materials that can be used for this purpose, including the following:

Putty or caulk: This type of fire-stopping material consists of a moldable putty that can be inserted into holes and sealed tightly around the base of a cable, conduit, or duct. Putty can be purchased as either curing or noncuring. Noncuring putty maintains its softness and can be removed or reshaped without significant effort at a later time. Curing putty hardens after it is set in place (air contact) and can be more difficult to remove. Curing putty may be desirable in situations where movement of the cable or distribution hardware may not be acceptable. Similarly, the noncuring putty allows slight movement of distribution hardware without breaking the seal or diminishing the effectiveness of the material. A disadvantage of curing putty is the fumes which are given off by some puttys during the hardening process (some putty simply gives off bad odors all the time).

Foam: Some of the more common fire-stopping products are foam-based. Foam systems rely upon a hardening foam which is inserted (via pressure) into a hole or breach. These types of systems can be very effective but tend not to allow for additional cable installation (the foam usually cracks or crumbles under stress). One strong disadvantage of foams is that they require special pressure application equipment and can sometimes deliver noxious fumes during the curing process.

Barriers: Physical barriers, constructed of fiberglass, ceramic, or other fireproof material, are often used to cover open breaches around conduit or cable. Often used in conjunction with foam or putty systems, a barrier system provides an outer protective shell which can enhance the appearance and improve the outer structure of such systems.

In addition to the fire-stopping materials listed above, each of which uses a foreign material, there are several "similar-construction" materials in use today, materials that match the content and structure of the wall or floor being fireproofed. For example, if a hole is drilled through a concrete floor, it is not uncommon to see additional concrete poured into the opening (around the cable or conduit) to provide a similar-construction fire-stop.

It is strongly advised that a registered professional familiar with your building environment, local regulations, and current product offerings be consulted before any decision is made regarding the breaching of a firewall or the selection and installation of fireproofing materials in any wall or floor structure (fire-rated or not).

8.4.3 Testing Activities

Whether you have just completed the installation of a horizontal distribution system or you are in the process of planning or installing the system, it is important to consider how it will be tested. This includes both the physical hardware and the telecommunications cable. The following is a list of suggested testing procedures that should be followed for inspecting and testing the distribution hardware:

Structural tests: The physical distribution system should be tested for structural integrity (support capabilities) prior to any cable being installed.

Proper labeling: An inspection should be performed to ensure that the physical hardware is clearly labeled as telecommunications-related.

Hardware inspection: The entire system, including junction and pull boxes, should be visually inspected for any of the following: environmental damage, rust, improper fittings, rough edges, excess moisture, and contact with other systems (electrical, security, etc.), where appropriate.

Grounding: The system should be checked for proper grounding, where appropriate.

Fire-stopping: Any fire-stopping seals should be carefully inspected. This includes any areas that have been sealed with any material for non-fire-stopping purposes as well.

SIO connection: The connection point between the SIO and the distribution hardware should be carefully inspected to ensure that the connection is both properly fitted and that the distribution hardware does not prevent the SIO outlet box from properly attaching to the supporting structure (wall, floor, ceiling, power pole, etc.).

The list does not include every possible testing procedure; it includes those that should be performed prior to installation of any cabling system. The following is a list of suggested procedures for inspecting and testing the horizontal distribution cable after installation:

Junction inspection: Each junction box should be inspected (where possible) to determine if cables have been scraped against any surface—look for sheath shavings.

SIO inspection: There should be ample spare cable (2 to 3 ft) beyond the end of the distribution hardware to allow for simplified installation and later retermination if required. The spare cable can be coiled within the distribution hardware or the housing structure (i.e., in the drywall).

Continuity testing: Cables should be tested end to end for signal continuity. This is accomplished with either a tone tester or a proper polarity tester. While these both check for signal continuity, the polarity test is preferred since the tone test can give misleading results, due to signal strength, between multiple pairs of cable. For example, a tone being sent down one conductor can often be detected on other conductors because of crosstalk signals. Another method of testing the cable involves the use of a time-domain-reflectometer (TDR) which sends a signal down the cable. By measuring the delay in the signal reflection (and the waveform of the reflection), the TDR can determine how long the cable is and if there are any short or open points along the circuit. This can also be used to test the electrical characteristics (including background EMI noise) of a copper cable. Fiber-optic cables should be tested for both signal integrity and signal strength (i.e., dB loss) with an optical TDR (OTDR).

The real test, of course, is whether or not your network system actually functions as it should. However, the tests and inspections listed above will help to ensure that the cabling plant is fundamentally sound at installation time, thus improving the chances of detecting and correcting installation and cabling flaws.

8.5 SUMMARY

The following is a summary of the key points contained within this chapter:

- The selection of a horizontal distribution cable should be carefully matched to existing and future requirements, including voice, data, and video equipment as well as building and environmental constraints.
- There are many different types of distribution hardware and layouts which can be custom-mixed to best suit the users' site-specific requirements.
- The complete horizontal distribution system should be properly inspected prior to acceptance and actual use.
- All building-related codes should be followed at all times during the design and installation process.

Chapter

9

The Distribution Frame

Definition: *Distribution frame*
A centralized facility from which telecommunications cable is distributed to other centralized locations within a campus, building, or floor or to end-user SIO units.

Just as the horizontal distribution system can be considered the spokes of a distributed star topology, the distribution frame (DF) can be considered the heart of the hub. It is from the DF that all cables emanate. This includes to other DFs, through vertical distribution, as well as to individual SIO units through horizontal distribution. Commonly referred to as a *telephone closet* or *cross-connect room*, there are four acceptable variations on the DF. These are as follows:

- *Campus distribution frame (CDF):* The centralized location in a campus facility where all distribution cables originate
- *Building distribution frame (BDF):* The centralized location within a single building where cable is distributed to intermediate distribution frames on each floor
- *Intermediate distribution frame (IDF):* The focal point on each floor where all horizontal distribution originates
- *Satellite distribution frame (SDF):* A small distribution point that serves end users beyond the distance limitations of an IDF

The function and placement of DFs throughout a building or campus are based on a hierarchical model similar to a star or tree distribution. All DF facilities originate from a large central point that feeds smaller decentralized DF points. See Fig. 9.1. Each of these DFs shares many common characteristics. However, there are some characteristics, such as location, layout, and functionality, that differ be-

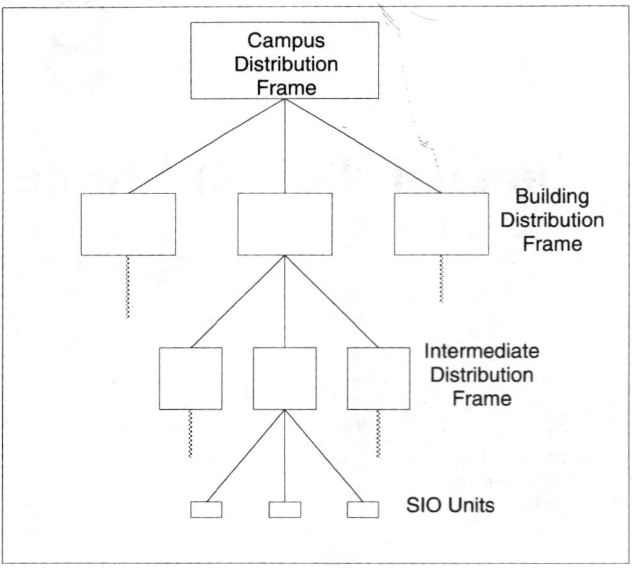

Figure 9.1 Centralized tree DF structure.

tween the four variations. Each of these similarities and differences are discussed in further detail in the following sections.

9.1 DISTRIBUTION FRAME CHARACTERISTICS

9.1.1 Distribution Frame Functionality

Any DF must perform a minimum set of functions that relate more to applications than to actual equipment, and may vary considerably between distribution frame types. They are:

Cross-connect access: The primary function of any DF must be to house appropriate cross-connect equipment supporting the building wiring system.

Communications equipment: A DF must be capable of supporting communications-related equipment, such as modems and communications servers.

Control equipment: A DF can include control equipment, such as security devices, if the devices use the telecommunications wiring system within the building.

Some of the functions that a DF must never be used for include the following:

Power equipment: The DF should never house power equipment unless it is directly related to the operation of the housed communications equipment.

Environmental equipment: Unless special A/C for communications equipment is required, there should never be any form of environmental control equipment present.

Storage facilities: Many DF rooms end up cluttered with excess equipment or non-communications-related items. A DF must be kept clean and orderly.

Nontelecommunications usage: A DF should never be used for nontelecommunications functions, such as the DF/office, the DF/coat room, and the ever-popular DF/rest room (which has happened before).

Please note that the above guidelines may be modified for special cases. However, if a room is shared between telecommunications and nontelecommunications functions, every effort should be made to provide as much physical separation as possible.

9.1.2 Distribution Frame Equipment

A DF will usually contain a fixed set of equipment. The amount of equipment may vary as the size of a DF increases. However, the following components can be expected to be associated with most DFs.

Access point: Each DF will have a single (usually) location where main distribution or telco cables enter the room.

Cross-connect blocks: There will be multiple sets of cross-connect blocks used to terminate incoming and outgoing cables.

Patch panels: Patch panels, providing modularity between cabling or communications systems, have become a mainstay within most DF rooms.

Distribution hardware: There will be miscellaneous equipment such as cable guides or raceways used to route cabling within the DF room.

Networking equipment: It is not uncommon to find networking-related equipment within a DF. This includes modems, multiplexers, and local area network (LAN) hubs.

Communications power supplies: Many high-usage communications devices will require an uninterrupted power supply (UPS) to filter power and provide emergency power in case of failure. Additionally, many newer communications devices require special power (i.e., non-110 V or conditioned) to operate.

Line protection: Most DF rooms will require line protection on all incoming circuits. This protection will help protect end-user and centralized equipment from power or lightning surges that could result in equipment loss or accidental fires.

9.1.3 Distribution Frame Room

While the rooms used to house a DF may vary considerably between installation size and location, there are certain parameters that should always apply:

Single-use facility: The DF room should be dedicated to communications functions. There should not be any equipment present that is not directly related to telecommunications (i.e., HVAC ducts, steam or water pipes, electrical cabling).

Security: Each DF room should be secure in nature with locking doors and limited access through windows (windows should never be put into a DF room). Security control will help to ensure that maintenance personnel or employees do not unintentionally, or purposefully, alter the cabling configuration.

Access: Each DF room should have adequate doorway access to enable the installation of communications equipment. Double removable doors (3 by 6.5 ft each) are usually adequate.

Nonfluorescent lighting: All DF rooms should have adequate standard incandescent lights installed around all entrances and work areas. Fluorescent lights should always be avoided because of their high electromagnetic interference (EMI) emanations.

Dedicated (clean) power: Each DF room should have dedicated power circuits unique to communications support. This will help to prolong the life of any power-drawing communications equipment placed within the DF.

Common power: In addition to a dedicated power supply, each DF room should have an adequate supply of common power outlets for use by maintenance personnel during equipment installation or maintenance (i.e., power tools, exhaust fans).

Grounding system: Each DF room should have an approved isolated building ground (common within a building). The ground must be easily accessible with 6AWG cable support.

Tiled flooring: A DF room should always have an unfinished concrete (painted) or tiled floor. Carpeted floors should never be used. Although some designers advocate the use of antistatic carpets, these carpets tend to be expensive (they have metal grounding coils running through them) and difficult to clean. DF rooms tend to have a great deal of cable clippings on the floor. It is much easier to sweep an unfinished floor than it is to vacuum a carpeted floor (not to mention the implications of vacuuming in a communications room—who vacuums, how much electromagnetic interference, or EMI, will the vacuum cause, etc.).

Structural support: A DF room should be able to support a minimum of 50 psi. This is very important for larger facilities that may house power or large private branch exchange (PBX) or computer equipment.

Nondrop ceiling: Drop-tile, or suspended, ceilings should not be used within the confines of a DF room. It is preferable to have direct floor-above access. Additionally, if the building has a plenum-rated ceiling, the plenum-rated area should be sealed from the DF room.

Fire protection: Any room used for housing DF facilities should have adequate fire prevention and protection systems in place. This will ensure that damage to telecommunications equipment during a fire will be kept to a minimum. Remember that the ability to communicate in any type of disaster can be crucial to minimizing the impact of the disruption. Also note that many control systems (security, fire alarms, etc.) that depend upon electronic communications may have cross-connects located within, or may pass through, a DF facility. As such, all DF rooms should have walls and doors with a minimum 1-hour (hr) fire-wall rating. Additionally, all finished walls within a DF room should be painted white with a flame-retardant paint. Every DF should have some form of fire-suppression device. Larger DF facilities, especially if they house computer or PBX equipment, should have a Halon or similar suppression system. Smaller DF rooms should have at a minimum one hand-held fire extinguisher, properly rated for electrical equipment and located near the entranceway.

Environmental control: Any DF room should maintain a minimum set of environmental parameters. Temperature should be maintained between 10 and 30°C (50 to 85°F) with a noncondensing humidity level of less than 65 to 70 percent. Higher humidity levels can result in corrosion of cross-connect equipment. Temperatures higher, or lower, than the acceptable range can result in improper operation of communications equipment.

9.1.4 Grounding

It is important for any DF facility to have access to a proper building ground. This will be used to ground communications cable and equipment within the DF room. There should be a single common ground, isolated from any other function, within each building dedicated to communications functions. The ground should support access via 6AWG properly bonded cable. There are many different types of grounds available for use today. The following is a list of the more acceptable building ground systems (a certified electrical engineer, with experience in building grounding, should always be consulted prior to a ground being designed, installed, or used for any purpose):

Cold water pipes: This provides the easiest, lowest-cost, and possibly most-effective source of building grounding. This requires access to a cold water pipe, constructed entirely out of steel, that has a minimum soil penetration, without breakage, of at least 3 m. The pipe should also be located at least 2 to 3 m below ground level.

Buried grounds: A buried ground is a large piece of steel that has been buried in open soil. This steel ground may consist of rods, plates, or coils. The size or length of the buried ground will depend upon the type, shape, and thickness of the metal being used (consult an appropriate electrical engineer for exact information on grounding requirements). These grounds must typically be placed at least 3 m below the surface and 3 m apart from any building or other ground source.

Building steel: It is possible, in some cases, to use the building steel infrastructure as an acceptable ground. This precludes the use of the building steel for any other grounding purpose. The use of the structural building steel as a ground may require special planning during the construction phase to ensure that the building steel meets the criteria for adequate soil penetration and electrical dissipation characteristics.

It is important for all communications facilities to be connected to the same ground. Using multiple grounds within a facility can result in ground differentials (resulting in electrical flow, or current) between hardware systems.

9.1.5 Multiuse Facilities

It is common in large buildings for multiple tenants to occupy portions of the same floor. More importantly, these tenants may be located within the same service area of a single DF. Tenants may require access to the IDF located both on their floor and in the BDF. Furthermore, tenants may wish to house equipment within these facilities and may be unwilling to allow open access to their facilities for both control and security reasons. When this occurs, there are two viable alternatives: a shared DF architecture and a multiple DF architecture.

The first alternative involves joint tenants sharing the use of a single DF. This can work only if the DF is structured in such a way as to offer some level of separation between tenant equipment. This requires providing a means of securing the cross-connect blocks associated with each tenant. Individual sets of cross-connect blocks should be established, if possible, and enclosed in lockable metal cabinets to restrict access (this does not prevent access to cables outside of the cabinets).

The second option involves the physical separation of DF facilities. In this scenario, individual organizations are given SDFs located in separate rooms adjacent to the serving IDF. The IDF serves as a distribution point only to the individual SDFs. All cross-connects are performed in the SDF rooms with distribution cables returning to the IDF for access to the building distribution system.

Tenants who occupy a significant portion of a building may need a similar joint usage arrangement with the vertical riser system and the BDF. Arrangements for the BDF can be similar to the IDF. The riser, however, is more difficult and will almost always require a dedicated cable for each user (see Chap. 10).

9.1.6 Secondary Distribution Frame

Many existing building wiring systems in use today use what is called a *secondary distribution frame* for horizontal distribution. The term *secondary* refers to the use of a DF method other than one of the acceptable primary methods (i.e., satellite, intermediate, building, campus). Secondary distribution frames usually involve one or two cross-connect blocks connecting a single multipair horizontal distribution cable (i.e., 25, 50, or 100 pair) to multiple three- or four-pair cables for connection to individual standard information outlet (SIO) units. By using a single high-pair cable for the majority of distribution, the amount of distribution hardware (size) can be reduced.

One of the more common applications for which secondary DF units are used involves modular furniture. A single bulk cable is pulled from the serving IDF to a modular furniture cluster. The distribution cable is cross-connected within the cluster (via a secondary DF) to short distribution cables that attach to the SIO units within each modular desk.

There are several big disadvantages to the use of secondary DFs in a standardized building wiring system, such as:

Location disadvantage: In order to realize the advantage of a single main distribution cable, the secondary DF must usually be located physically within a workgroup. This results in the cross-connect equipment being located under desks or in other unsuitable locations which can be hard to access and difficult to secure.

Line loss disadvantage: The additional cross-connect point within the secondary DF adds to the dB or signal loss of the distribution cable. This additional loss can be disruptive to high-speed networking systems that are extremely line-loss sensitive.

Failure point disadvantage: The secondary DF adds a potential failure point through the requirement for an additional cross-connect or splice.

Management difficulties: Adding an additional cross-connect point increases the difficulty of properly tracking all cross-connects through the horizontal distribution system. This can also make fault isolation more difficult because of the number of potential failure points.

Security disadvantage: Since these secondary DF units are very rarely located within a secure facility or cabinet, they are subject to all manners of accidental abuse.

Cable incompatibilities: Any electrical characteristic discrepancy between cable types can result in an impedance mismatch. This mismatch can contribute to line loss and is very common between bulk distribution and SIO access cables.

The above disadvantages are usually significant enough to effectively prohibit the use of secondary DF units in most installations. If a secondary DF must be used, perhaps because of extremely diverse physical constraints or limited distribution facilities, precautions should be taken to minimize any of the above potential failure points.

9.2 CROSS-CONNECT BLOCKS

The cross-connect architecture implemented within a building wiring system can literally make or break the success of most future networking implementations. While the SIO unit and the horizontal and vertical distribution systems remain relatively static after installations, the cross-connect is in constant use as new connections are made and old ones are torn down in support of ever-changing voice and data network requirements.

9.2.1 Block Placement

The placement of cross-connect blocks within a DF must be both organized and consistent throughout the DF room and the entire building or campus cabling system. A structured approach to cross-connect block placement will allow simplified installation of cable and facilitate management throughout the system. For example, if all cross-connect blocks are installed in the same manner throughout multiple DF rooms, a cable installer or maintenance person will not require special instruction in equipment layout. The areas that, if addressed in the initial design, can provide this structure include mounting equipment, voice and data alignment, and block placement within the DF room.

9.2.1.1 Mounting equipment. All cross-connect equipment should be either wall- or rack-mounted. Wall-mounted equipment should be mounted on 3/4-in plywood backboards measuring 4 by 8 ft. The plywood should be horizontally hung with the bottom of the plywood approximately 0.5 m from the ground and to within 0.5 m of every wall (it is good practice to place plywood backboards on every wall within a DF room). If the walls of the DF room extend above 3 m, the backboards may be hung vertically; however, no cross-connect or communications equipment should ever be placed higher than 2 m off the ground.

The plywood should always be painted with a flame-retardant paint; no backboard should ever be left untreated. In no case should any cross-connect block or communications component be installed directly onto a DF room wall, including drywall, concrete, or plywood.

Wall mounting can prove suitable for most types of equipment, including cross-connect blocks, fiber-termination equipment, splice cases, patch panels, and most other types of communications support equipment. Larger equipment, such as multiplexers and LAN hubs, can be mounted on plywood backboards through the use of wall-mounted shelves (no equipment should ever be placed directly on the floor of any DF room unless the room has been specifically designed for such activities, such as a computer room).

Rack-mounted equipment should use standard 19-in equipment racks. Cross-connect blocks, best suited to wall mounting, can be mounted on either plywood or sheet metal attached to the equipment rack if wall space is limited (plywood is preferable). The rack equipment should be at least 2 m in height and should be firmly bolted to the ground. There should be at least a 0.5-m clearance on either side of the rack and at least a 1-m clearance on the front and rear of the rack.

9.2.1.2 Block alignment. A structured approach to cross-connect block alignment is required in all DF rooms. This requires both vertical and lateral alignment. By convention, voice blocks are mounted above all other blocks, including data and video equipment. Mounting should be consistent in this fashion. While the vertical alignment separates cross-connect blocks by voice and data affiliation, the lateral configuration separates blocks from incoming and outgoing distribution systems. See Fig. 9.2.

This vertical and horizontal separation of equipment may vary depending upon the type of media being used. For example, fiber-optic equipment—usually dedicated to data applications—is typically housed alongside of twisted-pair voice applications and not underneath as twisted-pair data cross-connect blocks would be located.

9.2.1.3 Equipment placement. Once an established pattern of cross-connect block alignment has been achieved, it is important to physically allocate space within the DF room, which will most certainly

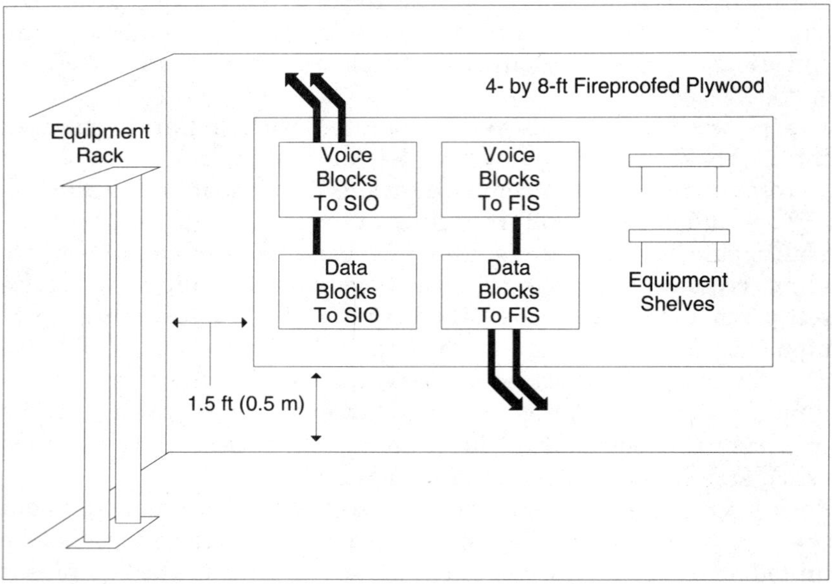

Figure 9.2 DF mounting.

vary depending upon the type of DF room and the equipment being used. There are, however, some general guidelines that will always apply:

Block placement: All wall-mounted cross-connects should be mounted on a single wall if possible. Multiple types of cross-connect blocks (i.e., coaxial, fiber optic) should be mounted on different walls prior to a single-block architecture (i.e., twisted-pair) being split between different walls. No cross-connect equipment should be placed within 1 m of a cable access point. Equipment racks should not be placed closer than 2 m of an access point.

Power: Power outlets should never be mounted in the middle of a backboard unit. There should be at least one duplex power outlet for each three backboards installed. The outlet, for maintenance usage, should not be more than 1 m from the centrally located backboard.

Communications equipment: Wall- or rack-mounted communications equipment should never be mounted higher than 2 m from the floor. This will allow ease of installation and access for equipment maintenance.

Cable access point: All riser cables should be located in one area, preferably in a corner near the entranceway on the shortest open wall. Horizontal distribution cable entrances should always be located in a corner, on the same wall as the riser entrance if possible. Cables should never enter the room in the middle of a wall; this will force the segmentation of the wall and make a structured approach to cross-connect block design difficult.

9.2.2 Cross-Connect Cable

Cross-connect cable is used to establish connections between incoming and outgoing cables attached to individual cross-connect equipment. It may consist of twisted-pair, coaxial cable, fiber-optic cable, or other specialty cable (such as a silver/satin patch cord). Any cable used to connect two different segments of cable together should match as closely as possible the electrical and optical characteristics of the cables being connected (some cable characteristics cannot be matched, such as shielding).

It is recommended that different colored cable be used for connecting different types of systems. For example, voice systems can use white and blue cable while data systems can use white and green; control systems that are integrated into the telecommunications wiring can use white and red. Patch cables, if they are being used, can also segment voice and data applications by cable color.

9.3 NETWORK ACCESS POINT

Within any building or DF facility there is usually a single location where cable enters the facility. This location is called the network access point (NAP) and may require special design, location, or equipment depending upon the type of facility being served. It is very important for proper design and management that there be only one NAP per DF room or building. The use of multiple access points within a single facility can make management difficult and add cost to the overall building wiring system through the requirement for additional cable, distribution hardware, and room equipment.

9.3.1 Building Access

The building access point (often referred to as the only real NAP) can be considered a special case for DF facilities since it often is not located within a true DF facility and may require special equipment. The NAP is usually located within the basement of the building where the telco or campus cables terminate. It is possible for the NAP to also be located above ground. However, this is usually only the case for campus-oriented buildings where distribution between buildings is served by aerial distribution or where there is no basement in the building. The method of access into a building is usually conduit or tunnel.

Recent rulings by the Federal Communications Commission (FCC) have proposed that the network termination point for telephone company demarcation (a.k.a. telco demarc) must be within 12 in of the point where the telco cable actually enters the building or property line (this is somewhat confusing for multitenant buildings. This is quite a bit more stringent than most current telco demarcation points. For the NAP, this requires that the user place distribution cross-connect equipment at the physical cable entrance. While the telco will usually provide a standard lug or wrap distribution block, the user should provide a cross-connect block consistent with the interior building distribution system (66, 110, etc.). The user block should also provide surge protection as detailed in Sec. 9.3.4 (note that electrical protection is required for any incoming cable terminated at a building NAP).

The NAP should be located, when possible, between 1 and 2 m from the base of the floor. For ease of access, the NAP should not be located within 30 cm of any wall corner. It is vital that the NAP never be located within any room or area where adverse environmental conditions may apply. This includes excess water, moisture, or diverse temperatures that may corrode the cross-connect equipment.

The ideal location for the building NAP is within a standard DF facility, usually the BDF. This requires that additional space be allocated within the BDF (or any DF) for NAP telco demarc equipment. The amount of space required for a telco demarc will vary depending upon the number of cables that must be terminated. However, between 1 and 2 m of horizontal wall space (floor to ceiling) is usually adequate for most buildings serving up to 500 users.

Collocation of the NAP and a DF may not always be possible since it requires the telco demarc point and the BDF to be collocated. When collocation is not possible, there should be a direct cable run between the NAP and the BDF (or CDF where applicable). Additionally, noncollocation of equipment may require a physical enclosure to house the demarc and cross-connect facilities.

9.3.2 Riser Access

Each IDF within a building (as well as the BDF) must have access to the building riser system. Since IDF facilities are usually vertically aligned, they rely upon riser systems, and the associated riser access methods, for the cable or network access point. There are three main types of hardware for riser access within an IDF:

Sleeves: This consists of short segments of plastic or metal tubes placed vertically within the floor or ceiling of the DF room. A sleeve is typically 10 to 20 cm in diameter and can protrude from the ceiling or floor from 10 cm up to a maximum of 1 m. The protrusion of the sleeve helps to prevent miscellaneous materials (such as cable trimmings, water, trash, etc.) from passing through the sleeve. The ends of a sleeve may be slightly flanged to prevent the cable from coming into contact with a sharp edge. See Fig. 9.3. Sleeves are usually installed during the pouring of concrete but may be added later (see slots and cores). This type of riser access hardware is ideal for installation in DF rooms that are vertically aligned.

Slots and cores: Slots and cores consist of square, rectangular, or circular holes cut into the ceiling or floor of the DF room. The process of installing a slot or a core is referred to as core drilling and is ideal for installing riser cable access points after initial floor or ceiling installation. There are, however, several disadvantages to this type of installation. First, core drilling may not be useful in all buildings depending upon the type of building material and the placement of vertical DF rooms (core drilling does not help if the two DF rooms being connected are not perfectly aligned on the ver-

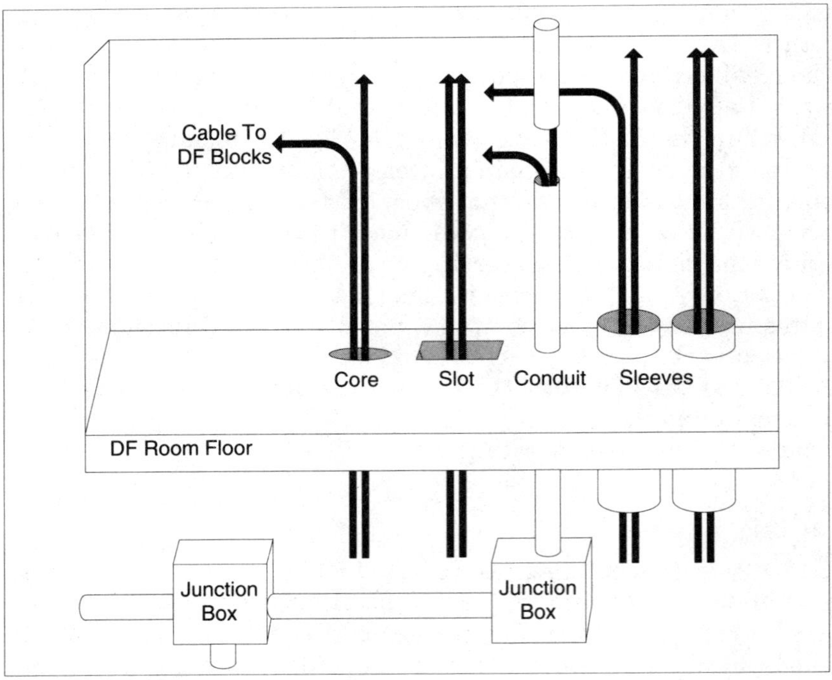

Figure 9.3 Riser access.

tical axis). Second, core drilling does not provide the physical protection afforded by sleeves (i.e., the flanged edges or the protrusion). Lastly, core drillings may not be permissible in some rooms because of the structural design of the building (slots or cores cannot be installed near structural supports).

Conduit: Conduit in the DF room consists of long segments of conduit that extend from floor to ceiling with openings for cable access. Conduit, when used in a DF room, is similar to an elongated version of a cable sleeve. The conduit, usually under 20 cm in diameter, is typically configured with a break midway between the ceiling and floor to allow for riser cables to bridge to the DF cross-connects. Like sleeves, conduit requires planning prior to the pouring of a concrete floor. Conduit is ideal for riser distribution between DF rooms that are not vertically aligned. In such a case, the conduit is extended between the two DF rooms through or directly under the ceiling or floor.

The location within the DF room of the riser access can be very important. Unlike the placement of cross-connect equipment, which is best located in a central manner, the best location for a riser access point is near a corner (not closer than 30 cm) where cross-connect blocks will not have to be placed on either side of the entrance.

9.3.3 Horizontal Access

Any DF room serving horizontal distribution must have an access point within the room where all horizontal distribution cables enter. The location of the access point will usually be directly related to the type and installation configuration of the horizontal distribution hardware. Like the riser, or vertical, access point, the location of the horizontal access point should be to one side of a room adjacent to the location where the cross-connect blocks will be installed. It is common to see the horizontal access point opposite the placement of the riser access point.

9.3.4 Surge Protection

Surge protection is a required element in many DF rooms. Electrical surges over any telecommunications cable can quickly destroy delicate electronic equipment, such as modems, LAN servers, and communications processors. Most end users think of surge protection as applying only to the power line connected to their computers or communications devices. The use of an electrical power line surge protector will protect your equipment from surges or drops in the power level which could result in damage to the connected equipment. This type of protector, however, does not offer any protection from increases in line voltage over communications circuits.

Communications equipment is very sensitive to changes in line voltage. Minor changes (i.e., a volt or two) can often disrupt many communications signals by making them unrecognizable to the destination device. In addition, frequent fluctuations can emulate some carrier signals, resulting in disrupted service (i.e., loss of sync). Larger changes in line voltage can just as easily destroy delicate communications equipment.

The communications line voltage can rise slowly as a result of induction (i.e., a sneak current resulting from too close a proximity to a power line) or very sharply (i.e., the result of direct contact with an electrical line or a lightning strike). Such changes can happen both within and outside a building. For this reason, it is often a good idea to have multiple surge protectors serving a single communications line.

The first location that should always be protected is the NAP to any building or campus facilities. The protection at this point should be immediately after the access to the building. There should be as little unprotected line within a building as possible; less than 1 m is not only recommended but is also required by many building codes. Any unprotected line (i.e., between the NAP and the protection units) offers a potential danger point. A power surge entering the building on a communications cable can cause fire or result in personal injury if a person is in close proximity to or working on the unprotected cable.

The second point that should be protected is the end-user equipment location. This involves the connection of an external surge protector to either a station outlet cable (apparatus cable) or to the cable connecting directly to a computer or communications device. The latter installation may require protection equipment to be located within an IDF or SDF. These types of devices (commonly grounded to the serving power outlet ground) offer additional protection for communications lines that may experience in-house surges. It is important to note that end-user protection devices are not a replacement for building NAP protectors.

9.3.4.1 Protection technology. The goal of a surge protector is to prevent an improper electrical current from reaching end-user equipment or transmission lines. This is accomplished by offering a shorter path to ground (electrical current is driven to ground in the fastest possible path that offers the least resistance). Most protection devices use either a carbon-block or gas-tube mechanism to provide proper grounding and protection. A carbon-block device allows a surging current to arc between two carbon components that are placed very close together and offer a faster (better) path to ground than the telecommunications cable. Carbon-block devices can also fuse the two carbon components together if the power surge is significant enough or lasts long enough (this offers a permanent path to ground and requires replacement of the surge protector). Gas-tube devices, while offering a fast path to ground like carbon-blocks, do not fuse permanently, thereby automatically resetting between power surges. This enables a communications line to be activated again without physical replacement (resetting) of the protection device. This is ideal for locations that are likely to incur repeated surges where an automatically recovering unit eliminates the requirement for on-site maintenance activity.

In addition to the above devices, there are fuse-based systems which offer a fuse as protection against rising currents. These systems have a fuse filament which burns out given a rise in current (similar to the fuse in an electronic device). This type of protection, mainly used for low-rise or sneak currents, is not widely implemented except for SIO unit protection and some add-on cross-connect protectors since it requires physical replacement after every surge and does not offer adequate protection for significant surges (such as those caused by a lightning strike).

9.3.4.2 Protection devices. Protection devices may be either integral or standalone components. Integral components function directly with special cross-connect blocks, such as the 110- or 66-blocks. The protection units generally consist of plug-in protectors that provide protection between incoming and outgoing cables. Some integral protectors function as add-on components to existing cross-connect blocks, allowing protection to be added after the installation of a system (such as a protector that fits over the insulation displacement clips of a 66-block and provides a connection between the clips and surge protection).

Standalone protectors are completely separate from any type of cross-connect equipment; they consist of a special unit that is usually designed to be connected directly to a bulk cable (at the NAP). While requiring more space than integral units, standalone units can offer "industrial strength" protection required for connection to outside lines which may be subjected to harsh conditions (lightning or power-line contact). Standalone units can connect via single pair connectors (punch or wire-wrap) or bulk connectors (large splice units or amphenol connectors).

Any type of protection device, whether integral or standalone, must be properly grounded to be fully functional. An improperly grounded protection unit will not offer the fast ground path (shortest path to ground) and thereby will not route surging signals from the communications line to the protector ground.

9.3.4.3 Cable requirements. Any cable which contains metal is a candidate for surge protection or grounding. This includes both in-building and outdoor cables. While this obviously pertains to copper-based cable such as twisted-pair or coax, it also applies to any fiber-optic cable that has any metal components. Additionally, any in-building metal-based cable that passes through harsh EMI areas or could potentially come in contact with power lines should be properly protected at either the DF or SIO location.

9.4 CAMPUS DISTRIBUTION FRAME

The CDF is the centralized point within a campus where all distribution cables originate. The function of the CDF is to feed BDFs located within each campus building. All cables from individual BDF facilities must pass through the CDF to reach any other BDF. In this manner, networks can be easily planned and maintained. Additionally, access to the campus from outside sources (telco and wide area and metropolitan networks) can be centrally controlled.

9.4.1 CDF Placement

The CDF, as the central focal point for all campus cable distribution, should be centrally located within the geographic area. Every effort should be made to locate the CDF such that all BDF facilities (or at least the major ones) are roughly equidistant from the CDF. Unfortunately, this may not always be possible because of campus layout. When center placement is not permissible for the CDF, it should be placed in the location that best serves the majority of BDF facilities without restricting access to any particular BDF.

A common design goal of many wiring system planners is the integration of the CDF with a centralized computer or data processing facility. This can help to provide the proper environment since many campus computer centers have well-trained staff housed in an environmentally controlled room. Additionally, it is not uncommon to see the CDF for a campus colocated with a BDF in one of the larger campus buildings. While this may simplify some of the room considerations and present easy access to the communications network for campus phone or computer systems, there is no fast requirement that the CDF be colocated with another communications facility. In fact, it is not unheard of to see a CDF facility located in a standalone building or in an underground distribution enclosure. The buried approach often allows centralized placement of the CDF when otherwise it would not be possible. It also offers easy access to many forms of campus distribution such as buried cable, underground conduit, or tunnels. See Fig. 9.4.

Many campus facilities are designed in such a configuration that a single CDF cannot serve the entire campus. It is not uncommon to see auxiliary CDF facilities installed in such situations. These secondary CDF sites are often standalone facilities that may be either above- or below-ground enclosures. Communications equipment, such as modems or computers, is usually not located in such facilities whose function is a simple distribution point.

9.4.2 CDF Room Characteristics

The room characteristics of the CDF may vary considerably depending upon the location of the CDF and the types of equipment that must be located there. Standalone CDF facilities, typically not capable of sup-

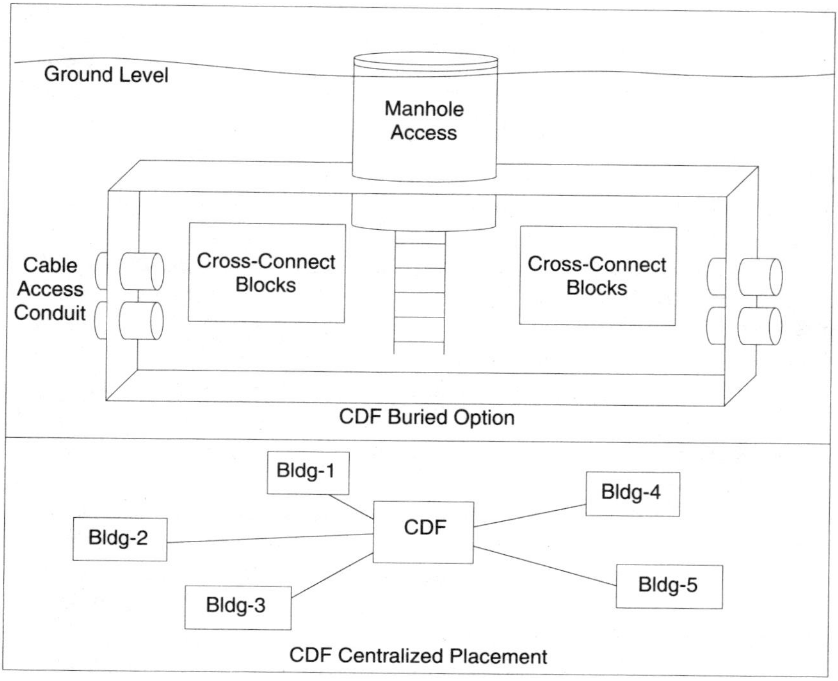

Figure 9.4 The CDF.

porting additional equipment (i.e., not cross-connect), will be fairly simple in construction and design. Such facilities, even if they are modular units purchased preconfigured for burial, should adhere to the specifications set forth for DF facilities in Sec. 9.1.3.

The CDF facility may require special design considerations if it is to house communications equipment. These special characteristics often include special access, power, air conditioning, or distribution hardware. It is unusual to see a standalone CDF that can properly support computer or communications gear. For this reason, if the CDF can reasonably be expected to house equipment other than cross-connect blocks and distribution panels, it is recommended that the CDF be placed within a building facility where colocation is possible.

9.5 BUILDING DISTRIBUTION FRAME

The BDF is the central point within each building where distribution cables originate and have access to the building NAP and campus distribution. The BDF serves multiple IDF facilities. It is commonly colocated with a PBX or computer room that can provide the room characteristics required of the BDF.

9.5.1 BDF Placement

The placement of the BDF is dependent upon several points. These include the location of the NAP, the design of the building, and the presence of large computer or PBX facilities. Each of these points should be considered since proper placement of the BDF will allow support for buildingwide networking systems, such as local area networks.

The NAP within a building must support both surge protectors and cross-connect equipment. Locating the BDF with the NAP will allow for solid NAP support (power, environmental, security, etc.). If the NAP and BDF cannot be colocated, an in-building distribution cable must be pulled between the two facilities.

Many networking systems in use today have very specific distance limitations for cable runs between devices. These distances are often limited to around 100 m. This contrasts markedly with the networking systems in use 10 or 15 years ago that could easily span several hundred meters without any difficulty. These new distance limitations have put design constraints on the placement or location of a BDF within a building. While the BDF was traditionally located in the basement of many buildings, where space was most likely available for the BDF, or in computer rooms which typically do not have windows, the newer trend has been to locate BDF facilities within the center of a building. It is not uncommon to see a BDF located on the fifth floor of a 10- or 15-floor building. See Fig. 9.5.

The center-located approach not only applies to vertical alignment but also to horizontal configuration. Locating the BDF within the center of a long building can help provide equidistant access to both ends of a facility. This can be critical in buildings which are in excess of 200 m in length or "L-shaped." Such buildings often require multiple riser systems (see Chap. 10) which should have equal access to the BDF.

Another consideration when determining the location of the BDF is the presence of large computer or PBX rooms within the building. Combining these two facilities into one larger facility can often save both time and money (and time is money) during installation and maintenance procedures. Regardless of what floor the BDF is located on within a building, it is very important that the BDF be centrally located around the building core and have easy access to the vertical (riser) distribution system.

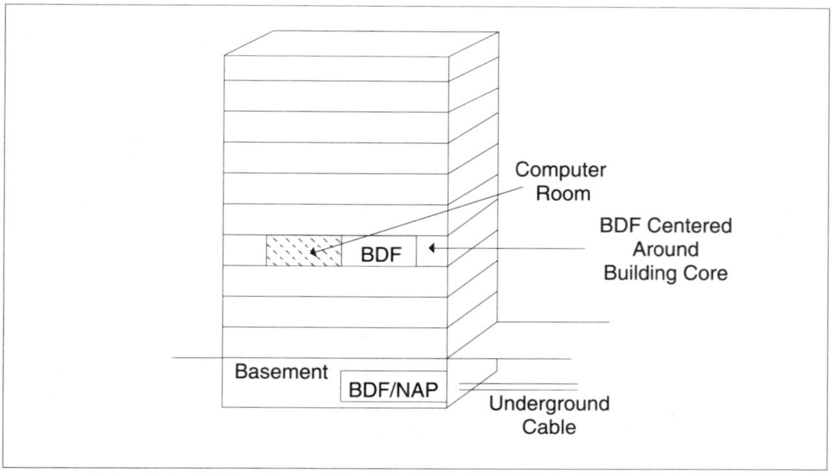

Figure 9.5 BDF locations.

9.5.2 BDF Room Characteristics

The BDF is the largest distribution frame within a building. Given the likely support for colocation of equipment, and the sheer volume of cross-connect-related material that must be supported, the BDF should meet all of the requirements listed in Sec. 9.1.3. BDF facilities are typically square or slightly rectangular in layout. Extreme shaped rooms (i.e., long but narrow) do not offer the floor space required for communications equipment support.

The layout of the BDF will usually be quite different from the other DF facilities in the building. In addition to a larger number of cross-connect blocks, the BDF must have ample room to enable a flexible configuration. There should be clearly defined areas for each type of equipment to be located within the BDF. Figure 9.6 shows a sample configuration for a BDF that contains computer and PBX equipment in addition to standard cross-connect blocks.

9.6 INTERMEDIATE DISTRIBUTION FRAME

The IDF is the central point on every floor from which all horizontal distribution cables originate. The IDF is fed directly by the BDF through the vertical (riser) distribution system and may be required to house limited communications equipment, such as file servers, LAN wiring hubs, and modems. Large equipment, such as T1 multi-

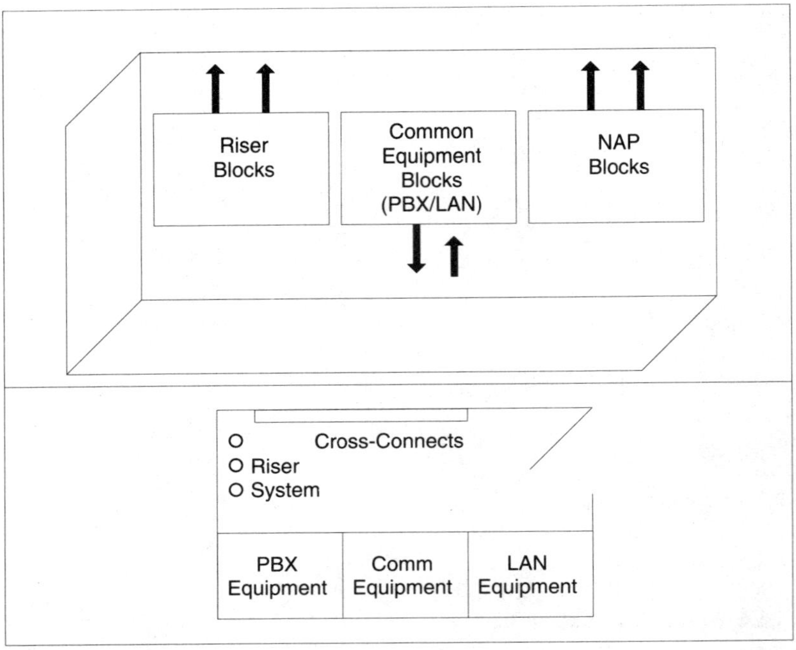

Figure 9.6 BDF layout.

plexers, large computers, and PBX/Key equipment, should be located within the BDF.

9.6.1 IDF Placement

The farther from the CDF or BDF you go, the more choices and flexibility you have in DF placement. This is necessary to support the varied requirements that arise in working office environments. The following guidelines apply to IDF placement and should be applied consistently throughout a building:

Vertical alignment: All IDF facilities within a building should be vertically aligned, preferably over the BDF location. Riser cable distribution between nonaligned IDF facilities often requires special distribution hardware (i.e., conduit) be installed within the ceiling, which can be costly and add to the overall length of cable required.

Core placement: Every building has at least one area designated as a building core. The building core is the location where all vertical components are installed (i.e., stairways, power distribution, elevators, etc.). Typically centrally located, the building core provides

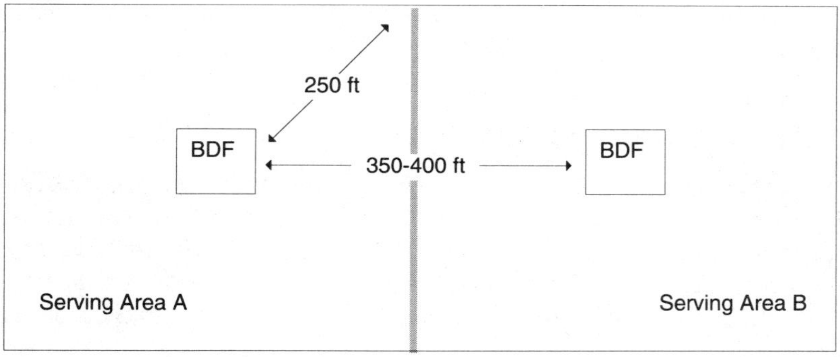

Figure 9.7 IDF serving area.

an ideal location for IDF placement. From a structural perspective, the building core must be capable of having distribution hardware installed to serve riser distribution (i.e., conduit, sleeves, slots).

Serving area: Unlike a BDF which serves an entire building, the service area of an IDF is fairly limited. The straight-line distance between an IDF and the farthest served SIO unit should not be more than 75 to 80 m (approximately 250 ft). This creates a serving area of approximately 20,000 m^2 per IDF (about 200,000 ft^2). See Fig. 9.7. This is the maximum serving size that an IDF should ever support. If all users on a single floor cannot be served by a single IDF, multiple IDF facilities or SDFs should be used. The use of multiple DF facilities per floor is common in large buildings that are more than 200 m long.

9.6.2 IDF Equipment Placement

The type of rooms used for IDF facilities may vary considerably depending upon the building configuration and the type of communications equipment that may be housed there. Since the primary function of an IDF is to house cross-connect equipment, with minimal emphasis on large communication components, a large segment of open wall space is more important than large quantities of open floor space. For this reason, it is permissible to have an IDF that is rectangular in shape (this contrasts to the BDF and CDF, which are best served by a square facility).

The most practical IDF layout will have all cross-connect blocks located along a single wall with communications equipment located on an adjacent or opposite wall (if the room is deep enough). IDF rooms that are less than 2 m deep cannot provide adequate working room for cross-connect installation or communications equipment colocation.

9.7 SATELLITE DISTRIBUTION FRAME

The SDF is the smallest distribution frame permissible within a building. Serving as an extension of an IDF, the SDF is used to reach small numbers of end users that are beyond the serving area of an IDF. The SDF, while extending the range of horizontal distribution, cannot typically offer the range of room characteristics that are required of an IDF.

9.7.1 SDF Placement

While BDF and IDF facilities are usually centered in building core areas, the SDF, through its role as an extension of an IDF, is usually located away from the building core areas. The SDF serves remote building areas or areas between multiple IDF serving areas. As such, the SDF is connected to the building wiring system by horizontal distribution only; vertical support for an SDF would classify the distribution point as an IDF.

Once it has been determined that a portion of a building floor cannot be served by an IDF, a location for an SDF must be determined. The SDF should be centrally located within the serving area (typically 7500 to 10,000 m^2 or 80,000 to 205,000 ft^2) and have access to a main distribution channel originating from the serving IDF. Figure 9.8 shows an example of where an SDF may be used to serve an area unreachable from an IDF.

The SDF should be treated as any other DF from a design and distribution perspective. This means that the SDF will have large multipair cables connecting it to its serving IDF, just as an IDF is connected to the BDF via large multipair cables.

9.7.2 SDF Room Characteristics

The SDF, being the smallest allowable frame, has the most flexibility in terms of room layout, location, and configuration. This flexibility is required since the SDF is often forced into locations where there is limited space and room configurations often take a secondary consideration to office design.

While it certainly is possible to locate communications equipment within an SDF, it is usually preferable to locate such equipment within an IDF where possible. This may not always work with some

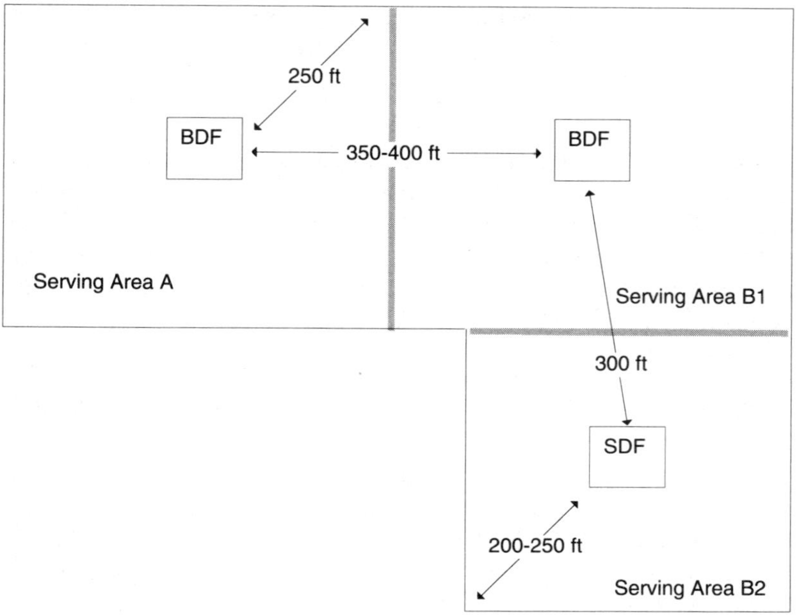

Figure 9.8 SDF serving area.

devices, such as LAN wiring hubs, that have strict distance limitations. It is recommended that the layout of the SDF be designed to accommodate a limited quantity of communications gear.

When space is available, the SDF should mirror the room characteristics of the IDF. When space is not available, it may be necessary to improvise. The most common configuration consists of a long and narrow room that emulates a modified coat closet.

This type of configuration, when given proper depth (at least 1 m), can allow for both ease of access and placement of limited communications equipment at one or both ends of the room. An SDF should never consist of a set of cross-connect blocks mounted in an open room or under a desk where control and functionality are severely limited.

9.8 DESIGN CRITERIA

The design criteria for any DF should mirror the goals set forth earlier in this chapter. These goals are designed to offer a maximum of functionality, providing adequate support for cross-connect and communications equipment. This support includes areas such as ease of access and maintenance, as well as security and flexibility for future growth. The most important design decisions that must be made (most other factors are the result of following basic guidelines) relate to the type of cross-connect block to be installed and the size of the DF facilities.

9.8.1 Block Selection

The decision to use one particular type of cross-connect over another must be made very carefully. The selection of an inappropriate block can result in the decreased ability to support high-speed networking applications and an increase in management overhead. The following criteria should apply to any cross-connect evaluation process (including twisted-pair, coaxial cable, and fiber-optic cable):

Applications support: The block must provide electrical and optical characteristics within the guidelines of the networking equipment being used.

Modularity and flexibility: The system should offer a high-density approach that can be added to in a modular fashion to accommodate future expansion. This includes the availability of a large number of add-on products that can provide limited customization of a standard block as well as support future growth.

Density: Wall space is typically at a premium within most office buildings. This is especially true given the current trend in the early 1990s toward a slower global economy in which cost savings have become the rule, not the exception. Being able to place a high number of cross-connects can help to reduce the amount of space that needs to be allocated to a DF. It can also allow future expansion of cross-connects without requiring additional allocation of space (expansion of the DF facility).

Availability: Sole-source cross-connect blocks should not be selected unless the source is an established organization and not likely to discontinue support for the product within the near term (±10 years).

Management: The cross-connect block must offer the ability to accommodate cable management information that indicates which incoming cables have been cross-connected to which outgoing cables. Additionally, the blocks should offer ease of installation and maintenance. The availability of trained personnel should also be considered.

Ease of installation: This criterion becomes very important when dealing with newer modular components that may not be completely field tested or have a limited number of qualified installation technicians available. A cross-connect block that is difficult to install or requires special mounting hardware can often increase the amount of time, and money, that must be spent on installation activities (including testing of the system and punching down distribution cables).

Cost: This should be the penultimate (if not the least important) factor in selecting a cross-connect block. Most blocks are within the same general price range (about $15 to $20 per 100 pairs for twisted-pair).

An additional consideration when selecting a cross-connect block architecture is the presence of an existing block installation, such as occurs during a building renovation or addition installation. It is very important that only one cross-connect block architecture be used for each type of media installed throughout a facility. This will help significantly with maintenance and communications system support. A mixed-block architecture can be difficult to integrate within DF facilities and can add to maintenance costs through the requirement for multiple sets of punch-down hardware (twisted-pair only), multiblock training for installers and maintenance personnel, and configuration management.

Additional information on cross-connect blocks and patch panels can be found in Chaps. 2, 3, and 4.

9.8.2 Sizing

Determining the amount of space that must be allocated to a particular DF facility can either ensure that the facility will provide adequate support for future applications or can guarantee that future expansion will be both painful and costly. Figure 9.9 presents minimum space guidelines (wall and floor) for the various forms of DF facilities. These guidelines are intended to serve as a minimum sizing configuration and can be exceeded if space is available (there are very few DF facilities that are too large, but plenty that are crowded beyond reasonable functionality).

Space	CDF (ft)	BDF (ft)	IDF (ft)	SDF (ft)
Floor	30 x 20	20 x 12	12 x 10	8 x 8
Wall	48 x 8	24 x 8	12 x 8	8 x 8

Figure 9.9 Minimum DF space allocations.

9.9 INSTALLATION PARAMETERS

Once a DF has been designed, it is important for the components to be installed in an orderly and scheduled fashion. This will help to reduce the total duration of time required to install a building wiring system as well as help to ensure the reliability of the system. Failure to properly schedule and test DF facilities can result in delays involving cable and communications equipment installation.

9.9.1 Installation Scheduling

The general construction of any DF must be accomplished in a timely manner consistent with the general components. For example, the riser system and any underfloor hardware should be installed during floor installation, lighting should be installed during electrical installation, etc. In this regard, a DF should not be given any special treatment unless it requires special structural or environmental components. However, even special components should be installed at such a time that is consistent with other such facilities within the building.

The real scheduling and installation of DF equipment must occur after the basic DF room has been completed and protected from any environmental disturbances (i.e., construction is completed). Once this occurs, the room can be painted with fire-resistant paint, cross-connect backboards can be installed, and any equipment racks can be bolted into the floor. After backboard installation, individual cross-connects and patch panels can be installed and prepared for wiring connection (which cannot occur until the wiring system is installed at a later time). Only after all necessary racks, blocks, and shelves have been installed should any cable or communications equipment be installed.

Within a campus or building, the installation of DF-specific hardware should occur in accordance with the basic tree distribution of the wiring system. For example, in a campus, the CDF or BDF should be designed and installed first since their final design will determine how IDF and SDF locations are designed and implemented.

It is not uncommon for renovations to pose some delicate planning problems. For example, when building in a new cross-connect block architecture, it is important to have a smooth transition period. This often means that the new system must be constructed while the old system is still in operation. This can be difficult in DF rooms where space is limited. One solution to this problem is a layered DF, which is one in which a new DF (cross-connect block system) is built on the top or in front of an existing cross-connect block installation. The new block system is built on a new plywood front that is mounted at least 10 cm from the face of the old cross-connect blocks. See Fig. 9.10.

Once the new cross-connect block system has been installed and tested, individual SIO or distribution connections can be made. This allows for partial cutover while retaining the old block structure as a

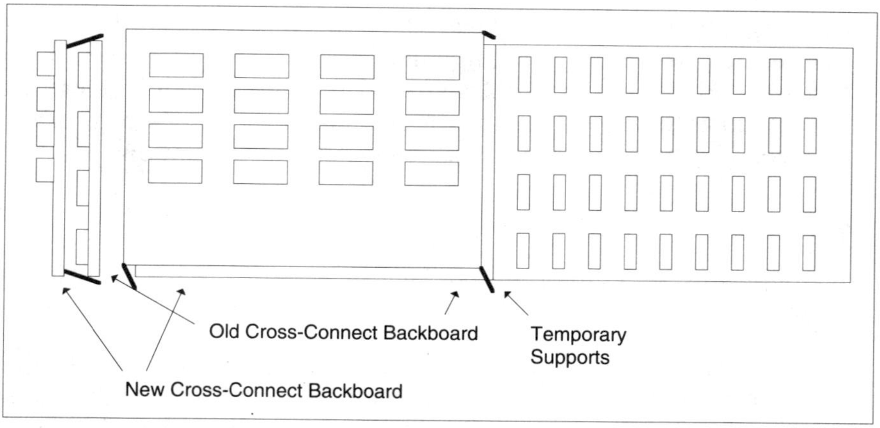

Figure 9.10 Layered DF (temporary).

backup mechanism. Once the new system has been approved for use, the old cross-connect blocks can be removed and the new cross-connect boards collapsed to fit against the wall. It is important to note that this type of layered installation is for temporary use only. The old cross-connect equipment should never be left in place.

9.9.2 Fire-stopping

Fire-stopping is one of the most commonly overlooked, and potentially the most deadly, items in DF planning. As discussed in terms of the horizontal and riser distribution systems, fire-stopping should occur any time a wall, ceiling, or floor is breached for either conduit, duct, or poke-through cable distribution. This includes underfloor distribution systems where flames or smoke could potentially travel via cables to or from remote parts of the building. While fire-stopping within conduit, duct, or sleeves can only take place after cable installation, any conduit or duct system should be properly fire-stopped (as outlined in Chap. 8) during installation prior to cable installation.

9.9.3 Configuration Testing

The components installed within a DF facility should be properly tested prior to cable and communications component installation. Any working components involved in the DF should be properly tested (air conditioning, humidity control, lighting, power, fire suppression, security and access control, etc.) prior to component installation. Additionally, any fixed component, such as cross-connect blocks or equipment racks, should be inspected and tested for structural integrity as well as quality of component. Cross-connect blocks must be able to support the weight of all installed cables, which in many cases can be greater than anticipated because of unplanned expansion or the retaining of obsolete cables while new cables are added. Checking equipment prior to cable installation can eliminate expensive recabling which can be required if components need replacement after installation.

9.10 SUMMARY

The following is a summary of the key points detailed within this chapter:

- There are four major types of distribution frame: campus distribution frame (CDF), building distribution frame (BDF), intermediate distribution frame (IDF), and satellite distribution frame (SDF).
- A DF will house cross-connect blocks, cable, and associated communications equipment.
- It is important to have a consistent cross-connect architecture throughout an entire building wiring system.
- The location of DF facilities is very important to the ability of the wiring system to adequately service user requirements.

Chapter

10

Frame Interconnection System

Definition: *Frame interconnection system* The system that links together, using cable and wiring, multiple distribution frames within a building or campus. For example, the vertical riser system within a building.

Within every building or campus wiring system, there must be a well-defined method for the interconnection of the various distribution frame (DF) facilities. A structured frame interconnection system (FIS) not only supports the other components of the wiring system but also can significantly add to the flexibility and manageability of the entire system. This includes connection between:

- *Campus distribution frame (CDF) to building distribution frame (BDF):* The campuswide distribution of cables between buildings, including the network access point (NAP).
- *BDF to intermediate distribution frame (IDF):* The vertical, or riser, distribution within a single building.
- *IDF to satellite distribution frame (SDF):* The distribution of cables between phone closets within a single floor of a building.

It should be noted that the FIS does not provide horizontal distribution of cables to standard information outlet (SIO) units, as discussed in Chap. 8. While cable used for horizontal distribution typically involves small-count cables (i.e., less than a dozen conductors per cable), the cables used for DF interconnection are high-count, potentially involving hundreds of conductors within a single cable sheath. Additionally, FIS cable systems often must be capable of operating in harsh environments encountered in locations such as outdoors, underground, and steam tunnels.

There are many different components and options that must be discussed when designing an FIS. An improper design can result in reduced ability to communicate throughout an organization. A lack of resources or flexibility in the FIS can be costly to correct, especially if the interconnection system involves buried or undergound distribution systems that may be difficult to expand or alter. Some of the decisions that must be made include the following:

Types of wiring and when each should be used

The quantity of cable to be installed

The distribution hardware required for system support

What design criteria should be applied to frame interconnection

How and when this type of system should be installed

It is not uncommon for multiple types of cable within a single FIS to support multiple applications. For example, the type of cable installed for voice system support is very often quite different from the cable installed for data support (voice and data cables should *always* be separate within an FIS).

The following sections address these concerns for campus (outdoor) and building (indoor) interconnection systems.

10.1 CABLE SELECTION

The type of cable selected for the FIS will vary greatly depending upon the characteristics of the building and campus environment in which it is being installed. Decisions regarding the type of cable media, shielding, plenum rating, and physical protection elements must be made to ensure that the cable properly matches both the user requirements and environmental constraints throughout the entire DF system. In many cases, this may require mixed-media solutions that have varying physical cable characteristics depending upon the serving area specifications. The types of cable used for frame interconnection must very often be able to support both indoor and outdoor installations.

There are many types of cable that can be used for DF interconnection within a building or campus. This includes twisted-pair, coaxial, and fiber as well as mixed-media combinations. It is important, in order to make an informed cable selection decision, to understand both the types of cables available and the forces which can shape the FIS cable selection process.

10.1.1 Media Selection

It is important that the FIS cable properly support the end-user applications within a facility. This includes support for any type of communications system such as voice, data, video, and control (security, paging, environmental) support systems, intended for support by the building wiring system. For this reason, the cable used within the FIS must directly support the type of cables selected for installation in the horizontal distribution system. For example, if twisted-pair and coaxial cable are both used for horizontal distribution, these same media must be used to support communications between various distribution points.

Matching the FIS cable to the horizontal distribution cable involves not only the selection of a similar media but also the matching of the specific electrical or optical cable characteristics. For example, if 24AWG cable with a nominal impedance of 100 Ω is used for horizontal distribution, a similar cable should be used for the FIS. Similarly, if 62.5-μ/125-μ optical fiber is used to support SIO unit connection, the same optical fiber should be present in the FIS as well.

In many cases, multiple media will be required. It is often recommended that certain types of cable be installed as a precaution against future, unplanned, requirements. The most common example of this is fiber-optic cable. It is recommended that most FISs include a minimum quantity of fiber-optic cable to support new technology applications as they emerge. This practice, commonly referred to as the installation of *dark fiber*, can often save time and money as user telecommunications become more bandwidth intensive and migrate to newer technology (dark fiber is a fiber-optic cable with no termination equipment attached; the term *dark* indicates that no light source is connected at time of installation). It is anticipated, for example, that many users will begin to interconnect local area networks (LANs) throughout a building with high-speed optical Fiber Distributed Data Interface (FDDI) networks. Additionally, high-speed circuit-switched technology (such as DS-3 or T3 networks operating at 45 Mbps) is unsuitable for connection over twisted-pair and requires either coaxial cable or (more commonly) fiber optics. Installing such surplus cable for future use at the same time as the FIS installation can help prepare for these emerging technologies.

The most common fiber-optic cable installed within buildings is 62.5-μ/125-μ fiber. This multimode cable has become a de facto industry and user standard for local and metropolitan area fiber networks.

It is uncommon for other media (i.e., coaxial cable) to be installed in FISs unless there is a clearly defined current or near-term application that cannot be accommodated by twisted-pair or optical fiber.

10.1.2 Cable Sizing

The number of conductors (or pairs of conductors) that are contained within a single sheath are often referred to as the "size" of the cable. For example, the size of a twisted-pair cable containing 50 individual twisted-pairs would be 50-pair. Coaxial cables are sized in terms of the number of solid inner conductors, such as a four-conductor cable, which indicates that four complete coaxial cables are contained within a single outer sheath. Fiber-optic cables are labeled according to the number of optical fibers (strands) within a single sheath. For example, a cable containing 12 optical fibers is considered a 12-strand fiber.

Cables used within an FIS are usually very large cables with high pair, conductor, and strand counts. This is a practical decision that involves both the simplicity of installing a smaller number of cables and the fact that since multiple conductors are placed within a single sheath, they occupy less space than the same number of conductors with a smaller in-sheath count. For example, 100 twisted-pairs within a single sheath occupy approximately the same volume (space) as 52 twisted-pairs encased in four-pair sheaths. This volume difference can be even more extreme as special sheaths, such as shielding, plenum, or strength casings, are placed on the cable.

Different size cables are commonly associated with different FIS segments. For example, in the twisted-pair arena, CDF-BDF segments typically consist of 100-, 300-, or 600-pair cables while BDF-IDF segments typically consist of 100- or 300-pair cables and IDF-SDF segments rarely exceed 25 or 100 pairs per cable.

In addition to the size of FIS cables, it is important that the proper number of conductors be present in each segment of the FIS to support the number of SIO units throughout the entire system. A general rule of networking is that large numbers of small systems are usually combined to form a smaller number of larger systems. For example, all personal computer users within a single floor of a building may be connected to a single local network. In turn, multiple local networks, on each floor of the building, may be connected together via a single building backbone network. Taking this another step, multiple building networks are commonly connected together via a single campus backbone network. What is demonstrated by this example is the idea that as you bring a wiring system together to a central focal point (away from the individual SIO unit) fewer cables are required for total system support (occasionally known as a cable compression ratio).

	FIS Pairs Per SIO Pairs	
	Voice	Data
SIO-SDF	1:1	1:1
SIO-IDF	1:1	1:1
SDF-IDF	1:1	1:1
IDF-BDF	3:4	1:2
BDF-CDF	1:2	1:4

Figure 10.1 Twisted-pair allocations.

This is ideally shown in the voice world by the telephone private branch exchange (PBX) which concentrates a large number of phone units (each connected to an SIO) into a relatively small number of actual connections to the local telephone company.

Given that the number of cables contained within an FIS is related directly to the number of SIO units being served and the level of concentration that is expected throughout the entire building wiring system, we are able to estimate the number of cables that are required at each step of an FIS installation. Figure 10.1 presents some general guidelines, based on the number of cables supporting SIO units, for the minimum number of cables that should be present in FIS segments. These cable sizes are for example purposes only and do not represent firm and fixed rules for cable installation. Other considerations, such as the number of SIO units that must be served by a wiring system, will help determine the number and size of cables that should be present within a building or campus FIS installation.

10.1.3 Plenum Rating

Plenum rating, most commonly associated with horizontal distribution only, can also be an important factor in FIS installations. The most common situation where plenum-rated cables may be required involve the FIS between IDF and SDF facilities. This will track almost item by item with standard horizontal distribution for SIO unit cables. Information on this topic can be found in Chap. 8.

The other common area for plenum-rated FIS cable involves the vertical distribution of cable between the BDF and IDF facilities. Any time an FIS cable passes between floors, or through a suspended ceiling, there is a chance that a plenum area may exist. For this reason, suspended ceilings should not be used in DF rooms (there are other reasons as well, discussed in Chap. 9).

There are two options if a cable must pass through a plenum-rated ceiling area. First, a form of distribution hardware (sleeves, conduit, duct) can be used to isolate the cable from the plenum area. Second, a plenum-rated cable can be used. This second option is greatly preferred unless there is some overriding factor (possibly cost) which prohibits the use of plenum cables in the FIS. It is recommended that plenum cables be used where possible to minimize the emission of toxic fumes from a standard plastic-sheathed cable during a fire.

With the possible exception of tunnel distribution systems, most outdoor distribution methods will never require plenum-rated cable (i.e., buried cable, cable in conduit, and aerial cable).

10.1.4 Shielding

Just as in horizontal distribution, shield cable may be required within an FIS for twisted-pair depending upon the end-user applications and the installation environment. Shielding can help guard an FIS cable against interference from electromechanical energy sources (EMI) such as power cables, machinery, and other telecommunications signals. It is recommended that shielded cables be used when possible for protective purposes, as described below.

Horizontal shielding: If shielded cable is being used for the horizontal distribution system, it is usually a good idea to use shielding for in-building frame connection as well.

EMI sources: Shielding should always be used within an FIS cable if the cable is installed within an environment subject to heavy EMI (i.e., power, machinery). A common source of EMI within vertical distribution systems is the building elevator. Vertical distribution systems (phone closets) are often installed in the building core next to elevator shafts. Telecommunications signal disruption has been known to occur when elevators pass by the cable.

Data support: While voice applications rarely require support of shielded cable, it is common for data applications to require shielding. This shielding may be required to either protect the cable signal from EMI or to prevent emanations from the cable itself (i.e., a data network operating at 16 Mbps will give off a 16-MHz signal, which exceeds FCC regulations).

System power: There are communications systems that send low-voltage power over twisted-pair (in excess of normal communications currents) for user devices. While this is generally not recommended unless specified by the manufacturer (even then an electrical engineer should be consulted first), it does happen and can interfere with data signals on adjacent copper cables.

The use of shielding for campus FIS cables (installed outdoors) can also be required in certain situations. For example, shielded cable should always be used for any cable distributed along aerial wires. This is also true for any telecommunications cable that could potentially come into contact with a power line or any outdoor cable that has even a remote chance of being a primary or secondary recipient of a lightning strike. This includes cable within underground structures (buried, conduit, tunnels) if it could potentially serve as a lightning or power path.

Any cable that is shielded should be properly grounded through either the ground braid or drain wire. For outdoor cables, where the primary concern is usually lightning strikes, the cable should be grounded periodically throughout the cable run (this contrasts with horizontal distribution cable which should only be grounded at one end of each cable). All segments of such a cable should be grounded together; the shield of a cable should never be severed or broken between two portions of the same cable installation. A qualified electrical engineer should be consulted concerning proper grounding techniques for any type of shielded FIS cable.

10.1.5 Protective Jackets

Many cables used for FIS installations, especially outdoors, have special protective sheaths that serve to protect the inner cable from damage. There are many sources of potential cable damage, such as moisture within the cable, cable nicks or cuts, and rodents (who actually eat through a surprising number of outdoor cables). Some of the more common forms of protective jackets include the following:

Pressurized sheaths: A pressurized sheath uses an air-pressure system at one, or both, ends of a cable to maintain a high-pressure level within the sheath. Pressurized cables are common in outdoor cable facilities and provide very good protection against moisture within a cable. A disadvantage is that they require special pressurization equipment and regular maintenance to maintain.

Gel-filled sheaths: These cables are similar to air-pressure cables; they are filled with gel that effectively seals off the inner conductors of a cable from any moisture external to the cable.

Steel strands: Used primarily to add tensile strength to a cable for installation purposes, steel strands are often placed within a cable sheath. These strands, typically 1 to 12 per sheath, are very common in aerial cable installations where a cable may be supported over a great distance by as few as two support points.

Cable armor: Consisting of a complete solid metal sheath, cable armor is very common in outdoor installations. The armor is designed to protect the cable from accidental cuts (i.e., digging backhoes) as well as to provide protection from hungry rodents (it is not uncommon to find some poor rodent that has tried to chew its way through what it undoubtedly thought was the world's longest edible tuber). Several disadvantages of cable armor include an increase in cable price and a decrease in cable flexibility.

Although protective sheaths are usually associated only with outdoor cable installations, such components can often be quite useful for in-building FIS installations. Cable armor is extremely valuable in installations where a great deal of heavy machinery is present (i.e., a maintenance garage).

10.2 DISTRIBUTION HARDWARE

The distribution hardware used for an FIS installation must be properly selected to suit the building and campus environment. Implementing the wrong type of hardware can lead to an FIS that is either unable to support the distribution wiring or is difficult to work with and maintain. Imagine how inappropriate it would be if a building designer installed nothing but stairs to distribute people in a 50-story building. Likewise, imagine the waste and overhead if the only distribution path between floors in a two-story building were elevators.

The following sections present a general overview of FIS distribution hardware and their respective attributes for typical building and campus cable distribution.

10.2.1 Horizontal Building Distribution

Given the similarity between horizontal station cables (between an IDF or SDF and the SIO) and horizontal FIS cables (between an IDF and an SDF), it is acceptable to apply the guidelines established in Chap. 8 to the distribution of horizontal FISs.

10.2.2 Vertical Building Distribution

Vertical building distribution provides hardware support facilities for FIS connection between BDF and IDF rooms. There are many types of components that are used to support vertical distribution, including slots, sleeves, conduit, and ladders. More detailed information on these distribution methods can be found in Sec. 9.3.2.

In addition to the "regular" types of vertical distribution hardware, there are several items required to support vertical cable distribution. These items enable vertically installed cable to be secured in place.

This can be very important given the sheer weight of a cable installed vertically for more than a few floors. For this reason, it is a good idea to anchor all vertical riser cables at each DF location.

There are several common approaches that can be used for anchoring cable, including clamps and mesh loops. Clamps used for cables are common U clamps with flat surfaces (not round) to spread out the pressure on the cable. Cables that pass through a DF (and do not terminate there) are occasionally clamped in a slight zig-zag configuration to help reduce vertical stress.

A second type of mounting hardware is the mesh loop, which involves a cylindrical mesh (synthetic or metal strands) that encases the cable. The top of the mesh is connected to a ring which can be affixed to a wall or other support structure. The mesh loop, similar to the Chinese finger trap, tightens around the cable when stretched lengthwise. In this fashion, the more vertical stress placed on the mesh loop the tighter it encases the cable.

It is very important to properly anchor the cable. A cable that is not capable of withstanding the vertical stress can dislodge, disrupting the cable and any communications systems using the cable at that time. Even a minor slip of a cable can be sufficient to separate a distribution cable from a cross-connect block.

10.2.3 Campus Distribution

The distribution of cable within a campus facility requires specialized components capable of providing a secure path between multiple buildings. Not only must a cable path be provided, it must be capable of housing cable in adverse environmental conditions. The most common methods of campus distribution are tunnels, underground conduit, buried cable, and aerial cable.

10.2.3.1 Tunnels. Tunnels within a campus environment provide a very suitable mechanism for cable distribution between buildings. Often used for utility distribution, such as electric, water, or steam, tunnels provide a well-protected environment with reasonable access for installation and maintenance (it is not uncommon for many tunnel systems to allow a person to stand upright—well, almost). Access to a tunnel system is accomplished through manholes or underground DF facilities. See Fig. 10.2.

There are many advantages to the use of tunnels for frame interconnection distribution systems. For example, most true tunnel systems (as opposed to large conduit systems) allow ample space for cable placement as well as installation activities. Cable facilities installed within a tunnel system are also less likely to be damaged by external

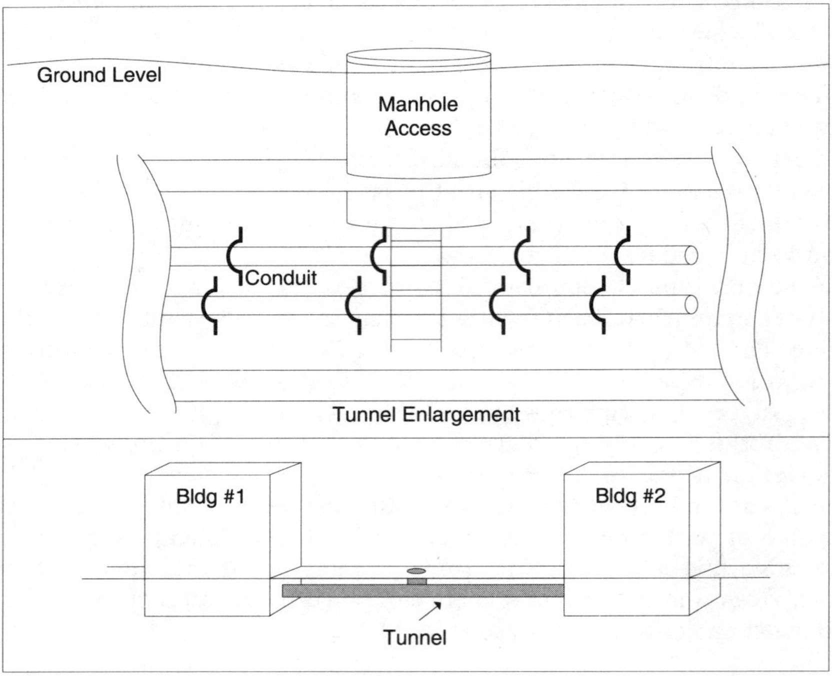

Figure 10.2 Tunnels.

activities (i.e., the masked construction crew and their faithful backhoe, Tonto, a process referred to as *backhoe fade*. Facilities management can also be aided through the use of tunnels since the layout of most tunnel systems is usually very well documented in a campus environment, which can enable cable maintenance personnel to better locate and repair faulty cable.

Prior to the selection of a tunnel system as a distribution method, the disadvantages must also be known. For example, tunnel systems can often flood, which, during winter, can lead to frozen cable and limited access. Also, if the tunnel is used for utility distribution, there is always the risk of cable coming into contact with electrical current or steam (should a steam pipe break). This exposure to extreme heat and cold or electrical current can require special cable.

Tunnel systems are also fairly inflexible by nature and cannot easily be modified to accommodate new building locations or expansion in capacity. However, when considered against the alternative methods, tunnel systems provide a very attractive distribution method.

10.2.3.2 Conduit. Conduit distribution in a multibuilding environment comprises segments of conduit buried underground. Usually

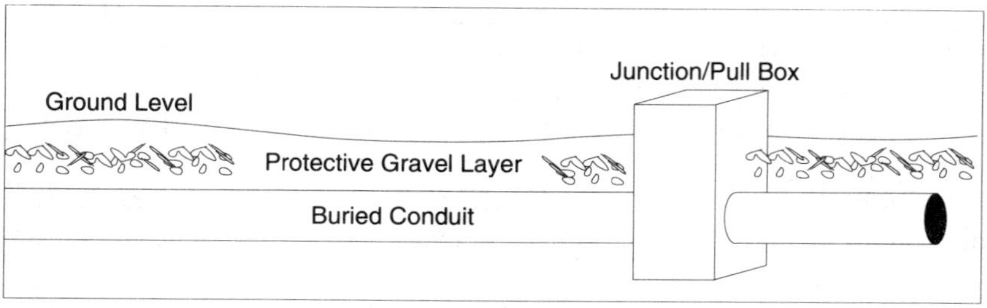

Figure 10.3 Buried conduit.

more than 4 to 6 in in diameter, the conduit is either encased in concrete or surrounded by a protective material (i.e., a layer of gravel). The conduit material may be either metal or synthetic. Access to buried conduit is accomplished through junction boxes (pull boxes). See Fig. 10.3.

The use of conduit is very common for campus distribution systems, especially in environments where tunnels do not exist (or are not available) or where distribution systems are installed after building completion (i.e., extending wiring facilities to a new or renovated building). Conduit is usually installed a minimum of 0.5 m below the ground surface to allow for minor groundwork (i.e., sprinklers, sidewalks, benches) without system disruption.

There are some strong advantages to the use of conduit for frame interconnection distribution systems. Conduit systems, especially when encased in concrete, offer very adequate physical protection against the environment and mechanical accidents, and they provide a good means of securing a cable from physical intrusion. If there is a reliable lock on any junction access point, it is fairly difficult to access installed cable. Buried conduit also provides protection against underground rodents.

Conduit systems can also offer moderate flexibility. Compared to tunnel systems, underground conduit is relatively easy to install. This is important for renovations or communications system expansions where tunnels may not exist or may be unsuitable for supporting communications cable.

As with any type of communications distribution system, there are some limitations to underground conduit systems, including a lack of reconfiguration options, limited growth capabilities, a susceptibility to flooding, and limited repair access.

10.2.3.3 Buried cable. One of the easiest systems to install, yet one of the least desirable, is underground buried cable. In this type of installation, cable is buried directly in the ground without the protection of a conduit or tunnel structure. Cable is buried (a minimum of 0.5 m deep) with a special mechanical tool that first digs a trough and then places a cable within the trough (this process can be performed separately for installations where multiple cables will be installed in a single trough).

Since cable is buried without the aid of solid conduit systems, buried cable must often contain special protective and strength components. These include steel strands (both for added strength and rodent protection) and weatherproof coatings. Additionally, it is a common practice to place buried cable in plastic sheaths, usually brightly colored, to help in cable location. At a minimum, a protective layer of fill, similar to that required for buried conduit, should be present (a minimum of 0.3 m deep or as required by local codes).

Despite the lack of solid distribution hardware, there are two strong advantages to the use of buried cable. First, buried cable is relatively inexpensive; it is extremely low in both installation costs and hardware costs compared to most other distribution systems. Second, the lack of physical distribution hardware allows buried cable to be installed relatively quickly.

Unfortunately, the advantages of using buried cable for campus distribution can be easily outweighed by the disadvantages of such a system. For example, although buried cable is usually covered by a gravel or concrete fill (to protect the cable and to serve as a warning track for persons digging in the area), it is very susceptible to accidental damage from digging equipment. The "fixed" underground aspect of buried cable means there is virtually no way to physically inspect or repair a cable short of digging it up. Buried cable is also extremely limited in growth. The only way to add additional capacity to a buried system is to bury a new cable. But, it can be risky to dig alongside installed cable to add new facilities (it can be done, but care must be taken).

10.2.3.4 Aerial cable. The term *aerial cable* refers to any cable installed above ground between raised distribution supports. An example of aerial cable is wires supported by telephone poles or strung between buildings. This type of cable often contains special steel strength components which are required to prevent damage or accidental breakage of the cable caused by stress between distribution points. As an alternative, many aerial cables can be supported by separate steel cables (attached to the cable with cable loops) in order to provide needed support between distribution points.

Aerial cable is by far the cheapest and most flexible of all cable distribution methods; however, it is one of the least preferred methods. The advantages of aerial cables include low cost, ease of reconfiguration, and total physical access. This is by far the most inexpensive method of cable distribution for small quantities of cable (the larger the number of cables, the less cost effective this method because of added physical support requirements).

Aerial systems can be reconfigured at any time without significant construction requirements provided support structures can be installed (something that is relatively simple compared to building a tunnel or underground conduit system). Aerial cable also allows physical inspection and repair to all portions of a cable (provided you can get access via a cherry picker or other lifting device).

The strengths of aerial cable make it well suited for temporary installations or situations where cable cannot be placed within the ground, perhaps because of rocky soil or limited right-of-way paths. The disadvantages of aerial cable, including a lack of physical protection, susceptibility to vandalism (planned or unplanned), and appearance, are considerable compared to the strengths.

Aerial cable is constantly exposed to the elements (i.e., rain, wind, snow, ice, heat). Extreme weather can damage the outer sheath, resulting in potential inner-conductor damage, and also can damage support structures. For example, extreme ice buildup on a cable can add enough weight to the cable to bring down the communications line. Additionally, strong winds can topple even large cable support structures or cause large trees to fall against cables or distribution systems. Aerial cable is also subject to direct lightning strikes (this includes fiber-optic cables which usually have steel support strands either external or integral to the cable). This requires special grounding and lightning protection systems (i.e., lightning rods) which can add to the complexity, planning, and cost requirements of a campus distribution system.

The lack of a protected environment makes aerial cable a perfect target of vandalism. Included in this category are accidental incidents, such as kites or other unintentional guests, that can require removal from communications cables. This can heighten requirements for frequent physical inspection and can be dangerous if communications cables are placed on the same support structures as electrical cables; something that is never recommended. If communications cables and power cables must be supported by the same structures, the two types of cables should be clearly marked, located a minimum of 2 m from each other, and placed in a configuration such that communications maintenance personnel never have to interact with any electrical cable or pass it in any way.

10.3 DESIGN CRITERIA

There are many different factors which can influence the design of an FIS. These factors cover a wide range of functions and may vary in importance depending upon the particular installation environment.

Some of the most universal design factors are presented, in alphabetical order only, within this section.

10.3.1 Aesthetics

Having an FIS that is visually appealing can promote user acceptance and facilitate proper management of the system. For example, users that see a well-organized (i.e., hidden) cable system are more likely to think that the system is being managed properly than if they are constantly seeing cables between buildings, in hallways, and on office walls. As an example, there is very little about aerial cable that denotes a "high-tech" image. Looking for nontelecom examples, would a clean automobile repair shop attract more business than one littered with spare parts and old automobiles? Most likely, yes (I will acknowledge, however, that the better service shop will attract more repeat business; an item discussed in Chap. 11).

For cabling distribution systems that must be installed within common view, such as molding, choosing an appealing system can make the difference between an acceptable system and a system that can be classified as an eyesore. It is recommended that distribution systems installed within common areas be located in areas that do not make the item a visual highlight of the room. Additionally, blending FIS hardware into the facility environment can also be used to make a distribution system more appealing. This is commonly seen with aerial systems where using a distribution pole that is similar to other poles or at a minimum painted to blend into the surrounding background, can improve the image of the system.

Visual acceptance can also play a role in the geographic placement of FIS hardware. Very few people would argue the aesthetic value of placing several large conduits within a suspended ceiling versus in the open down a main office hallway. The same holds true for external installations. Placing an aerial system through the center of an office park is sure to gather at least passing attention.

While most people subscribe to the old adage "you can't judge a book by its cover" or "looks are not everything," most would also be willing to agree that aerial cable simply does not have that aesthetic appeal sought after in many of today's campus installations. In fact, it is not unheard of for local municipalities, associations, or business parks to prohibit the installation of aerial cable.

10.3.2 Distribution Frame Configuration

The location of both in-building and campus DFs can significantly affect the design of an FIS installation. Looking at a building configuration, if the IDFs are vertically aligned, the FIS configuration should follow a vertical path using some form of slot or sleeve device. For frames that are not vertically aligned, some form of horizontal distribution hardware must be used. The horizontal FIS distribution hardware will be subject to the environmental and plenum constraints that would affect regular horizontal distribution of IDF-SIO cables.

If a building is extremely diverse in its DF alignment, it may be best served by multiple FIS paths. These paths typically originate at the BDF and may follow the same route to a convenient divergence point. It is not required that multiple FIS paths converge a second time. See Fig. 10.4. The design of a campus FIS depends on the availability of established distribution paths, the location of the individual campus buildings (including the CDF), and the type of terrain connecting the buildings.

Two important questions that need to be asked during the design of a campus FIS are whether there is an existing distribution path through the campus and whether or not this path can support the requirements of the telecommunications system. This pertains to both renovations and new building construction within an existing campus environment. It also applies, to a limited extent, to new campus construction where distribution systems for other utility services may be used for telecommunications purposes (note that all utility and support systems should be designed at the same time to work in a cohesive manner).

When evaluating an existing distribution system, it is important to make sure that the system not only meets the distribution requirements but also offers the support required for a telecommunications system. For example, placing FIS cables within an existing conduit system is not acceptable if there are existing electrical cables present. Looking at existing tunnels, a tunnel system is not very suitable if it does not offer adequate protection from water or other physical elements (it is very difficult to perform fault isolation on a telecommuni-

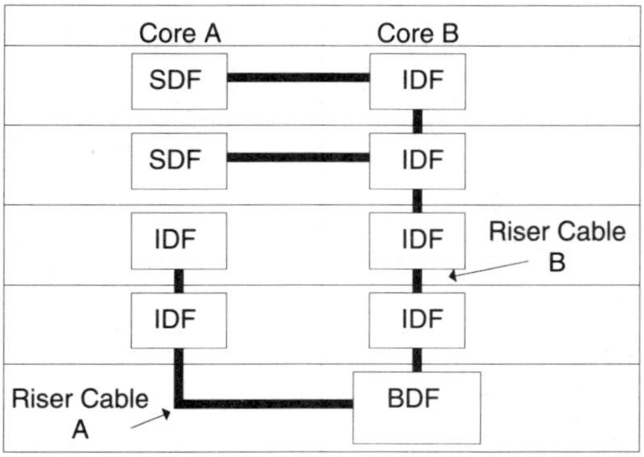

Figure 10.4 Dual FIS configuration.

cations cable that is either submerged in water or does not allow maintenance staff to enter because of high water levels).

When planning the layout of a new distribution system, the location of each building and the proximity to the CDF are very important. The goal should be to design a wiring distribution system that allows central concentration of cables at the CDF while maintaining a minimum distance between any two buildings. This is usually achieved through the use of a star-shaped FIS configuration. See Fig. 10.5.

In many campus environments, however, it may not be feasible to have the only means of interconnection through a single star configuration. Factors such as extreme distance or extremely high levels of communications between two adjacent buildings may require a modification to the star shape. This type of installation, a modified or hybrid star, involves the installation of redundant connections between facilities that may bypass the central CDF (note that this is not the same as an alternate or redundant path between a BDF and a CDF).

One approach that is generally not recommended for the FIS configuration is the loop or ring. In this type of configuration, a single distribution path is used to connect buildings together in a serial fashion. While a secondary loop may often be used as a backup to a primary

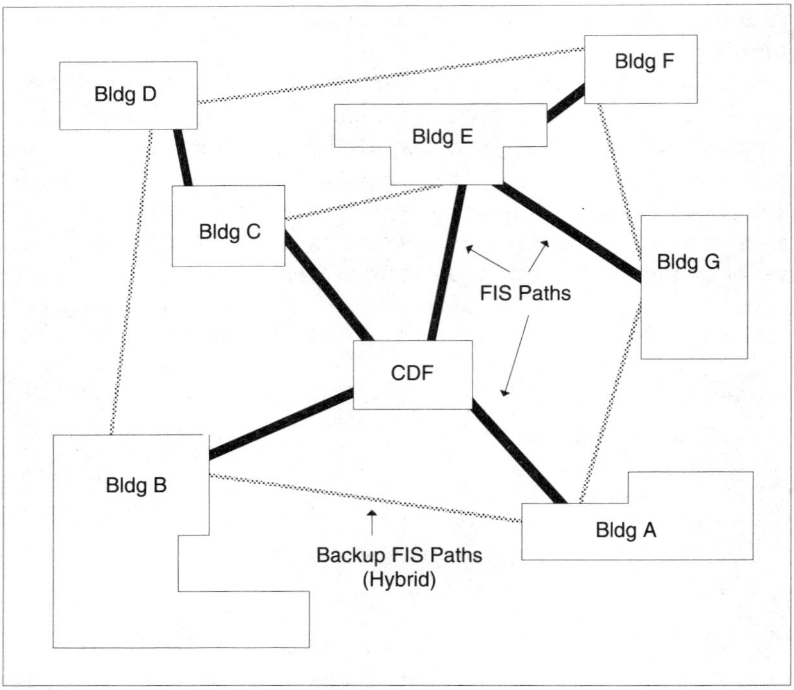

Figure 10.5 Campus FIS configuration.

star-shaped configuration, it does not provide the centralized support and minimum intrabuilding distances required for most campus communications systems.

An additional factor that must be considered when designing a campus FIS installation is the terrain between campus buildings. It is not uncommon for many types of distribution to be greatly restricted by rocky soil, lakes, or portions of unowned land between buildings. These types of environmental conditions should be considered both during the FIS design and during the selection of the CDF site.

10.3.3 Flexibility

The flexibility of an FIS can be key to the system being able to support additional growth or applications in the future. There are several key areas to be considered for FIS flexibility:

Capacity: The FIS should always be able to support an additional quantity of cable equal to 50 percent of the base requirement (100 percent or more is certainly acceptable).

Configuration: The FIS design should take into account any future growth in areas not initially directly served by the FIS. For example, FIS hardware to future installations should be built during the initial installation, where possible and not overly cost inefficient.

Modularity: The FIS should allow minor reconfiguration without too much difficulty. This will permit additional buildings to be added to the campus without having to completely reinstall all distribution hardware. Modularity should be used in installing the distribution system so that portions can be added and connected to new segments without affecting more than a small portion of the FIS installation.

A system that is rigid in both capacity and configuration can extremely limit the ability of the system to support future installations without significant cost or compounded management (i.e., trying to manage multiple distribution systems between two buildings).

10.3.4 Installation and Management

The design of an FIS must take into account the resources required to install and maintain such a system. Some of the issues that must be addressed include the following:

Installation costs: How much money will the installation require? This includes not only equipment and personnel costs but also the costs of resources not being available (i.e., users not being able to access a computer system because of FIS installation activities).

Installation duration: The amount of time required to install an FIS must take into account the time of year the system is being installed and the urgency of the installation. Failure to consider these points together can result in almost certain delays. For example, if a system must be installed within 2 months (such as might be required on a college campus), a tunnel system that requires 3 or more months may not be an effective solution. Additionally, the weather must be considered as well. Installing underground conduit can be difficult if not impossible during winter months in many northern locations. Such circumstances can often require an initial temporary system to provide service during the main system installation (such systems can also be useful when cutting over a major new system).

Maintenance costs: The costs associated with maintenance of an FIS installation include three important factors, personnel, equipment, and time, all of which relate directly to money. The number of people, and their qualifications, can vary from system to system. For example, an aerial distribution system that uses pressurized cable requires the talents of a maintenance person trained in pressurization systems. The type of cable used within an FIS can often require different types of testing equipment. Also different types of FIS hardware and cable can take different amounts of time to maintain. For example, a buried conduit system can be more difficult and require more time to perform fault isolation activities than an aerial cable.

Taking the above issues into consideration prior to the selection and design of an FIS can save time and also help to conserve financial and personnel resources.

10.4 INSTALLATION GUIDELINES

The installation of an FIS is both complex and time consuming. In addition, there are many factors outside of the normal FIS hardware and cable that must be considered. Factors such as when to install an FIS, potential interaction points with other utility systems, the testing of an FIS (including the end-to-end applications), and the management of the system must all be considered.

10.4.1 Installation Scheduling

The FIS, unlike most other wiring components, can include an extremely wide range of both equipment and cable types. Additionally, a single FIS may involve many different buildings and a diverse campus environment. This requires that the installation of the FIS be accomplished according to multiple schedules. For example, the schedule for in-building installation can be separate from the campus installation. Individual sections of the campus FIS may be installed at various times depending upon building completion and campus growth plans.

This divergence of schedules also affects when cable is installed. While all cables are usually installed within a building at the same time, there may be multiple schedules for installing campus wiring depending upon building requirements and the completion of portions of the campus FIS.

10.4.1.1 In-building installation. Basic guidelines for in-building FIS installations are very similar to the guidelines for most other types of in-building distribution systems, especially where physical components are concerned. The timing of the installation should be as follows:

Hardware installation: The installation of in-building FIS hardware often must take place during building construction. This is especially true of integral hardware items such as the NAP or other related building access points and any component that breaches a wall or floor, such as a slot or sleeve. This type of hardware must be installed during the main building installation (i.e., a sleeve is placed into the floor of a building as it is being poured). It is normal for the installation of FIS hardware to occur as soon as the building structure can support it, provided it is not in risk of damage from continued construction. Equipment such as ladders, ducts, and conduit should only be installed after the building exterior has been completed and the interior areas protected from outside elements. The nonintegral hardware is often installed concurrently with horizontal distribution systems. It is common for in-building FIS installations to mirror the installation activities of electrical hardware system installation.

Cable installation: As with any type of cable installation, there should be no FIS cable installed prior to the completion and satisfactory testing of the FIS hardware system. This must not occur before the building interior has been completed (basic structures only). This mirrors the installation guidelines of any in-building cable system. Given that telecommunications cable is much more sensitive than electrical cable systems, it is often installed well after the electrical grid is in place.

Building renovations may require special time-planning activities to accommodate the "must have it now" approach to many renovations. It is not uncommon for building renovations (or telecommunications system renovations) to take place while a building is being occupied. This usually requires the installation of the renovated components in a modular fashion so as to provide some level of working service immediately.

An entire FIS does not need to be installed at one time. The basic components that must be present to make any system function include the NAP-BDF connection and any connections between the BDF and "must have it now" floors. For example, if a 12-story building only requires immediate use of the first floor, the remaining floors do not need to be completely renovated immediately. The FIS equipment required for each floor can be installed in sequence as the next level (floor) is required. This is, obviously, an inclusive approach to multiple floors. If the third floor of a building requires immediate occupancy, the FIS for any other floor between the third and the BDF location must also be completed at the same time. Failure to perform concurrent renovations will prohibit the FIS cable from passing through intermediate floors to the served floor. Note that the facilities required to support the horizontal installation or cable termination (i.e., cross-connect blocks) do not need to be installed in IDF locations for pass-through purposes.

10.4.1.2 Campus installation. The installation of any campus FIS hardware must take place prior to the cutover of any served building. This often requires that FIS hardware be installed during building construction. Installation after building completion can make NAP installation difficult and often takes up too much time (digging tunnels or trenches is never a expeditious activity), thereby delaying the use of the facility. At a minimum, the entire hardware installation must be completely planned out prior to building construction. Failure to do so can put cable installations off and can also make NAP location and installation activities very difficult.

One of the advantages of a properly designed campus FIS is the modularity with which it can be implemented. This often translates into the requirement that only those portions of an FIS configuration required to provide immediate service to a particular building need be completed (unless of course a portion of an FIS is being used as a pass-through point for reaching other buildings). However, given the disruptive nature of FIS hardware installations, it is generally recommended that the entire FIS hardware system (within reason) be installed at one time. Not only will this save time, but it can also be less expensive and less disruptive than multiple installations occurring over a greater time period.

The installation of cable within a campus FIS should not be undertaken too early in the construction cycle. It is generally acceptable to install cable only after the relevant FIS hardware and serving building locations have been completely finished. Installation too early in the cycle can result in cable being exposed to environmental or construction abuses. Additionally, it is not a good idea to "prewire" future locations unless there is compelling financial justification (i.e., if we don't do it this year, it will never be funded again). The installation of cable to a future building site (provided the FIS hardware has been installed) can often result in disaster. First, the cable is often unmanaged during this time period and may be subject to abuse or adverse environmental conditions without fast problem detection (and, as we all know, an undetected problem grows very rapidly in magnitude). Second, given that buildings are constructed relatively slowly, the requirements for the FIS cable may change drastically during the months or years between FIS installation and building completion.

10.4.2 Danger Points

There are many potential danger points within any FIS installation. These points should be clearly discussed during the design and installation phases. Some of the more common danger points are as follows:

Power line interaction: Unknown to most people, the eighth deadly sin is the placing of telecommunications cable alongside power cable. Put aside, for a moment, any technical issues regarding collocation of equipment. The predominant concern is the accidental interaction between the two systems from a maintenance perspective. A telecommunications technician should never have to navigate around or come into contact with any type of power-bearing circuit. Even close placement of the two systems is not very acceptable since most contact will be accidental. Given the potentially fatal nature of a worker cutting into or otherwise coming into contact with a power line, the two types of cable should not be buried together or placed alongside each other in tunnels or on aerial systems. It is important to make sure prior to approving installation plans and during the field inspection that all power and communications cables have been properly designed and separated.

Steam pipes: The ability of steam to burn and melt cable is well documented. Steam can quickly damage the outer sheath and interior insulation of many different types of cable. No type of communications cable should be placed near a steam pipe. If this is required, the cable installed should be environmentally rated to handle not only the heat but also the moisture associated with a steam accident.

Flooding: The addition of moisture into a cable can alter the electrical properties of the insulation material. This can change the velocity at which an electrical signal will propagate through a cable, potentially resulting in an open or short in the cable. Flooding can occur in any cable environment, including buried cable, cable in tunnels or conduit, and aerial cable (which can have water enter through exposed contacts and "flood" the interior of the cable). All cable systems should be properly designed to use waterproof cable (i.e., sealed, pressurized) and/or have sufficient drainage capabilities for tunnels, conduit, or buried cable. Buried cable drainage requires sufficient drainage capabilities in the soil surrounding the cable, such as gravel fill. This will help to keep the water table low in the vicinity of the cable. Care should also be taken not to install cable in areas with high water tables that are likely to be flooded or moist for extended periods.

Cable supports: A common failure point within building wiring systems involves improper cable support. This includes the supporting of horizontal as well as vertical cable (both internal and external cables). Horizontal cables must have support systems that can adequately support the current cable weight and the total possible weight of expansion cable. A major weakness of cable support systems is the inability to handle major additions of cable, which although unplanned for do occur in the majority of cable installations. Vertical support (and aerial support) requires that the cable be properly supported both at the highest point of the cable run (i.e., the top of a vertical riser) and at intermediate points throughout the cable run (i.e., within individual intermediate DFs) to help offset the strain on the top cable support. Cables are often added to systems after initial installation and, unfortunately, a large number of new cables are often lashed to existing cables for support. This type of installation is not acceptable. However, given that it is bound to happen sooner or later, it is wise to plan in advance for this and build additional support into FIS cable systems.

Lack of access points: A common mistake when planning or installing an FIS is a lack of proper access points. It is imperative that maintenance personnel be able to access cable and related facilities in a fast and convenient manner. This will not only facilitate routine maintenance but will also help to promote expeditious fault isolation and resolution. This is an issue for both campus and in-building cable facilities and should be a consideration in the type of distribution hardware selected since some types of hardware offer better access properties than others.

Utility proximity: It is often thought that by placing telecommunications cables within a relatively close proximity to other utility system components, there is an advantage to be gained in structural support and access capabilities (i.e., reduced costs). While this is true to a certain extent, there is the risk of a utility maintenance worker (or utility malfunction) damaging a telecommunications cable. For example, a water utility excavation operation should not have to worry about encountering telecommunications cable during digging or repair activities. Complete diagrams of any FIS configuration should be supplied to any maintenance worker who may be working on any type of system in the general area. Additionally, many localities have special hot-line phone numbers that must be contacted prior to any digging activities. These numbers are staffed with personnel who can readily tell you if there are any buried cables or utilities within your immediate area and in many cases can provide on-site support or inspections.

Cable protection: There are many components to cable protection, and each of them may be a potential danger point. Cable must be properly secured from an access perspective. It must also be protected from environmental elements, such as lightning, water, or hot and cold. An additional concern is vandalism. Any activities by people can be partially hampered by proper access security. Accidental animal vandalism, however, requires much more than access control. Special cables are often required for both in-building and campus installations. Using special hardened plastic sheaths or metal rods, these cables provide a deterrent to rodents and burrowing animals from chewing into or through communications cables.

Third-party contact: An extreme danger point in any FIS installation, especially outdoor installations, is accidental interaction between the FIS and a noncommunications or utility person. For example, the child who gets a kite tangled in your aerial wires or who plays detective in your underground tunnel system. Care must be taken to ensure that accidental contact is minimized as much as possible. This includes posting warning signs (i.e., Do Not Fly Kites Near Wires) and installing physical barriers such as steel bars and/or secure entranceways over potential access points. This will not only improve the up-time of your system but will also help to ensure that accidental injury to inquisitive third parties does not occur. There should also be an emergency plan in place to deal with any type of unplanned interruption of service or personal injury (falling trees, flooding, etc.).

The above list is not intended to be all-inclusive. There are many other potential points that may vary in scope or severity according to the environment of a particular installation. Whenever an area of potential concern arises, a qualified person (i.e., certified relative to the potential danger point) should be consulted.

10.4.3 Configuration Testing

There are two methods for testing the viability or soundness of an FIS installation. The first is by far the easiest; you cut over to the new system and see if it works. The second is a bit more complex and costly, although it will help with job security. Any type of cable installation must be properly inspected and tested prior to use. There are three main areas of testing activities: hardware, cabling, and communications.

The testing of hardware (support) systems involves a thorough physical inspection. The system should not only be tested for physical soundness but also for its finish, which refers to the absence of any elements that may damage cable during installation or maintenance activities (rough edges, undersized junction boxes, etc.). This includes the interaction (or preferable lack of) with other utility systems. Areas such as grounding systems, cross-connect block mountings, and SIO installations are some of the critical areas often overlooked.

The testing and inspection of telecommunications cable should take place both during and after system installation. During the installation process, the work activities should be monitored to ensure that the cable is being installed in such a manner as to reduce the chances of accidental damage to the cable (i.e., too much cable bending or scraping of cable against rough edges). After the installation process, the cable should be checked to determine if the final installation is both terminated properly (cross-connects and SIO unit punch-downs) and meets the electrical or optical requirements of your communications applications. The electrical or optical testing will involve specialized test equipment, with the final test being the ability of the cable to support a working installation (see the chapters dealing with specific cable media for more testing information). Basic testing activities, such as passing a telephone tone down a twisted-pair or checking for shorts and opens, are not enough.

Once a system has been physically installed, it is generally a good idea to stage a small trial run of some of your more critical communications systems. These do not have to be full-blown installations. Small test-bed installations, or network simulations, can often be used for this purpose. These types of tests, generally reserved for the more sensitive high-speed data applications, help to serve as the final OK and can often pick up minor cable problems that would go undetected by standard cable testing activities. This type of check can also help to determine if the requirements given to the vendor actually meet the system requirements (occasionally an end user has not fully explained the system's requirements to a vendor).

Of course, all the criteria listed above are targeted to ensuring that the system meets your basic requirements. All too often systems are installed and tested for stability without being tested to determine that they actually meet the requirements set forth in the requirements analysis portion of the design cycle, in other words, testing to see whether or not the vendor has installed a system that meets all of the specifications as provided to the vendor. This area should not ever be overlooked, especially when the work is being performed by an outside organization.

10.4.4 System Marking and Identification

An important component of the FIS is the system marking and identification element. The proper marking of all telecommunications distribution components will help to not only facilitate proper management but also to reduce the accidental contact that can often disrupt or damage communications facilities. For example, the proper marking of individual cables within a conduit can help a maintenance person quickly locate a particular cable. Likewise, the proper labeling of aerial cables can be used to visually separate telecommunications cables from other types of aerial wires. Taking this one step further, proper marking can help prevent digging accidents.

There are several very important components that must be clearly labeled within an FIS, as follows:

Riser cables: These cables should be clearly labeled at both intermediate and end locations. Intermediate locations are any DF facility. Individual pairs or cables within a multiunit sheath should also be clearly labeled within the distribution block system. Riser cables can be marked with either tape markers or marker plates. A tape marker consists of a piece of white (or other light color) tape upon which a cable identification is marked. This method is generally suitable for indoor use only. A marker plate is a flat metal or plastic plate, containing identification information, that is fastened to the cable. Marker plates, given their durability, are best suited for outdoor installations.

Campus cables: Cables installed within the campus portion of an FIS should be clearly marked. The most common method is with marker plates because of their physical strength. Tape markers generally do not stand up to the environmental rigors, although this is changing as techniques are improving. Problems such as smeared letters or fraying from wear and tear are common with tape markers.

DF facilities: Any component within a DF that supports an FIS connection should be clearly labeled as an FIS support system. This includes NAP facilities and any supporting hardware. It is not sufficient to mark a component as an FIS facility; it must also be marked to indicate what DF locations are being supported.

External support hardware: Campus FIS installations are notorious for not properly marking facilities. Items such as aerial support poles, exposed conduit, and buried cable should all be properly marked as supporting telecommunications cable. Marker plates should be used for exposed components while flag markers should be used for any type of buried cable or conduit. This will help prevent accidental damage and also facilitate faster reporting of damaged components. For example, suppose a passerby reports that a telephone pole has been severely damaged and appears in danger of falling over. If the pole clearly is marked as supporting telecommunications cable, the telecommunications maintenance personnel can be quickly notified of the impending fault and potentially bypass the system before the component failure.

Additional information about types of labeling schemes can be found in Chap. 11.

10.5 SUMMARY

The following is a summary of the key points of this chapter:

- The FIS supports campuswide communications and in-building communications between distribution frames.
- The FIS may include a wide range of transmission media, including twisted-pair, coax, fiber-optic, microwave, and infrared, based upon end-user requirements.
- The FIS within a building closely mirrors the cable and distribution hardware of horizontal distribution media.
- Campus FISs that use aerial and underground distribution may require specialized cable to withstand outdoor installation.
- Design, installation, and management of the FIS should promote flexibility in DF service.

Chapter

11

Cable Management

Definition: *Cable management* The act of maintaining and administrating a building wiring system in a sound working order to allow simplified operations of telecommunications systems. Note that this must be accomplished without the use of telepathic or astrological agents.

An important, yet often underemphasized, component of any building wiring system is the cable management system. Cable management is the element that binds together the individual components of the overall system (i.e., the "cosmic glue"). It is a required element in any successful wiring system. Unfortunately, since it is often the most nebulous aspect of building wiring, it is often overlooked or underdeveloped. This chapter discusses the various components of building wiring management and presents some guidelines for developing a comprehensive support and maintenance capability.

11.1 THE ENTROPY FACTOR

There is a well-known law within the universe that governs the order of all elements. Commonly referred to as the *entropy factor*, it simply states that all elements will eventually strive to reach a chaotic state of affairs. In other words, a state of disarray. Randomness. Confusion. Pandemonium. The list goes on and on and it applies to every building wiring system ever installed.

The entropy factor can often manifest itself in the form of unmarked cables connected to distribution frames (DFs), messy DFs, spare cable with no purpose, or, even worse, connected cable with no known purpose. It is not uncommon to discover cables in older wiring systems that have no known endpoints, yet plenty of data traffic is present. Another common chaotic scene involves old unused cables that cannot be removed from the wiring system because nobody is quite sure of the cable path through the building or campus (these are usually in distribution hardware that is so choked with cable that it would be impossible to remove them without damaging all other cables). These scenarios are very common in point-to-point wiring systems that do not follow a structured hierarchical approach. However, the latter type of wiring system can also be just as dysfunctional given a lack of proper design or cable management practices.

The advantages of a good cable management system can be found in the goals of building wiring cable management (which are designed to promote a sense of order and well-being into a wiring system):

- Increased cable action audit trail capabilities
- Fast isolation of cable faults
- Increased system flexibility for future expansion
- Reduced recabling costs for new applications
- Increased user satisfaction through improved management support

These goals generally cover most aspects of the advantages of cable management and, when taken into consideration, can make the difference between a successful or a failing cable system.

There are many individual components of a complete wiring management system, each of which must work directly with the other components or the system will not be manageable in and of itself. When used together, they can help prevent the chaotic dark side of building wiring from taking over your building wiring system.

Cable labels: These clearly label the identity of every cable within the building wiring system. This includes all horizontal, vertical, and campus cables (but not their individual pairs; strands of fiber are included here when possible).

Component labels: Each cross-connect block, patch panel, splice point, and standard information outlet (SIO) unit should be clearly marked with a unique identifier. This includes distribution hardware such as conduit and ducts.

Hardware configuration records: The general layout of all hardware should be clearly documented. This information should include other utility system information where applicable (power lines, sprinkler systems, air ducts, etc.).

Installation records: The installation dates and accompanying notes can be vital to maintaining an aging wiring system. In addition, having a record of any difficulties during an installation process can help in fault isolation activities at a later time. Installation records are not only concerned with wiring and support hardware but also must take into account the installation of end-user applications that result in cross-connects or patch-panel connections (or in any way alter the state of a wiring system).

Change records: Any change made to a building wiring system should be clearly recorded. This not only helps to make an audit trail of changes but can also be used to monitor work activities for action trends (i.e., the frequency of certain management activities or the failure rate of certain cables or cross-connects).

Circuit records: Multiple individual cable segments within a building wiring system comprise a complete end-to-end circuit connection. There must be a system for correlation of individual cable segments to an end-to-end circuit which serves a single network application [modem line, local area network (LAN) device, private branch exchange (PBX) handset, computer-computer connection, etc.]. Such a record is immensely useful when trying to isolate networking problems within a cabling system.

These components can be implemented using many different methods. However, they must all be present regardless of the implementation structure. Without such components in place, the remaining components become ineffective and will eventually be discarded by maintenance personnel (people in general tend to not continue the use of any system that takes up time but does not give anything of appreciable nature in return).

11.2 NATIVE RECORD SYSTEMS

The most basic components that make up a cable management system are those that involve the labeling and configuration of a building wiring system. This includes the labeling of cable and hardware and are made possible by proper wiring design. These components, when properly implemented, provide for a built-in, or native, form of cable management. Often referred to as a *living record system*, such a system that is properly labeled can allow a maintenance person to quickly ascertain the location and status of any cable within the system. In effect, it is a system with built-in tracking mechanisms.

While such a native system does not provide for chronological (or any form of) audit trail tracking, it can be useful in many installations. More importantly, without such a system in place, most other forms of cable management (i.e., installation or change records) are bound to fail as strategic components of a building wiring management system.

An important preparatory action to facilitate a native record system is a thorough building wiring design following a structured approach. This allows native systems to be implemented to their fullest potential. Only with a building wiring plan that is orderly and hierarchical in nature can the numbering and labeling of such a system be of benefit because of the structured nature of any management system. While an unstructured building wiring system can be supported by native record management, it does not allow the structured management principle to fully integrate with the wiring system.

11.2.1 Numbering Systems

The numbering scheme used within a building wiring system must provide a structured plan for relating individual identifiers to segments of cable and also for relating an overall circuit number to a group of individual cable connections (an end-to-end circuit). Such a numbering scheme (which may consist of both numbers and letters, i.e., alphanumeric) must begin with the overall picture in mind. This requires that each building within a campus be given a unique identifier, which can be either an alphanumeric or a short code representing the name of the building (FR for Franklin, HE for Health, etc.). Within each building, each building distribution frame (BDF) and intermediate distribution frame (IDF) is assigned an alphanumeric representing the floor and the sequential DF within that floor. Each satellite distribution frame (SDF) is in turn given a sequential identification associated with the serving IDF. In this manner, we can create a logical identifier representing each DF within any campus environment.

Looking at this type of hierarchical numbering scheme, we can see that each DF will have a unique identifier that cannot be duplicated. For example, suppose that we have a campus consisting of four buildings: North, South, East, and West. All DFs associated with each building will start with either an N, S, E, or W. Looking at the IDF on the fourth floor of the North building, we can derive a name of N4. If there are multiple IDFs within the fourth floor, we end up with N4A (where A represents the sequential IDF). When multiple IDFs are present, there are usually multiple riser columns (vertical distribution). It is recommended that all IDFs served by a single riser system have the same sequential identifier. When SDF units are served by an IDF, we can extend the hierarchical identification to include a number or lowercase letter to represent each SDF. For example, the third SDF on the fourth floor of the North building is identified as the N4A3 DF.

Once each DF within a campus has been identified and labeled, an identification code can be placed on each SIO unit. There are two numbering recommendations that are popular today. The first relies upon a flat numbering scheme within a building or floor. For example, the thirty-fifth SIO within a floor is numbered as unit 35. There is no attempt in this system to relate any type of ordered identification.

The second approach follows the form of the DF numbering scheme where each SIO unit is given a sequential number based upon the number of units served by an individual SDF or IDF. For example, the same thirty-fifth SIO served directly by the second IDF on the third floor of the South building is known as S3B#35. The advantage in using the hierarchical numbering scheme throughout the DF and SIO identification process is that the addressing scheme allows a technician to clearly identify the DF path of any SIO serving cable without having to refer to records or database files.

After each SIO and DF have been given an identification code, the cables connecting each location can be identified as well. There are many different methods for creating the cable numbering scheme. Fortunately, there are no right or wrong systems here; telecom managers can use what they feel the most comfortable with depending upon the size and complexity of their building wiring system.

A common method of labeling cables that pass between DF locations is to list the endpoint locations with the addition of an application specifier (i.e., *V*oice, *D*ata, vide*O*, etc.). In addition, the prefix C is often used to make clear that the following nomenclature refers to a specific cable and not just a DF or SIO. Using this numbering scheme, we would label a voice cable between the BDF and the fourth floor IDF within the North building as CV-N:N4V (the use of the N without any number refers to the BDF which is the prime DF within any building). Likewise, a cable serving an SDF on the same floor could be represented as CV-N4:N4A. The CV indicates that this is a voice cable while the N4:N4A indicates the two points connected by the cable. In addition, a sequential number is appended to indicate multiple cables between locations. Occasionally, an additional number can be appended to indicate the number of twisted-pairs or fiber strands within a single cable sheath. It is recommended that this number appear in parentheses to clearly differentiate it from any type of sequential identification number. For example, the code CD-N:N5A(100) refers to a 100-pair cable used for data traffic between the BDF and fifth floor (A) DF. Individual number codes can also be appended to represent an individual pair or strand within a sheath. See Fig. 11.1.

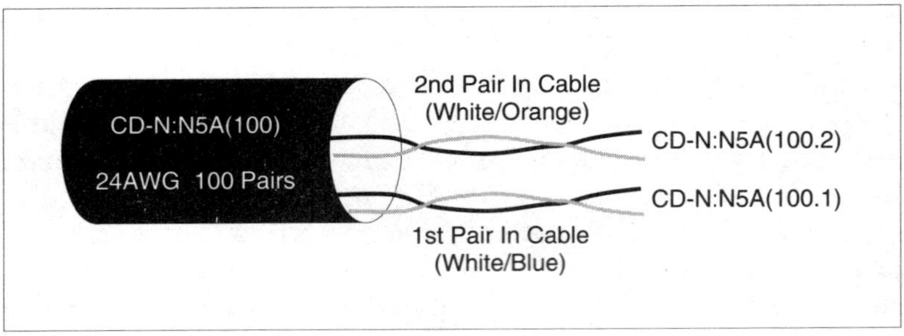

Figure 11.1 Numbering scheme.

The use of a unique identifier can also be applied to complete *circuit connections* within a building or campus wiring system. A circuit connection links two remote sites together and is made up of individual point-to-point links (typically between DF facilities). An example of a circuit connection would be a circuit between two computers located in different buildings. For example, the first computer may be attached to the fifth SIO unit on the fourth floor of the North building (N5A#5) while the second may be attached to the sixth SIO unit served by the second SDF attached to the third IDF on the third floor of the South building (S3C2#6). While these may seem complicated, they can be easily used with a CC prefix for circuit connection to represent a complete path between two endpoints: CC-N5A#5:S3C2#6.

The use of a circuit connection identifier only applies to a logical connection since the identifier does not specify the path of the circuit, only the end locations. However, this unique code can have individual circuit connections assigned against it within a record tracking system. Such a system helps to eliminate the alternate path code of N5A#5:N5A:N:S:S3C:S3C2:S3C2#6, which, although long and complex, does indicate a complete path from SIO to SIO (such a number could be even more complex if the individual link codes for each cable segment and pair and strand number were used).

11.2.2 Color Coding

The use of color as an identifier of different types of elements has long been used within both technical and nontechnical trades. There are numerous examples of color being used to differentiate similar, yet different, components. So let's apply color coding to individual components within a building wiring system. There are many different candidates for color coding:

Station outlet cables: It is common to see differently colored cable used for identifying voice, data, and video applications. For example, while silver/satin cables are extremely common for voice, many users are making use of black or brown cables for data applications.

Standard information outlets: Many modular outlet vendors allow different colors to be used for voice, data, and video outlet jacks within a single SIO. This goes one step further than simply labeling a jack as V or D (especially since labels can be accidentally removed or can fade with time).

Distribution cables: While less common than SIO coloring, the use of differently colored cables to represent voice, data, and video types is growing. Even the frame interconnection system (FIS) can benefit from a colored cable approach, allowing multiple applications (voice, data, video) as well as cable type (coax, fiber, twisted-pair) to be identifiable by color coding.

Cross-connect blocks: Many of the more advanced cross-connect block systems offer color coding that matches the band-marked color code used throughout the industry for coding the sequence of twisted-pairs within a multipair cable.

Patch cables: The cable used to connect cross-connect blocks and patch panels is often color coded. A common coding scheme uses white-based cross-connect cable for voice applications (i.e., white/blue and blue/white) and red-based cable for data applications (i.e., red/blue and blue/red).

The use of color-coded components may cost a bit more in some installations. However, most mid- to large-sized users can usually find a cable vendor willing to accommodate color coding without significant charge. Remember that any slight increase in cable cost can usually be quickly offset by reduced fault-isolation and repair times. Coded components can also reduce the instance of mismatched connections of end-user equipment (not surprising, a large number of "network" problems result from the network plugs being inserted into the wrong wiring jack, i.e., voice into data).

Color-coded components used for cross-connect equipment are generally standardized around the band-marked color code. However, the use of color for other components is generally up to the user to specify an appropriate color scheme. Additionally, the use of a color code does not mean that all cables need to be completely colored. For example, the use of colored labels or periodic painted or taped markers can work equally as well in many installations (especially FIS cables installed in open areas).

11.3 HARD-COPY RECORDS

Each building wiring system should have a complete set of hard-copy records (read this as documentation) that details both the current status and past history of the system. This is required in addition to the use of "native" numbering schemes which help significantly with the creation of hard-copy records. While a hierarchical numbering system throughout a wiring system will help with cable and circuit identification, the hard-copy system is what provides the true record of events, which can be vital in understanding the changing nature of a wiring system.

11.3.1 The Role of Hard-Copy Records

The hard-copy record serves as the mark in history of each installation or modification made to a wiring system. This type of recordkeeping has several useful functions:

Component status: It provides an accurate means by which the status of any component can be easily brought to light. Any action taking place involving a wiring component should have either an installation record or a change and action record. This record should indicate the original and the new configurations.

Audit trail: Good records can provide the manager with an audit trail of events that can be used to follow the progression of a wiring system. The ability to look at a chronological series of events can be beneficial in assessing the failure rate of a component, the types of repairs or changes that are required, or (perhaps the most important) who has been making changes to the system.

System heartbeat: If a hard-copy record system is accurately used to record all relevant wiring system action, it can often be used to provide a pulse or health check of the overall system, since it records the type and frequency of actions being performed.

11.3.2 Information Content

To provide truly useful information, the hard-copy record must be able to accommodate a detailed list of actions. This includes:

- *Work actions:* Repair activities, such as cable or SIO replacement
- *Installation records:* New cable or hardware installations
- *Changes:* New cross-connects or cable reconfiguration

The tracking of the above information may require separate forms, depending upon the complexity of information being gathered. Combined forms can be used; however, they will add to the complexity of the form. Regardless of the type of action being performed, there will be certain bits of information that must be tracked and included on the form. This common information includes the following items:

- *Who:* The name(s) of the persons involved
- *What:* The type of action performed
- *When:* The date and time that the action took place
- *Where:* The location of all components involved
- *Why:* What work order or activity prompted this action
- *Testing:* What type of test was performed, the results of the test, and who was responsible for certifying the work as satisfactory

Developing a form or set of forms that allow for the above information to be recorded can be increasingly difficult in large, complex building wiring systems. Specialization of forms, where certain information areas are customized to meet the exact physical layout or configuration of a wiring system, can often make the task of recording information much easier. As an example, if your wiring system is properly structured in such a manner that all DFs are arranged in the same basic configuration, a diagram of that configuration may be useful on the hard-copy record. Wiring technicians can use the diagram on the form to mark changes or special features of the performed work.

11.3.3 Administration

An information tracking system can only be useful if the system is properly implemented. This requires that any hard-copy record system be strictly administered and followed at all times. Failure to include information on a form can result in inaccuracies later on since current information is not available or is not correct. Even worse, failure to use the forms for all activities can result in a lack of confidence in the entire system (of what use is a file system that only contains half the required records?).

To ensure that the appropriate forms are used properly, it is necessary to make the form as easy to use as possible. The required information fields should be structured in such a way as to encourage their use; unorganized forms that do not follow the flow of work performed will not be used properly since they will require too much time to fill out. All forms should also be collected daily and filed within a central storage area. Access to these files should be strictly controlled and only copies of individual records should be allowed to be removed. The file system should also provide some type of cross-index that allows records to be located by several different criteria, such as the day work was performed, the person performing the work, or the type of work performed.

11.4 AUTOMATED RECORD SYSTEMS

One of the best methods available to encourage the use of a cable management system is to automate the system on a computer. Computerized record systems can allow fast retrieval of information. They can also be used to prepare detailed reports pertaining to cable management activities. However, they do require special development efforts and administrative support. Before an automated record system can be implemented, there are several important factors that must be understood. These factors, such as the role of an automated system in relation to a hard-copy system, how it will be used and maintained (not to mention created), and what type of information can be retrieved, must be understood by all involved parties for the system to be successful.

11.4.1 The Role of Automated Records

It is often mistakenly thought that an automated record system will negate the requirement for a hard-copy record system. The concept of a paperless office is often directly linked with computer automation. Unfortunately, this link is not valid. The primary function of any automated record system, including the most sophisticated systems available, should be to augment the capabilities of a well-designed hard-copy record system.

The logging and assignment of incoming fault reports and change requests are well suited to a computerized tracking system. Information can be entered directly upon receipt of the request and work orders can be generated on a computer printer. Major changes to the system, however, are typically not well supported through this type of computerized process; details of major renovative changes are usually decided by committee and implemented in large, sweeping projects. Such major projects do not usually mesh well with automated trouble response systems, although if they can be integrated (i.e., the major plan is accomplished through much smaller work orders), they should be. Nonetheless, the most effective method of conveying information to technical staff is through the use of hard-copy work orders.

An alternative to the hard-copy work order is to have each technician use some type of electronic device to access the master database.

In this system, each technician has a personal computer or other such device that can communicate with the main automated record system. This communication can often be bidirectional where the technician cannot only receive information but can pass back information on work activities. While this type of system can often be effective for large implementations where technicians must cover large areas (large campuses or metropolitan areas), they are not usually effective for smaller organizations confined to only a few buildings.

Even when an automated system is in use, there should be hard-copy records available. These records act as a backup of the electronic version, although they often serve as the first-line technician interface. This is required since computer systems are subject to periodic failure. Such failures should not disrupt management activities altogether (backup computer systems are strongly recommended).

11.4.2 System Design

The capabilities of an automated system can vary greatly from system to system. It may be a simple database allowing electronic storage or it may be a complicated system that graphically displays cable paths throughout a building or campus. Some of the more complex systems treat the wiring system as an extension of a telecommunications network and even allow individual cables to be allocated against an end-to-end connection (of course, the actual work must still be performed by a technician). In this regard, the software and hardware required to support an automated record system can differ considerably between simple and complex systems.

The simplest automated record system can usually function on any standard personal computer using a commercially available database package. The programming required to set up such a system can often be performed by in-house programming staff in a matter of weeks. The simple system usually mirrors the information contained within the hard-copy record system but does not go any further in terms of additional capabilities (i.e., graphical design and layout display). Any automated system should include the ability to print standard forms and reports based upon previous activities.

The more complex systems, however, are usually based on computer-aided design (CAD) programs and can be much more complex. Permitting the graphic representation of complete buildings, campuses, and wiring systems (down to the individual cross-connect pair), these systems often require powerful computers such as high-end personal computers and UNIX workstations. This type of system can be extremely costly, not only for the workstation but for the software development as well. However, there are a few commercially available packages which can be easily customized to meet most end-user wiring requirements.

While there are some systems available off-the-shelf today, many organizations often create their own customized systems. This can be especially common in large organizations where the automated wiring record system may be integrated into a larger management information system (MIS). Such systems often integrate the wiring and network systems into one overall system. This can make sense since the wiring within a building is often viewed as merely the physical layer of the much larger network. The network, which can include voice, data, and video applications, often involves mixed fault-isolation and resolution activities. For example, the addition of a new user onto a LAN involves not only computer and LAN support but building wiring support as well. The same is true for voice and video applications; each uses the building wiring system as a subcomponent. The process of fault isolation (i.e., an end user cannot access the network) can often involve both computer network and building wiring components.

When deciding on the type of automated tracking system to be implemented, it is important to consider many factors:

System complexity: *How complex a system is required to augment the hard-copy record system? Will the new system be a simple electronic version of paper records or will it have value-added features?* Generally speaking, the value-added approach often affords a greater degree of flexibility and growth. Even if additional functions are not needed immediately, a system that has the capability of providing additional functionality will usually have a better structural design and be capable of growing as your network needs expand.

System integration: *Is it advantageous to integrate the wiring record system into a larger or preexisting MIS system?* Unfortunately, the marriage of wiring and MIS management systems often requires significant custom development work because most MIS systems are mini- or mainframe computer-based while most network or cabling management systems are based on personal computers (PCs). Cross-development of applications between the high- and low-end computer types is not that common for most of the newer software that is powerful enough to manage entire building wiring or network computer systems. Additionally, most MIS systems tend to be very oriented toward the management and accounting aspects of business, often overlooking the operations aspect common in a network management system.

Computer support: *What type of computer will be required to sup-*

port the system? Is a multiuser system required so that more than one person can access the system at one time? What type of storage and printing capabilities are required? The type of computer used to support a wiring management system can vary considerably depending upon the size and complexity of your building environment and the level of detail supported by the system. Low-end PCs, either DOS or OS/2 based, can usually support most management systems provided extensive detail and graphics are not required. Most systems that provide CAD-like drawing and tracing of cable circuits usually require a high-end machine, based on an Intel 80386, Motorola 68030, or SPARC technology.

Technician support: How will the technicians access the information? Printed hard copy? Remote terminal or computer? The ability of a wiring management system to print information is essential. This should include not only replications of hard-copy forms but summary reports as well. Additionally, the ability of a system to support local or remote terminal access can be quite beneficial to the support technicians who can query or input information in a real-time manner.

System capacity: How many records can the system support? How many types of actions and what type of information can be stored? The obvious answer is that a good wiring management system can store the records of all actions that take place within a wiring system for instant retrieval. However, this is somewhat unpractical. More realistic is a system that allows older or less-current information to be archived. Archived information, often stored on a backup computer, disk, or tape, can be accessed for summary information searches or for tracing the life history of a wiring system. The only information that needs to be stored in the real-time management database is information that is relatively recent (i.e., less than 12 months old) or pertains to the current or just prior status of any wiring component.

System display: Is it acceptable to view information in a textual form only or is a graphic display required? If graphics are required, how complex a display is needed (the display of a route, a piece of distribution hardware, a cross-connect block, an individual twisted-pair, etc.)? Basic low-end management systems usually provide a text display concerning actions and status events. However, this does nothing to simplify management actions, such as tracing a circuit path through a campus. The ability to visually display a wiring system can be of significant benefit to both wiring system planners and maintenance staff. The most current systems on the market use CAD/CAM display technology to literally create a blueprint of a building or campus. Individual cables can be traced and allocated against end-user SIO units or applications. Distribution hardware, as well as electrical or utility information, can also be displayed to help present a realistic view of the wiring system. Perhaps the best example of a cable management display is the COMMAND system by ISICAD, which provides a CAD-like display of wiring, patch panels, floor plans; and communications equipment.

System cost: How much money and time can be expended to develop the system? This can range from a few thousand dollars to tens of thousands of dollars depending upon the complexity of your system. It is important to note here that a good management system is an investment that will help reduce costs in other areas, such as fault isolation, system configuration, and system planning. It is not uncommon to spend $5000 to $10,000 on a hardware platform and an additional $5000 to $10,000 on software for a fairly complete off-the-shelf wiring management package. Customized packages can be less expensive but usually end up costing much more because of the value-added or tailored capabilities.

Once these questions have been addressed, the decision to either purchase or develop a system can be made. If the decision is to purchase a system, it is important that the system be judged fairly prior to selection. This includes comparing a minimum of two or three different packages. Some of the more important off-the-shelf evaluation criteria include the following:

Documentation: The program should be well documented, including information on installation, system configuration, operations, management, and error control. The documentation should be clear, easy to follow, and, above all, complete.

Installation activities: The installation process for any computer software package can often be difficult and time consuming. A good package will allow system configuration in a fast and modular fashion. This includes support for any networking connections as well as peripheral devices.

User interface: One of the most important aspects of any management system is the user interface. A graphical interface, menu and mouse driven, is generally accepted as the easiest form of user interface to work with. The interface applies to all aspects of information input and output routines. Users should not have to go through multiple manual (i.e., keyboard) commands to enter or retrieve information, unless they want to. The processing speed of the system also affects the user interface. Beware of systems that require very much time to process simple commands; they tend to get very slow for large reports and information queries.

Expandability: The system should allow for expansion of both software and hardware capabilities. For example, you may want to add additional functionality to the system. A good system will allow functions to be added in a modular fashion without affecting the overall system. From a hardware perspective, the system should allow increases in the number of peripherals or support the ability to increase the size of the database without having to discard the entire current system.

Maintenance and support: The vendor should offer some type of service and support contract (not to mention a basic system warranty). On-site installation and training support is generally very helpful. Telephone support for the software is also a good idea for small problems while on-site support is almost mandatory for major catastrophes such as hardware or software failure. Looking at the inevitable disaster, the system should allow complete restoration (either within the existing platform or the migration to a new computer). This includes the ability to backup and restore information in a timely manner (all-night backups are not acceptable).

Installed base: As with any major purchase, it is recommended that the selected vendor have a large installed user base. Speaking with several customers with similar building wiring configurations can usually give you a good idea of how the vendor will live up to the above items.

An additional item for comparing the merits of several management packages is the cost of the system. The dollar amount of the initial purchase is not very important here, provided all systems are in the same range (within 25 percent of each other). The items of real interest are the cost of installing the system, maintenance costs, and the relative price/performance ratio.

Above all, any automated record system should meet your requirements, be easy to use, and be flexible enough to grow as your networking and wiring requirements evolve. The system should accept records in an easy, straightforward manner. Report generation should include the ability to print copies of the original hard-copy records as well as to collate information into statistics and overall "snapshot" pictures of the complete wiring system. And, as always, backup or redundant systems are preferred.

11.4.3 Administration

Installing a building wiring system is only half the battle. Maintaining the automated system can often be more costly and complicated than the initial implementation. There must be a well-established process for entering information into the system. This should occur in a timely manner to keep the system accurate and up to date. The entering of information should be handled by a select group of individuals to ensure that the information is entered in a manner consistent with prior record entries.

Another nonnegotiable aspect of automated record system administration is the backup and archiving of information. All automated systems should be backed up (copied) at least once a day. In addition, old records may become obsolete for daily use as new records and changes are made. An archive can be created for these records. The archive (i.e., a backup tape or disk) allows the old records to be reentered into the system for special reports or queries; however, they are not present for day-to-day operations. This archiving process can reduce the operational size of the automated system, which can increase the speed of database file operations, such as reports or information queries.

The overall administration of an automated building wiring tracking system should be handled in much the same fashion as any complex database system. A systems administrator should be appointed to control the daily operations of the database system. This includes control of the information input-output process, the management of system database elements, the backing up of the system, and maintaining the computer system that houses the automated tracking system.

The difficulty of administering a large database system should not be overlooked or underestimated. This is clearly a task that should be either controlled by an MIS organization or managed by a person trained in database management techniques. This is an important point since this type of administration often results in unplanned expenses after system implementation. Of course, smaller automated systems that are nothing more than document storage facilities are relatively easy to maintain. However, large systems, either in size or complexity, should all adhere to the following summary guidelines:

- *Administration:* There should be a single person responsible for maintaining the automated tracking system.
- *System training:* There should be adequate training for all persons involved in the operations or use of the system.
- *Access procedures:* All access to input or output functions should be tightly controlled and documented.
- *System archive:* The system should be properly backed up on a daily basis and old records placed in archive.
- *System restoral:* There should be a plan in place for the emergency restoral of the tracking system. This should include the ability to replace any component (hardware or software) as well as to relocate

the system to a backup or auxiliary facility in case of damage to the entire facility (fire, flooding, etc.).

11.5 TROUBLESHOOTING

Every building wiring system, no matter how well engineered and installed, is going to have a problem sooner or later. Somewhere, sometime, a user is going to realize that the connection to the network is not working properly. Enter the *network fault*. There are many potential failure points within a wiring system. While it is true that prevention is the best medicine, it is often very difficult to keep every potential fault from becoming a kinetic, or active, fault. Every cable connection has the inherent ability to disconnect. Every cross-connect can become uncrossed. And every cable is just waiting to be severed by an innocent building maintenance person or interfered with by EMI or environmental elements. When these situations do occur, and they will, it is important to be prepared and to react in a fast and efficient manner to quickly identify and resolve the problem.

There are three main tasks involved in the troubleshooting arena: fault recognition, isolation, and resolution. Each of these tasks involves a different set of functions and equipment. They are often performed independently of each other but work best when approached as a system-oriented solution to cable and wiring problems.

11.5.1 Fault Recognition

The first important step in properly handling a wire or cable problem is the acceptance of the fact that there is a problem. Denial only makes the problem hurt worse since an untreated problem never goes away by itself. An untreated problem means that an end user somewhere is not able to properly use the communications system.

The first person to usually come to the conclusion that there is a problem with the communications network is the end user. This usually occurs in the form of a partial or total disruption in service to the communications device. It is vital that the end user have an established means by which to convey to the appropriate organization that there is a possible problem. This usually requires the establishment of a help desk or at a minimum a designated "help" person who users can contact to report a problem with wiring systems. The help desk (or surrogate) should be reachable directly during normal business hours. For after-hours problems, an answering machine is quite common with some form of override for emergency situations (i.e., our mainframe link to Japan is down and we cannot transfer our banking funds by the bank deadline).

The above-illustrated problem brings up an interesting point; how do users know if the problem is with a computer, communications device, or building wiring? Unfortunately, in most cases the answer is that they don't know the specifics of the problem. For example, in the above case, the problem may be with the software in the computer that establishes the connection with the remote computer. It may also be a problem with the modems or other communications devices that enable the computer to communicate to a remote device. And, of course, there is the chance (however slight) that there is a problem with the building or campus wiring. This A or B or C situation requires that the help desk function cover many different aspects of computing and communications, such as MIS, telecommunications, and building wiring.

It is the responsibility of the help desk to try to qualify the source of the problem so that it may be directed to the appropriate service organization. If this is not possible, there should be a "try A then B then C" approach taken where a single group is assigned all "origin unknown" problems until the most likely source can be ascertained. It is not at all uncommon to see a single organization responsible for all fault isolation activities within an organization. Taking this one step further, many managers are actually combining all service organizations into a single group (this is discussed further in the next section).

Regardless of the type of service organization you have, there are several pieces of information that need to be gathered during the fault recognition process:

- Who is reporting the problem (and when)?
- What are the symptoms of the problem? Is there a visible cable problem? A problem communicating with remote devices?
- How long has the problem been going on? Weeks, days, hours, intermittent, etc.
- How urgent is this? Does the problem need to be fixed immediately?
- Has this problem been reported or occurred before? If so, when and was the problem ever corrected?

An additional function of the help desk is to perform some elementary sleuthing concerning the problem. For example, in response to a complaint that an end user cannot access the remote file server, the help desk should ask what functions can still be performed by the system. For example, can the user still access local file or print servers? The help desk can also be used to provide basic suggestions, such as "has the user checked that the device is plugged into the network?" (This one happens often.)

One of the easiest methods for gathering and storing this, and other, information is the use of preset checklist forms. These forms can be quickly "checked" concerning the nature of the problem and activities that must be undertaken to isolate the exact source of the problem. Depending upon the size of the help desk (and the associated communications system), the use of an automated or computerized entry system can reduce the amount of time required to take action to resolve the problem. Additionally, the use of a computer to enter information can help to quickly determine whether this problem has occurred previously or whether it could be a result of an ongoing action (i.e., new cable installation or renovation) taking place. The use of rule-based expert systems can also be beneficial in determining the most likely source of the problem and the best course of action to take (this is discussed further in the following sections).

11.5.2 Fault Isolation

Once a problem has been identified, the process of isolating the source of the problem can begin. This involves a detailed search from the top down to determine the exact location and scope of the problem. The determination of the precise problem can sometimes be difficult and involve electronic line-testing equipment.

This process should start as a series of broad questions designed to determine if the problem is with a cabling component or with another noncable system. Each subsequent question is designed to locate the physical problem within the cable system. For example, a communications failure between two sites may be a communications device problem. The first question that should be asked is "Have the communications units (noncable components) been checked for normal operations?" Once the status of electronic devices has been ascertained (this includes items such as network hubs, file servers, and bridge and router devices), the question of checking the physical connections can begin.

Checking a cable for nonvisible faults is usually best accomplished through the use of an electronic cable tester. These devices come in many different forms, each offering a varying degree of sophistication and error detection. On the high end, devices such as time-domain reflectometers (TDRs) send a pulse down a cable and measure the types and frequency of any signal reflections that bounce back to the device. By measuring the strength of the reflection and the amount of time required for the reflection to arrive back at the device, the TDR can usually indicate the length of the cable as well as the location of any cable faults (shorts, opens, crosstalk areas, etc.). A TDR can also be used to measure the characteristic impedance of a cable (or light characteristics of a fiber-optic segment). Additionally, taps or cross-connects onto the cable can be sensed.

There are many different types of TDR devices on the market. Most involve a relatively large device (10 by 10 by 6 in). These large TDRs usually include a large diagnostic screen and integral printing capabilities. There are, however, a growing number of small mini-TDR devices appearing on the market today, which are usually hand-held and perform a subset of the functions found in a real TDR. For example, they may provide TDR diagnostics but not include the precision or the display or print capabilities. These hand-held devices usually provide a small liquid crystal display (LCD) screen with text display only. However, some do include some traditional wire tests that TDRs tend to overlook. For example, many of these small devices can test pair continuity and/or send a "tone" signal along a copper cable that can be monitored at the remote end (this helps for remote cable locations). Some of the newer devices are also able to check cable for use by specific networking systems, such as those devices designed to check cable against the electrical specifications for 10BaseT Ethernet systems.

The use of electronic components may also help to isolate non-cable-related problems. For example, a protocol analyzer may be used to help troubleshoot intermittent problems, such as crosstalk, which result in large numbers of incorrect bits of information being received by an end user from a remote site.

The important point to fault isolation is that the maintenance worker should have the proper equipment, have access to the appropriate information (trouble report and current wiring system status), and have a well-defined procedure defining the steps necessary to locate a cable fault within a particular building environment. The creation of a step-oriented process (check SIO interface, check horizontal cable for shorts or opens, check cross-connect point, etc.) will help to ensure that all possible fault locations are checked.

Once the exact problem has been isolated, the process of fixing the problem and restoring service to the user can begin.

11.5.3 Fault Resolution

The fault resolution process is fairly straightforward. Once the problem has been isolated, you fix it. Simple. Well, not really. There are a few small items that must be taken care of to ensure that the repair is properly implemented:

User notification: Unless the problem is a fairly small one, it is usually beneficial to inform the user that the problem has been isolated and will now be resolved. The user should be made aware of the anticipated amount of time required to fix the problem (although this does not fix the problem any faster, it is a good political move since it tells users you are working on the problem and prevents them from reporting the problem again). User notification is also critical if any portion of the network, not directly affected by the failure, will be temporarily taken out of service during the repair process. This also includes notifying any utility organizations where there is any risk of system interaction (digging, working near power lines, etc.).

System testing: Any repair should be properly tested and certified prior to the circuit being declared operational. This means that the communications service being supported by the system should be tested over the repaired circuit. Passing an electronic tone or checking for pair continuity is not sufficient test activity. Failure to test the system with the user's application can often result in a reoccurrence of the problem for two reasons. First, the repaired portion of the network may not have been fixed sufficiently to allow the user's application to run (testing with a tone or continuity tester does not rate the quality of the repair). Second, the repaired portion of the network may not have been the only reason for the failure of the end-user's application.

Action documentation: After the fault has been resolved, the cable facilities should be appropriately labeled in accord with the labeling procedures set forth for the entire wiring system. Additionally, any action taken during the repair process should be so noted and stored in the manual or automated wiring tracking system. Failure to properly document the repair actions can cause confusion at a later time when additional maintenance people try to perform maintenance or fault isolation activities and find themselves working with outdated cable records.

Performing the above tasks with every fault resolution action will help to ensure that the entire building wiring system continues to perform at its optimal level. However, in the event that the problem cannot be easily resolved, there must be a firm plan of action for problem escalation. A problem escalation plan sets forth a set pattern of elevating a problem to the next higher level if the problem is either major or cannot be resolved within a preset period of time. For example, if a technician cannot resolve a problem within a predetermined period of time (i.e., 2 hr), the technician's immediate superior should be informed and brought into the resolution process. This process should continue until the problem is either resolved or all available managers have been exhausted (figuratively and literally). While the escalation process may not always result in a quick resolution of the problem, it will result in the appropriate attention of management to the problems that do occur and the type and quantity of resources required to support a building wiring system. Such a plan will also provide the user community with the feeling that management is concerned and working to provide a solid building resource.

11.6 ORGANIZATIONAL REQUIREMENTS

It is important at this point to present a brief word or two on the organizational requirements of a building wiring system. How your installation and troubleshooting organization is structured can make a difference in the effectiveness of cable support activities. An improperly structured organization can result is miscommunication between various organizations, often resulting in the proverbial finger pointing when problems arise. In contrast, an organization that promotes communications between installers and maintenance and MIS personnel can often work in tandem to quickly resolve problems.

The traditional telecommunications service structure within most large organizations is one of functional specialization. There is a different group dedicated to performing each specialized task. For example, MIS technicians (software and end-user hardware) are often in a separate group from the telecommunications staff (networks, modems, multiplexers, etc.). Additionally, the building wiring technicians may be separate from the telecommunications group. This type of segmented architecture can lead to gross miscommunication and finger pointing tactics.

While it may make valid sense to maintain separate organizations that have the specific skills necessary to resolve faults in specific types of equipment, it does not make sense to have them work independently of each other. Network problems can often consist of several different elements that cross the traditional service group structures.

There are some structural configurations that can help to improve communications between these diverse groups. First, the different groups should all come under a single upper-management entity. This provides a place where "the buck can stop" and definitive actions taken. Second, there should be a single help desk where all communications-related problems can be focused. Where possible, the fault isolation functions can also be integrated to a certain extent. Third, there should be a high degree of cross-training between the various groups. This will promote a high level of job-related understanding between the various groups and eliminate the "fear of the unknown" that often results in finger pointing between groups.

An extension of the cross-training approach is to rotate qualified personnel between groups on a regular basis. As an example, a service technician whose primary function involves telecommunications equipment support may benefit from spending 1 or 2 months a year in an MIS or wiring support group. This not only helps to promote job understanding, but it also helps to improve the value and troubleshooting capabilities of the person.

A second area where organization boundaries can cause problems is within the individual service groups. It is common to see groups structured such that installation and maintenance activities are undertaken by two separate groups of people. This often results in the classic double-action scenario. For example, first, the installation crew goes to a site and installs a circuit to an end-user's computer. The circuit is tested and the crew leaves. Second, the user calls in a trouble report immediately (or soon) thereafter complaining that the service does not function properly (note that good test procedures here would help a lot). In response to this trouble call, a maintenance technician is sent to the site; the second trip in n hours. This second person on site is at an immediate disadvantage. First, the technician does not have any first-hand knowledge of what took place during the installation. Second, if the installation was relatively recent, there may not be adequate documentation available describing the installation activities. Third, the end user is likely to be a bit irate over having to wait an additional period of time after the initial installation for an operational network connection.

Aside from proper testing activities, one of the best methods of preventing this double-trip situation is to combine the functions of the installer and the maintainer into a single position or group. Too many groups are structured such that the more experienced person is involved in the maintenance function while the "simple task of installing a cable" is left to the relatively unexperienced. Combining the two functions provides a good cross-training environment and also builds more path development opportunities since the combined staff can usually more easily support the creation of trainee, junior, and senior technician levels.

These simple structural approaches can benefit many service organizations. However, it is important to point out that this may not work in all situations. As with any group, the structure must be fine-tuned and oriented toward the individual peculiarities within that particular working environment.

11.7 EXPERT SYSTEMS

One of the fastest growing trends today in networking management is the use of expert systems in conjunction with automated management systems. This approach can be costly and may not provide all of the "pie in the sky" capabilities that are being discussed. Before implementing such a system, it is important to understand just what an expert system is and how it can realistically be expected to benefit a building wiring tracking system.

11.7.1 Expert System Definition

Often touted as a form of artificial intelligence, an expert system is much better described as a set of rules that assist a person in making a well-informed decision. While artificial intelligence is the ability of a computer to learn and grow from experiences (and apply this information to future events), an expert system can be thought of as a series of established rules, programmed by a person into a computer, to allow the computer to quickly analyze information and make a decision based upon preprogrammed logic.

Before looking at an example of how an expert system is created, we first must see how a person makes a decision. We will use a very simple example, the turning on of a computer terminal. When we first turn on the switch, we ask ourselves the question "Did the computer turn on?" If the computer did, we proceed with our work. However, if the computer did not turn on, we must try to determine why this event did not take place as planned. We know there are several possibilities: bad switch, bad power supply, power cord not attached to computer, power cord not attached to outlet, faulty power cord, faulty outlet, lack of power to the outlet, and of course, my favorite, improper operation of the on/off switch. Once we have determined what the possible set of problems could be, we go about testing our theories. Of course, we try the most likely ones first (such as checking to make sure the power cord is properly plugged in). The least likely items are usually checked last (i.e., faulty on/off switch). Eventually, we will eliminate possibilities until there is only one possible (and correct) solution to our little problem. We then take corrective action (usually making sure that we flip the switch all the way this time) and proceed with our working activities.

The process outlined above is all based on simple rules known to most people. We know the possible failure locations. We can also, to a high degree through intuition and experience, rank the most likely causes of the failure. Additionally, the more we know about other components, the more certain we can be about predicting the most likely failure point. This process can be defined in a set of simple rules. For example, rule 1 can be, If the on/off button has been switched and the terminal does not activate, there is a problem with the power system. Rule 2 could be, If the power cord is properly plugged into the terminal and the outlet, then the cord may be bad. Rule 3, If the power cord is good, there may be a problem with the power supply or the outlet supply. This process can continue until all the intuitive actions that a person would take are classified as if this, then that rules.

While a person can learn from experience and make judgment calls based upon "feelings," a computer is incapable of performing such a feat (at least with today's technology). To make a computer mimic our thought process, we must take the set of rules that we have defined for a problem and input them into the computer. Like any computer program, after feeding in information, such as, The terminal did not turn on when the on/off switch was switched, we can run this problem through our user-programmed set of rules for determining the potential problem. In addition, we can associate suggested actions with each set of rules. For example, instead of a simple if this, then that, we can program the computer to respond with if this, then try this. This is the essence of expert system technology today. Using predetermined rules to allow a computer to make a decision for us.

11.7.2 Advantages

The use of an expert system in the above example (to determine why a computer terminal did not turn on properly) is a bit too simple for practical applications today. Most people could make the same assessment in half the time it would take to input the information into the expert system computer. The reason for this is that there is a fairly small number of possible fault locations and we are fairly familiar with these possibilities. Where expert systems are advantageous, however, is when there are an extremely large number of input variables or potential solutions. Additionally, we may benefit through the use of an expert system if we, as the user of the system, are not very knowledgeable of the system. It is not uncommon for the rules of an expert system to have been created by a true "expert" in that particular area. These expert rules can then be used to assist less experienced personnel, in effect giving them the benefit of someone else's experience.

How does all this apply to the world of cable and wiring? The most practical application today is in the fault isolation activity. By programming an expert system with all of the possible fault locations, the characteristics that they exhibit, and the potential fixes, we can use a computer to quickly make a rule-based recommendation. This recommendation can be part of the work order generated in response to an end-user service call. A good expert system can function in the following manner:

Work Order #234

Problem:	Phone doesn't work
Suggest:	Check phone cable
Suggest:	Check nearby phones
Suggest:	Check phone handset and device

By analyzing the history of past problems and their faults, we can add to the expert system (reprogram) to create a true probability rating for each rule. Such a probability rating system might come in handy with the following example:

Work Order #234	
Problem:	Phone doesn't work
Based on past events:	77 percent likelihood of bad phone cable
Based on past events:	11 percent likelihood of bad handset or device
Based on past events:	10 percent likelihood of bad wiring
Based on past events:	2 percent likelihood of other problem

This second example illustrates how we can use a rule-based expert system to not only suggest possible faults but to assign them a probability ranking. This allows the service technician to immediately investigate the most likely source of the problem. The savings here is in the amount of time required to isolate and repair the problem. Taking this a few steps further, the use of an ongoing expert system (which constantly sifts input through a set of rules) could eventually be developed to enable preventive actions to be requested prior to the realization of a fault. For example, the system could have a rule that states, If a cross-connect has required retermination over 10 times in the past month, consider replacing the cross-connect block. Such an action is preemptive in nature, based upon past experiences. Of course, such capabilities would require an extremely complex system that maintains a database and has the ability to identify and classify many different types of actions. Nonetheless, it is possible that such a system could be designed, perhaps in conjunction with a much larger overall expert system designed to support an entire MIS function (including telecommunications equipment and building wiring systems).

It is important to remember that an expert system is only as good as the rules programmed into the system. An expert system cannot "learn" on its own. Since it is only running through a set of if this, then that rules that we program ourselves, we must provide additional, updated, information if the system is to evolve into a truly useful system.

11.7.3 Implementation Issues

The implementation of an expert system in conjunction with an automated building wiring tracking system is not an easy task. Fortunately, there are many inexpensive expert system packages available today (under $1000) that allow a good deal of customization and external database interfaces. The real expense comes in providing programming and MIS-type resources to define the rules, build the expert system, and then integrate it with the automated wiring tracking system.

Since the use of expert systems is relatively new in the building wiring arena, it is suggested that an automated system be first implemented without the expert system functionality, which can be added at a later time. This will allow the initial system to be developed in a more timely manner; expert systems are still a relatively new technology and can be difficult to master and implement. However, don't forget to design your automated system in a manner that provides detailed tracking of information and the potential ability to interface to additional programs, such as a custom rule-based assistant.

11.8 SUMMARY

The following is a summary of the key points of this chapter:

- Proper cable management is an essential component of any building wiring system.
- A cable management system can provide audit capabilities, promote fault isolation, increase future flexibility and support, reduce cable management costs, and increase end-user satisfaction.
- Cable management includes system documentation, cable labeling, and activity records.
- The use of automated record and tracking systems can enhance cable management.
- A cable management system should be tied directly into a help-desk function, perhaps integrated with overall MIS help-desk support.

Chapter

12

Design Principles

Definition: *Design principles* The guidelines that shape and drive the development of a telecommunications wiring system.

The foundation of any good building wiring plan is in the design. A good design can result in a wiring system that meets not only the current requirements but also provides the capacity for expansion to accommodate user growth. With a good design, new users can be added without expensive new installations, and emerging networking technologies, such as Fiber Distributed Data Interface (FDDI) and local area networks (LANs), can be accommodated without complete system reconfiguration. This chapter presents an explanation of the design process, including a set of guidelines to assist in the development of a solid wiring plan.

12.1 DESIGN PRINCIPLE GOALS

Every component within a building wiring system must have a predetermined function. The system must be designed to support this function in cooperation with the other components of the wiring system. A proper design process can help to ensure that each component fulfills this requirement. When managed properly, a good wiring design process can result in additional benefits. Some of the long-term benefits of a proper wiring design include the following:

- *Ease of procurement:* Enabled through the design of a wiring system that uses standards-based off-the-shelf equipment.
- *Reduced cost:* The selection of cable and components that are available from multiple vendors can help to get the best price.

- *Simplified installation:* A plan that is well organized supports an orderly installation of components.
- *Conservation of capital:* In a building wiring system, there are many functions that can be combined into synergistic services. Proper design and planning helps to identify these common services and to eliminate expensive wiring scenarios.
- *Quality of wiring system:* This is the generic benefit of a good wiring design: solid management capabilities, ease of system reconfiguration, and built-in expansion capabilities.

In contrast to a proper wiring system design, a poor design, regardless of the quality of the installation activities, can result in a wiring plan that does not support the communications requirements of the user base. For example, new user installations may require the installation of new cable, changing user locations within a building may be time consuming and costly because of overwhelming reconfiguration of cabling equipment, and the management and maintenance of the complete system may be difficult and incomplete.

It is important that a building wiring design take into account the many different factors which may influence a wiring design, such as the following items:

- *User requirements:* This pertains directly to the types of technology (and applications, since an application can often be satisfied by several different types of technology) that are likely to be employed by all user groups within the user organization.
- *Building wiring codes:* Each government, federal, state, or local, has certain types of codes and regulations in place that dictate the installation parameters for power and communications-related cable.
- *Design process:* Understanding the basic concepts that drive the planning and design of a wiring system can help to facilitate a proper design the first time around. A $10 to 20 million dollar office building is not the proper place to learn building wiring system design by trial and error.
- *System procurement:* The design must take into account any outstanding issues that may affect the timely procurement and installation of the wiring system.

The following sections within this chapter are intended to help explain and provide guidance concerning the design of a proper building wiring system.

12.2 REQUIREMENTS-DRIVEN DESIGN

One of the fundamental factors driving the design of any building wiring system is the set of requirements arising from the telecommunications needs of the user base. This includes not only telecommunications requirements but also building space considerations, technological level of management capabilities, and political situations. Telecommunications requirements and the communications systems in use will determine the type of cable and SIO systems to be used. Building and space considerations will determine the configuration and layout of the system. Political considerations, although unfortunate but common in every organization, can preclude the use of certain technologies that users may feel are not suitable for the strategic long haul. As an example, while multimode fiber can support many different types of data applications, some users feel very strongly that single-mode fiber (while costly today) will be a commercially viable alternative within the next 5 years. Based on this theme, users may not support any wiring initiative that involves multimode fiber (this is a very mild example of politicing; most involve control of resources). Each of these aspects must be taken into consideration when designing a building wiring system.

12.2.1 Developing a Requirements Specification

The process of designing a solid building wiring plan must be based upon a firm understanding of the complete requirements list. The only way to be sure that all requirements have been considered is to perform an in-depth requirements analysis. This requires that all organizations and concerned parties have an opportunity to present their current and planned telecommunications requirements. This is usually accomplished through either personal interviews or survey forms (in fact, both are recommended for large organizations).

The goal of the requirements analysis is to develop a comprehensive listing of communications requirements. The important information to gather here is not the specific end-to-end applications that are being run but the supporting information that may affect a wiring system design. This includes information concerning the following items:

Overall applications: What voice, data, video, security, and control applications will be in place at the time of initial installation? What systems are planned for future implementation? This information will help to gauge the scope of the wiring plan as well as the future growth potential of the user base. This can also help to focus attention on trends concerning the ratio of voice, data, and video applications, which may be influential in the selection of new and emerging technology for cabling systems. Additionally, it is important to note which user applications are planned for replacement within the next few years. It may be an option to not fully support an application if it requires special cable and is planned to be taken out of service shortly [i.e., it may be unwise to install coaxial cable to every desk to support a small ARCnet local area network (LAN) if the system will be replaced with an Ethernet over unshielded twisted-pair network within the next year].

Media requirements: Once the type of applications have been listed, it is important to determine what types of media are supported. This is usually hardware dependent. Although LANs can very often be supported by multiple types of media, there are some networking systems (vendor specific) that only function on a limited set of media. For example, although IEEE 802.5 Token-Ring networks can operate over shielded and unshielded twisted-pair, some vendors do not support the unshielded twisted-pair option in their equipment.

Geographic installation: Where will the users be located? Where will common equipment reside (file servers, communications servers, etc.)? This information will help to determine if the user requirements need to apply across the entire wiring system or just within a local environment. For example, installing fiber-optic cable to the desktop throughout a campus may be unjustifiable from a cost perspective if the only real fiber requirement is for a group of CAD/CAM workstations located within a single building.

Multiple installations: Will there be multiple types of applications present at user locations? Will users require support from two or more data applications (i.e., 3270 terminal and Ethernet PC) on a single desktop? This can directly affect the number of cables that must be installed at each SIO location. Where multiple applications are present, the opportunity to use multiplexing equipment or to combine applications into a single networking system should be investigated.

User mobility: How often will users be changing locations? Can user moves be predicted in advance? This information is vital to incorporating proper management principles into the wiring design. It also can be a good indication of the size of the management staff required to support the user base.

User sizing: How much is the user base expected to grow by organization and application? This will help to determine to what extent the wiring system must accommodate additional growth. It will also help to rank the relative importance (strength and pervasiveness) of each application and media requirement.

The information gathering process can take place through personal interviews or information survey forms. Personal interviews, with departmental MIS staff, often provide the most detailed source of information. In extremely large organizations, however, they may not be possible (this is also true for new buildings where the end users and/or their applications may not be known in advance). The use of survey forms can provide a fast and efficient means by which to gather information from a large group. A standardized form can also help to ensure that information gathered from different sources is provided in a consistent manner.

Once the applications and their associated media are understood, the process of analyzing the information can take place. It is important to look for any potential danger points between different applications. For example, a radio system may emit high levels of radio frequency interference (RFI) that may disrupt other systems (i.e., high-speed unshielded twisted-pair data applications). A ranking of the requirement can also take place after the end-user information gathering process has been completed. This involves the relative ranking of requirements based upon the size and importance of their installation. For example, a LAN application that serves the entire organization would receive a higher ranking than a data terminal or mainframe system that supports only a portion of the user community. This ranking will be beneficial when determining which of the multiple types of cable required for end-user support will become part of the overall wiring system and which ones will either not be supported or will become exceptions to the standardized plan.

The information gathering process can also provide insight into situations where applications common to multiple groups can be consolidated (i.e., the use of a large campuswide backbone network to replace multiple smaller networks, each of which requires an independent management staff). Unfortunately, many organizations succumb to the "What, Me Worry?" school of planning where many departments do not know (or care) what other departments are doing in the telecommunications arena. Cross-knowledge should be considered as basic a requirement as hand-eye coordination. This is an area where information gathered for one purpose (building wiring requirements) can be used to help benefit the overall MIS structure.

The overall process of gathering information must be fully supported by top-level management. The end-user process of gathering and preparing information to present in response to a requirements analysis request (either personal interview or survey form) can often be very time consuming. Unfortunately, it can often be difficult to ensure the cooperation of the end-user community unless they are aware that (1) top management has ordered the gathering of information, and (2) their requirements will not be supported by the system unless they provide adequate information.

The use of a firm timetable for the completion of the requirements analysis can be beneficial. Establishing set dates for the completion of the survey forms (at least 3 weeks) and for the personal interviews (with backup dates) can help to ensure that information is gathered in a timely manner.

An often overlooked area of the requirements analysis is the vendor connection. Meeting with the major telecommunications and system vendors that deal with your organization can often provide in-depth information concerning the direction the vendor is taking in regard to cabling support. For example, knowing that a LAN vendor is planning support for unshielded twisted-pair for a system that currently requires shielded cable can be advantageous in planning for cabling support. A working example of this is the current support of the Fiber Distributed Data Interface (FDDI) network that operates at 100 Mbps. While FDDI is available from many vendors over multimode fiber-optic cable, there are several vendors who have announced support for FDDI to the desktop over shielded twisted-pair (STP-FDDI). For those users who plan to migrate to FDDI within the next couple of years, the use of shielded twisted-pair (or perhaps even unshielded twisted-pair within the next 5 to 10 years) can be far less expensive than the use of fiber-optic cable.

The network management aspect of a wiring resource analysis is also often overlooked during the information gathering process. It is important to design a system that can be managed. This requires that input from the management and maintenance staff organizations, concerning how they manage their wiring systems, be collected and taken into account during the requirements analysis.

12.2.2 Building Wiring Consultants

A very good way to augment the requirements analysis process is to enlist the aid of a building wiring or telecommunications consultant. An experienced consultant can often provide numerous services that extend the capabilities of an existing in-house staff, including the following:

- Requirements analysis
- Evaluation of wiring, telecommunications, and computer trends

- Building wiring design
- Wiring vendor interface and selection
- Project management (including installation and testing)

One of the most important tasks involved in the use of a consultant is the selection process. There are many different types of consultants, many with high levels of specialization. Choosing the right consultant depends upon the type of resources that you need to supplement your internal capabilities. It is advisable that a minimum of three consultants be investigated prior to any final selection.

There are several key requirements that any consulting organization should meet prior to selection:

Established firm: The consultant should be an established entity with a history of successful contract completions. You do not want a firm that offers a low fee but cannot complete the contract successfully (the classic "fly-by-night" organization).

Wiring knowledge: The consultant should have prior experience involving the work you are contracting out. Additionally, the work should involve an organization that matches your organization's size, requirements, and communications architecture. References should be checked closely. The past experience of any person included in a bid on the project by a consultant should be properly evaluated. Hiring an experienced consultant organization that uses inexperienced staff on your project can spell disaster.

Certification: Perhaps the most important certification within the building wiring arena is that offered by the Building Industry Consulting Services International (BICSI) organization. BICSI offers certification for Registered Communications Distribution Designers (RCDDs). The RCDD certification relates to an in-depth understanding of the building wiring design process. An RCDD must be recertified (tested) every 3 years and must show an ongoing educational process. Additionally, certification as a professional engineer or as an electrician can be extremely beneficial when electrical components are involved.

Vendor independence: A consultant is not a true consultant if he or she is affiliated with a communications or wiring vendor. The job of a consultant is to provide impartial advice and services. This cannot be guaranteed if there is any possibility of a conflict of interests.

Industry presence: A consultant may often be called upon to act as an intermediary between an end user and a vendor. The use of a consultant who not only knows the industry but is known within the industry can help to facilitate a solid vendor relationship.

Once a consultant has been selected, a set of performance guidelines and payment schedules must be established. It is important that the end-user organization control, and not be controlled by, the consultant. Individual tasks, such as information gathering, information analysis, and wiring system design, should be clearly defined in terms of working activities and completed deliverables. Consultant reviews, to be completed after each task, should be performed. Where possible, each task should have a fixed duration and monetary value associated with it. Having a contract with a consultant that allows you to use or not use the consultant on a task-by-task basis can help to ensure that the work quality remains high throughout the project.

12.3 WIRING CODES

A quick word on building wiring codes. Know them and follow them. The National Electrical Code (NEC) is a set of codes and guidelines established by the National Fire Protection Association (NFPA). The NEC governs the use of communications and electrical cabling within a building and campus environment. The NEC itself has no jurisdiction over building wiring systems. However, many state and local governments have adopted portions of the NEC for their own legally binding codes.

Another set of codes developed by the NFPA and the IEEE is the National Electrical Safety Code (NESC). Adopted by the American National Standards Institute (ANSI), the NESC complements the NEC for campus-oriented communications facilities.

The purpose of the NEC, and other codes, is to ensure a safe working environment. For example, the use of plenum-rated cable is one of the more important regulations set forth in the NEC. Another involves the proximity of current-bearing cables (power and control) to communications facilities.

There are many organizations that either promote, certify, or set standards for telecommunications-related devices and cabling systems. These organizations include the following:

- *National Fire Protection Association (NFPA):* Developers of codes pertaining to fire prevention, cable flammability, lightning protection, and other cable (power and communications) guidelines.
- *Underwriters Laboratories (UL):* Cable and device ratings associated with electrical and flame characteristics, such as CL2(P) for

horizontal wiring (plenum), and the CM, CMR, and CMP ratings for vertical cable. The UL also has specified flammability tests for cable, such as the VW-1 flame test.

- *Electronic Industries Association (EIA):* Wiring and connector standards.
- *American National Standards Institute (ANSI):* Networking and wiring standards.
- *Canadian Standards Association (CSA):* Similar to the UL in the United States, responsible for testing and certification of cables and communications devices.
- *Intelligent Building Institute (IBI):* Promoting the development of the intelligent building concept.
- *Institute of Electrical and Electronics Engineers (IEEE):* The specification of telecommunications networks and cabling systems.
- *Construction Specifications Institute (CSI):* The development of a set format for specifying telecommunications facilities in construction bids and specifications.

The important point to note concerning codes and regulations is that each state and local government has a different code structure. Completely independent from building and safety codes [such as the Occupational Safety Health Agency (OSHA), Building Officials and Code Administrators (BOCA), Standard Building Code (SBC), etc.], government codes directly affect the design of telecommunications wiring. It is imperative that any person involved in the design of a building wiring system be familiar with the appropriate codes for the local vicinity. Information regarding specific building and wiring safety codes can be obtained directly from the particular local or state government.

12.4 PROJECT PLANNING

An important part of the building wiring design process is the initial project planning phase. Project planning refers to the development of the initial concept surrounding the decision to build or renovate a building. Usually considered to be similar to a *building proposal*, the project plan specifies the generics of a building (such as where the building will be located, how big it will be, who will occupy the building, space considerations and allocations, cost allocations for telecommunications wiring, etc.). The project plan does not involve detailed information on telecommunications requirements; it merely provides the outline for further development.

The project plan is developed at a very high level, such as the building developer or organizational director level. The information included in the project plan will set the tone for the development of the entire building or renovation. For this reason, it is important that it include some level of basic information concerning the building wiring process.

12.4.1 The Initial Concept

The project plan will be used during the design program phase leading to the actual design of the building wiring plan. The project plan can be either simple or complex depending upon the scope of the building development process. The creation of a new high-rise building in a downtown area can require an extremely complex plan. In contrast, the renovation of a wing of a building in a small campus can be relatively simple.

The project plan should clearly identify information concerning the function and scope of the building project. This includes the following pieces of telecommunications-related information:

Building location: Where will the building be located? Will there be access to established communications systems in a campus environment? If part of a multibuilding organization, is there microwave line of sight between existing buildings?

Function: Who will occupy the building? Will it be office personnel, maintenance, medical, security, etc.? This information will help to determine what level of telecommunications support system is required.

Communications requirements: Given the type of building occupants, will there be any specific communications requirements? There can be a vast difference in communications requirements between a manufacturing organization and a service organization. Special communications systems, such as radio, microwave, or satellite, can require specific space or structural modifications. An organization with a large computer center will probably require a large computer room with special power, environmental control, security, and fire suppression systems. These items can add significant cost to a building.

12.4.2 Space Planning

The amount of space required for a building wiring system can vary greatly depending upon the communications requirements. Areas of concern during the project planning phase include the following:

- *Network access point (NAP):* There should be a dedicated room (a minimum of 3 m by 3 m) for the NAP. This is separate from any other facilities.
- *Distribution frames (DFs):* There must be adequate room designated for the placement of DF facilities. There should be at least one DF per floor occupying a minimum of 10 m^2 of floor space for every 20,000 m^2 of office or work space. The building distribution frame (BDF) where all communications cables originate can occupy an equivalent space ratio. All DFs should be located adjacent to a building core, a process which can affect the development of "innovative" architectural designs.
- *Distribution systems:* In addition to floor space, there must be adequate space within the distribution mechanism for proper cable dispersement (in the ceiling, underfloor, etc.).
- *Computer facilities:* The recent increase in the use of personal computers and local network systems has only added to the space requirement for computer rooms. These rooms, which have controlled environments, are used to service mainframe computers, private branch exchange (PBX) and voice systems, and local networking file servers.

It is not uncommon to see up to 5 percent of the usable space within a building allocated to communications-related functions (i.e., distribution frames or computer or PBX facilities).

12.4.3 Installation Duration

The addition of telecommunications wiring and associated facilities within a building can only add to the amount of time required to complete a new or renovation project. While the amount of time is not very large, considering the length of most building projects, it does need to be taken into account. A single controlled computer room, for example, can take up to a month to construct and properly outfit. Likewise, the installation of DF equipment (cross-connect blocks, backboards, etc.) can take a week or more depending upon the size and complexity. The installation of the actual wiring elements, such as distribution hardware and cable, can also add weeks or months to the completion of a building. Fortunately, these activities can be performed in parallel with other activities once the building exterior has been completed. This does, however, require additional and specialized resources since the equipment is unique to the telecommunications system. A typical example would be a five-story building (approximately 100,000 ft^2) which may require 1 to 2 weeks to install all cable and standard information outlet (SIO) units. This does not include the installation of the distribution hardware, which may take a much longer duration (again, while other activities are taking place).

From an operational perspective, most large city areas are highly unionized. The presence of a strong union can require the use of union workers who may be subject to working hours or conditions that should be taken into consideration. The most difficult task is the sequential coordination of the different unions involved in the process. For example, it is not uncommon to find one group responsible for installing the distribution hardware while another installs the cable and wiring. This may require additional time for inspection between the installation of different components since different parties will be responsible for different implementation activities.

12.4.4 Estimating Costs

As part of the project planning process, costs must be estimated for various building components. This includes an estimate for the cost of the building wiring system. In preparing the type of broad estimate required for a project plan, there are several variables that can be evaluated to produce a rough estimate of cabling costs:

- *Cable types:* The type and quantity of cable installed will directly change the cost of the system. For example, the installation of unshielded twisted-pair to each SIO unit will invariably be less expensive than the installation of fiber-optic cable.
- *Labor rates:* Labor rates will vary considerably among different regions. Labor is often considered to be one of the most expensive components in the installation of a building wiring system.
- *Distribution method:* The cost of different distribution systems can vary greatly depending upon the type of hardware used. Underfloor distribution systems tend to be more expensive than ceiling distribution.
- *Communications facilities:* The cost of adding specialized computer and communications rooms can be significant. Requiring special power and environmental control, these rooms can range from \$20,000 for a small 20- by 20-ft room to over \$100,000 for a larger room capable of supporting multiple mainframes or PBX systems. In addition to this cost, there is the simple fact that space used for communications equipment cannot be used for office occupancy.

When considering the cost range for all items, including installation and hardware, it is common to spend anywhere from \$500 to \$1000 per SIO unit on building wiring-related items. This can translate into over $500,000 for a small building supporting only 500 users without any large computer rooms.

Looking at the highend of telecommunications wiring systems, there are those who feel that spending $1000 to $2000 per user on the building wiring and telecommunications system is not out of line.

When determining rough pricing estimates, contacting a local cabling providor or consultant can often give you a good idea of what will be required.

12.5 DESIGN PROGRAMMING

The design programming process involves the specification of building wiring goals and objectives to the architectural design team for inclusion in the building or renovation design. Unlike the project plan which involves the broad estimation of requirements and funding levels, the design program is directly based on the requirements analysis process. Based upon the detailed requirements (type of cabling required, distribution plans, DF space and configurations, SIO locations, etc.), the design programmer will create a preliminary blueprint for the building wiring system. This blueprint is incorporated into the working blueprint for the entire building.

12.5.1 Design Programming Goals

The overriding goal of the design program is to provide information concerning the building wiring plan to the architectural design team. This major goal, however, includes several smaller goals, each representing a portion of the overall system. They are all intended to ensure that (1) the building wiring plan is properly integrated with other building resources, (2) all users have equal access to wiring facilities, (3) the wiring system is ergonomically integrated into the building, and (4) the building support systems (power, air conditioning, fire protection, security, etc.) adequately support the building wiring system. These four overriding goals help to ensure that the many minor goals of the design program work together to truly form a complete system that is in itself a building wiring system. The goals of the design program each pertain to the definition of a basic component of the building wiring system, including the following:

- SIO units and geographic distribution
- The types and quantity of horizontal distribution cable
- The location and general size and configuration information for DF facilities
- The type and configuration of cross-connect blocks and support equipment

- The type and quantity of vertical distribution cable
- Information regarding the planned horizontal and vertical distribution hardware (layouts, type of equipment, etc.)
- Information on campus distribution systems (network access points, cable types, planned distribution methods, etc.)
- Special requirements, such as microwave or radio towers, that involve structural support or line of sight
- Communications room requirements (location, size, power, environmental control, fire suppression, etc.)

The more information that can be incorporated into the overall design program, the better the final building and wiring design will be structured. The information should be presented in a structured and hierarchical fashion, with information on each component of the building wiring plan clearly detailed. It is important to properly use this information in designing a wiring system that meets these user requirements. This is best accomplished through a very basic concept, the top-down design. The term *top down* refers to the application of end-user requirements as the base for all wiring planning. The end user, at the top of the support structure, dictates the underlying support mechanism.

Perhaps the most important element within the building wiring system is the SIO, which is the single component, at the top of the building wiring system, that will drive the majority of additional decisions, such as the type of horizontal distribution media, the placement of DFs, and the sizing of the overall system. Working up from this single component will help to ensure a structured design program. This may involve working closely with the architectural team to help collate the information (based upon the requirements analysis) into the best format for the individual design team.

12.5.2 Monitoring the Design

The process of developing the final building blueprints should be carefully monitored throughout the entire process. This not only involves working with the architectural team during the design programming process but also involves regular meetings with the team during the actual design development. It is important that the design accurately reflect the intent of the goals set forth in the design program. The monitoring of the design process will also allow minor changes to the building wiring system which may come about as a result of external factors or which may arise as a result of the design process itself.

12.6 PROCUREMENT ISSUES

As part of the entire building wiring design process, issues such as equipment procurement, system installation, and testing procedures must be considered. Each of these activities is both separate from and integral to the building wiring design. While not technically part of a design, they all can determine the effectiveness of the wiring design. Failure to properly acquire, install, and test the installed equipment can result in faulty or inadequate components being forced upon the end-user community.

12.6.1 Competitive Bidding

Perhaps the single easiest method to procuring a building wiring system (design and all) is to call one of the local equipment vendors, or perhaps one of the modular equipment vendors, and invite them to have a go at your system. Unfortunately, this often results in a system implementation that does not address your specific needs. Additionally, just like in the car business, nobody pays list price for wiring components and installation. A better method is to solicit a number of bids. There are enough cable vendors in the market today to realize a fairly good competitive bid process.

The selection of a cable and wiring vendor follows much the same guidelines as the selection process for a building wiring consultant:

Established firm: The vendor should be an established firm with a proven record of supplying and installing building wiring systems.

Installation experience: The vendor should have prior experience providing and installing the exact (or extremely similar) wiring system that you are planning to install. At least two, possibly three, references should be checked for prior installations into environments similar to your own.

Certification: The members of the installation should be appropriately certified for construction and installation activities (these may vary between different localities).

Vendor independence: An organization that provides and installs wire and cable components from several different vendors can often give you a better selection of manufacturers' hardware (i.e., there are many different manufacturers of SIO units, each with their own definition of quality).

Service and support: The vendor should have a local presence to provide support for your wiring system after installation. Components are bound to fail sooner or later and having to deal with a support organization thousands (or hundreds) of miles away can be extremely difficult.

Installation flexibility: It is important for the vendor to be able to conform to your schedule of events. You do not want to delay any portion of the installation, which could result in occupancy delays, because of a lack of vendor support. Additionally, the vendor should be able to accommodate minor changes (within reason) to the building wiring design after installation has begun.

At least three potential vendors should be involved in any competitive bid process. The process can begin with a general request for information (RFI) that may be sent to over a dozen vendor organizations. The purpose of the RFI is to identify qualified sources of cable and wire products and installation services, not to solicit detailed bid responses. The information included in the RFI should be fairly complete but does not require an exact configuration of your wiring requirements.

Based upon the responses to the RFI, which is more of an information gathering tool than a real bid, a request for quote (RFQ) or request for proposal (RFP) can be issued to a selected group of RFI respondents. The RFQ is where the detailed competition begins. The information included in the RFQ should be detailed and specific concerning the entire building wiring plan. Interested vendors should be given the opportunity to inspect the building site or plans so that they may provide the best possible solution.

Once a vendor has been selected, based upon the criteria above, the installation process can begin.

12.6.2 System Installation

The ability to coordinate multiple activities at a single time is perhaps the most required talent for managing a building wiring installation. The process is both complex and lengthy. Beginning with the basic installation of a NAP, this process continues through the installation of DF facilities, distribution hardware, cable and wiring, and SIO units.

One of the first decisions regarding the installation process is when to install the various components. In an average installation, which may vary depending upon the type of construction or renovation activities taking place, the components are usually installed in the following order:

- NAP
- DF rooms (basic flooring)
- Distribution hardware (vertical and horizontal unless underfloor

distribution is used, in which case it is installed during the pouring of the floors)

- DF hardware (cross-connect equipment)
- Cabling
- SIOs
- Communications equipment

It is critical to note that the order of the equipment listed above can vary from building to building, especially during renovations. The underlying format of the installation order, however, generally applies to any installation; components should always be installed in a support-oriented fashion, from the ground up, so that each installed component allows the installation of the next higher-level item. For example, cable should never be installed prior to the completion of all DF facilities, including the installation of all cross-connect and cable termination equipment. Additionally, installing an SIO faceplate prior to the installation of the horizontal distribution cable will only result in the SIO faceplate having to be removed again prior to cable installation. This entire stepping process is based on commonsense support (i.e., walk before you run).

Another fundamental question pertaining to system installation involves who should perform the installation. The best possible solution is to have the provider of the equipment responsible for the complete installation and initial testing of the equipment. Only in this manner will there be a single point at which the "buck can stop." Using different organizations for providing and installing the equipment exposes the end-user organization to potential disaster. For example, suppose a system does not pass testing. Is it the result of faulty equipment or faulty installation? Each organization is bound to blame the other. Suppose the equipment does not arrive on time to meet the installer's schedule. Who pays for the lost labor time? While you may be able to eventually work these problems out, the resolution process takes time (and money) and can unduly complicate the installation process. Having a single point of contact within an organization that is responsible for the entire process can help to make the entire installation process a smooth operation.

One aspect often overlooked during the installation process is the area of client-vendor contact. During any major installation activity, there should be a representative of the end-user client available to the installing contractor to answer questions about the installation configuration. Additionally, the installation team should provide complete access to the client for purposes of inspecting the ongoing installation. Finding out that the SIO units in your building have been wired improperly after the fact can be costly and time consuming for both the installer and the end user. Common mistakes like this, which do occur on a fairly regular basis, can be easily corrected by on-site inspections during the installation process. On-site access to installation activities can also be used to monitor adherence to building, safety, and wiring codes that may be present in your locality. Although the majority of vendors are good and responsible organizations, there are bound to be a few here and there that will try to cut a corner or two when nobody is looking.

12.6.3 Testing and Acceptance Procedures

An integral part of any building design is the testing and acceptance procedures that must take place during and after the initial installation. While this is clearly part of the installation process, it should be well established as part of the wiring design. This means scheduling installation activities in such a manner as to allow sufficient time for unit inspections.

The test and acceptance process must be custom designed to fit your particular equipment installation. Each component installed should be inspected in some form. This may involve a visual inspection or a physical test of the system. This includes all wiring-related components: SIO units, all cable and wiring, cross-connect equipment, cross-connections, DF facilities, and distribution hardware (more information on testing procedures can be found in Chaps. 7, 8, 9, and 10).

It is recommended that each major component be checked to make sure it matches the specifications prior to installation. Additionally, each major component should be tested before the next major component is installed (i.e., physically examine the distribution hardware prior to the installation of cable). The inspection process should be designed to spot mechanical as well as functional problems.

The final test, which should be performed before the complete system is accepted and paid for, is the actual operation of a working network component, such as a LAN or other data device. This is the only true test of whether the system meets end-user requirements. Unfortunately, this test may fail if the overall system has been improperly designed. For this reason, it is important that the methods used to test any component of the system be established and agreed upon by all involved parties prior to the beginning of the installation process. The equipment provider and installation vendor have every right to know in advance how the quality of their equipment and services will be measured. Establishing these benchmarks in advance will also make the testing and acceptance process easier for the end-user organization.

12.7 SUMMARY

The following is a summary of key points outlined in this chapter:

- The quality of a wiring design can either make or break a building wiring system.
- A design should take into account all current and projected requirements.
- The development of a project plan and a design program can help ensure a smooth design and implementation process.
- Wiring codes, such as the NEC and local codes, should be known and adhered to in all cases.
- The use of consultants and the system procurement process should be carefully planned and monitored throughout the entire project.
- A sound test plan should be part of any wiring system design.

Chapter 13

System Integration

Definition: *System integration* The process of designing and integrating a telecommunications network (voice PBX, local network, etc.) with a building wiring system. Typically involving timing, planning, and luck.

The function of a building wiring system is to support end-user system implementations. This includes a broad array of networking systems, including the following:

Voice private branch exchanges (PBXs)

Data PBXs (data switches)

Local area networks (LANs)

Facsimile machines

Host and terminal systems

Modem and dial devices

The process of integrating these communications systems with a building wiring system can often be a complex process. Each networking system carries with it a multitude of configuration and implementation options. This chapter deals with the system integration and implementation process. It discusses the basic technology behind networking systems today and presents network system design options that can help to take advantage of a properly designed building wiring plan.

13.1 THE SYSTEM INTEGRATION PROCESS

The system integration process is very similar to the process of designing and installing a building wiring system. Both processes have as their basic foundation the requirement to support communications between end users, computers, and control equipment. As such, the development steps are very similar.

Requirements acknowledgment: The recognition and determination that there is a requirement for an end-user network. This is usually the result of some unfulfilled requirement, such as a lack of information exchange between office workers.

Requirements analysis: The networking requirements of the end users are gathered and used to form a list of function specifications of the networking system. These requirements may include a wide range of items, such as remote access, distance capabilities, geographic presence, and specific terminal or computer support.

System design: Based upon the requirements analysis, the system design presents a complete configuration layout of both hardware and software required to support the network system. An important part of the design process is to structure the networking system such that it takes advantage of the building wiring system as a resource and does not require extensive custom cabling efforts.

System procurement: This process involves the creation of a request for proposal (RFP), or similar device, designed to solicit information or competitive bids from potential vendors. Based upon the strength of the vendors' responses, a single vendor (or group of vendors) is selected to provide the networking equipment. This process is very similar to the procurement of the building wiring system components.

System installation: The communications equipment is brought on site and integrated into the existing building wiring system. This involves both hardware and software installations.

System testing: Any new system must be properly tested prior to the system being accepted by the end user. Not only should end-user software be tested, but the ability of the network equipment to properly integrate with the wiring structure should be determined as well.

System maintenance: Similar to building wiring, a communications network or system must have a set process for management in place prior to the installation of the network. This process should be integrated, where practical, with the management and help-desk functions of the overall management information system (MIS) and building wiring functions.

The system integration process can begin at various times during the building wiring development cycle. However, it cannot be completed until the building wiring system has itself been installed, tested, and declared operational.

System integration steps that can begin prior to the completion of the building wiring system include those that do not involve specific installation activities, such as the requirements acknowledgment, requirements analysis, system design, and system procurement. For large installations it may be advisable to begin the system integration process either during or prior to the design of the building wiring plan. This is a result of two factors. First, the overall system integration process can be extremely lengthy in duration, especially for large campus- or organizationwide systems. Second, determining the requirement and general design constraints of a networking system can help to shape the development of the building wiring system. Although the system integration process cannot be completed until well after the completion of the building wiring system, it can begin well in advance of the building wiring activities.

The system-level requirements uncovered during the system integration requirements analysis may even be the impetus for a building wiring installation (especially in existing buildings that are found to have insufficient cabling systems to support end-user networks). System integration activities that cannot take place prior to the complete installation and certification of a building wiring system include the installation, testing, and maintenance steps. Undertaking any of these activities during the installation of the building wiring system can be disastrous. For example, suppose that an end-user network has been installed while portions of the building wiring system are either under installation or have not been tested and certified. The ongoing installation activities may disrupt the operations of the network. Furthermore, should there be a problem with the operation of the network, it may be difficult to determine if the problem lies with the network equipment or the building wiring system. The completion of the test and certification process of the wiring system will help to determine if there are any problems with the wiring and, it is hoped, will demonstrate the ability of the wiring to fully support the end-user networking application.

As with a building wiring system, the duration and actual time required to complete the system integration process will vary depending upon the complexity of the system. Small networking systems, or those such as Centrex with little hardware, can take as little as a month. Larger systems, such as a campuswide PBX or LAN, can take up to a year to fully plan, procure, and install.

The system integration and building wiring development processes often follow parallel courses. Unfortunately, a delay in the building wiring process can create a delay in the system integration process. This often necessitates the linking of the two procedures through a single status tracking system. Under such a system, milestone events in the building wiring process can be used to trigger the occurrence of events in the system integration process. For example, the completion of the wiring installation activities indicates that the procurement (actual delivery, not product or vendor evaluation) of networking equipment can take place. This equipment procurement, usually taking several weeks at a minimum, allows ample time for the building wiring to be tested and certified operational. Likewise, the certification of the wire systems signals that the installation of the networking equipment can begin.

Linking the wiring and equipment cycles can be difficult. Not only must these two areas be incorporated into a single tracking element, there must be sufficient coordination between the two management groups to ensure that the operation flows in a smooth manner in which each group fully understands the progress of the overall system, what type of delays can be expected, and how each delay will be overcome. It is recommended that where possible, the two groups be combined into a single structure. When this is not possible, each group should have a person appointed as the interface to the other group. This interface person will be responsible for attending meetings of both organizations and ensuring that communication between groups takes place.

13.2 OSI MODEL

When seeking to understand the system integration and networking concepts, it is important to understand how networking systems are structured. The Open Systems Interconnection (OSI) model is a layered construction in which the many functional components of a network are grouped into seven functional areas, called *protocol levels*. These areas address all networking components, from the building wiring all the way to the end-user application software. The OSI networking model defines a structured approach to communications systems in which each layer (defined as a protocol layer) has a specific function within the overall system.

The OSI seven-layer model was developed and promoted by the International Standards Organization (ISO) to facilitate standardization within the networking community. The theory is that if a vendor designs a networking system or application that follows the structure of the OSI model, that system should interoperate with other vendor equipment that follows the same specification (provided their level-specific implementation follows a common standard). Additionally, with a layered approach, it is intended that software and hardware components from one layer can be mixed and matched with compo-

Layer	Function
7 Application	End-user Application Software
6 Presentation	Presentation of Information to End-user Applications (Data Formats, Screen Maps)
5 Session	End-to-End Node Session Establishment
4 Transport	End-to-End Node Communications (i.e., TCP, TP4)
3 Network	Addressing Between Network Nodes and Logical Network Structures (i.e., IP, XNS, IPX, DECnet)
2 Datalink	Communications Formats, Access Control, Addressing (i.e., Ethernet, Token-Ring, FDDI)
1 Physical	Physical Signalling and Media Interfaces, Including Multiplexing and Building Wiring Systems

Figure 13.1 OSI network model.

nents from other higher or lower layers without compatibility problems. This lets end users choose the best components for each function. The individual seven layers are presented in Fig. 13.1.

Each of these layers represents a different level of networking and communications functionality. For example, the first layer (physical) relates specifically to the cable and wiring used for communications networks while the seventh layer (application) relates to the software used on end-user computers. Different vendors' protocols can each be supported through this modularity. For example, many different equipment vendors offer networking systems that can operate over the same Layer 1 physical media. Likewise, many different applications (spreadsheets, word processors, etc.) can operate over many different and diverse networks. As the protocol level rises, the level of interaction, especially from a building wiring perspective, will decrease. For example, designing an Ethernet LAN involves direct interaction with the building wiring system. However, deciding to use a specific application on an Ethernet LAN does not involve any building wiring issues (except where network configuration and the location of end-user devices are concerned).

13.2.1 OSI Layer 1—Physical

The OSI Physical Layer 1 defines the electrical characteristics of the networking transmission media. These characteristics include the type of signaling used to represent information and the physical interface specifications that are supported at Layer 1. Examples of Layer 1 networking include interfaces such as V.35, RS-232, and X.21. In addition, devices such as modems and multiplexers operate at this layer. For example, the T1 (DS-1) transmission service that operates at

1.544 Mbps uses time division multiplexing at Layer 1 to transfer information between network communications devices. The physical layer can include characteristics for many different types of transmission media, such as twisted-pair, coaxial cable, fiber-optic cable, microwave, and infrared.

13.2.2 OSI Layer 2—Data Link

The OSI Data Link Layer 2 defines the framing structure of information in a network including the determination of how two or more nodes within a network will communicate. Details such as when a node can send or receive information, what format that information is sent in, and how nodes are physically addressed take place within Layer 2. This layer is often called the MAC Layer since it is responsible for the Media Access Control functions, such as Token-Passing, for networks. Examples of Layer 2 protocols include High-Level Data Link Control (HDLC), Synchronous Data Link Control (SDLC), and Frame Relay. Additionally, Layer 2 is responsible for the structure of LAN protocols, such as Ethernet IEEE 802.3, Token-Ring IEEE 802.5, Token-Bus IEEE 802.4, Metropolitan Area Network IEEE 802.6, and FDDI ANSI X3T9.5. Using these LAN protocols as an example of the OSI modularity, each of these networking systems can support numerous upper-level protocols (such as TCP and IP) and applications (word processing, database, file transfer) without any network reconfiguration.

All physical addressing within a network takes place at this level. For example, all Ethernet interface cards (the card that connects a user device to an Ethernet network) have a 48-bit address physically embedded in their circuitry. Any device wishing to communicate with this node issues a message addressed to that specific 48-bit address. The bridging of networks takes place based upon the Layer 2 physical address.

13.2.3 OSI Layer 3—Network

The OSI Network Layer 3 is responsible for the logical networking of information between end-user nodes. The logical networking refers to the ability to pass information between multiple networks based upon a logical addressing scheme. This implies that the Layer 3 protocol has the ability to route or switch information through multiple switches and intermediate networks that may be located between end-user nodes. Protocols found within the Layer 3 segment are usually designed to operate over multiple lower-level protocols (such as HDLC, Asynch, Token-Ring, Ethernet, etc.). Examples of Level 3 networking protocols include the following:

Internet Protocol (IP): Currently one of the more widely used protocols for the networking of user devices. The IP protocol is widely used by the Internet community (academic and research institutions) where it was developed. It has recently found its way into the corporate environment as one of the more reliable, vendor-independent, networking protocols. IP is often coupled with the Layer 4 Transmission Control Protocol (TCP).

Xerox Network System (XNS): Developed by Xerox as the "Greybook" specification, XNS is currently used by a wide variety of LAN vendors, such as Banyan, Apollo (Domain), 3Com, and Novell (IPX). Most vendors who implement XNS use their own proprietary version of the protocol, so there is not a great deal of vendor compatibility between all XNS nodes.

Appletalk (Datagram Delivery Protocol, DDP): This protocol, a remote derivative of the XNS protocol, is used by Apple for their Appletalk network. Appletalk should not be confused with Localtalk, which is Apple's Layer 2 LAN that operates at 230 kbps over a single shielded twisted-pair in a bus configuration.

DECnet: This is a proprietary protocol developed by Digital Equipment Corporation (DEC) as part of their overall networking strategy. The current release of DECnet is called Phase IV. The next version, Phase V, is intended to be an OSI-compliant Level 3 protocol.

OSI Connection-Less Network Service (CLNS): This is the standard protocol for OSI-compliant networking that provides a connection-less network connection.

Level 3 is where network "routing" functions take place based on logical network addresses (i.e., where a user is located within a network) rather than on Layer 2 physical node addresses.

13.2.4 OSI Layer 4—Transport

The OSI Transport Level 4 protocol is responsible for maintaining the integrity of information on an end user to end user basis. This protocol must determine if the information received by an end user is valid (i.e., all bits are correct and in the proper order). The TCP is the most common example of this protocol in public networking today.

13.2.5 OSI Layers 5, 6, and 7

The OSI Layers 5, 6, and 7 are associated with end-user systems. They provide for user sessions (i.e., log-on), presentation (i.e., screen imaging and formats), and applications (i.e., word processors, spreadsheets, databases). These protocol layers are usually not associated with networking systems; rather they are associated with applications, such as the X.400 standard for electronic mail systems.

13.3 VOICE SYSTEMS

The most common application that will be seen in almost any building environment is the support of voice communications. This typically involves the connection of telephones to either a public or private switching system. Supporting this type of application is a relatively easy task given the flexibility of most voice systems, the unbelievable error correction of the end-user devices (the human ear), and the extreme distance that most telephone systems can transmit a voice-grade signal.

13.3.1 Physical Requirements

There are three main types of voice support systems available today: Centrex, key systems, and PBXs. Each of these systems offers several strengths and weaknesses. More important, however, is that their requirement for building wiring support is virtually identical.

Centrex is a service offered by the local telephone company (telco) that allows the telco to provide virtual switching for voice calls within an organization. Every telephone within an organization is connected directly to the telco central office. Within the central office, the telco uses their own telephone switch (such as #5ESS) to connect end-user telephones together (i.e., complete a call). With Centrex, there is no requirement for the customer to purchase any switching equipment. Additionally, there is no telco equipment or switches located on the customer premises.

A key system is an enhancement to the Centrex offering that allows multiple telephones within an organization to share a smaller number of incoming telephone lines connected to the central office. A key system does not provide any switching of calls or connection between telephones. All connection and switching services are still provided by the telco switch. Older key systems used a 25-pair twisted-pair cable to connect telephones to the main key unit. Newer systems, however, usually require only two twisted-pairs per telephone. A key system is typically a small unit that can be wall-mounted.

The PBX is a telephone circuit switch that is located on a customer premises. This allows the end user to control all functions associated with telephone service, such as the assignment of telephone numbers, call blocking systems, call forwarding systems, and voice mail. All telephones are connected directly to the PBX. All outside connections (to telephones outside of the organization) are routed by the PBX to the central office switch over a small number of circuits (there typically does not need to be a circuit for every telephone unless all telephones will be connected at all times to outside connections). While there is no equipment required for Centrex and only a wall-mounted unit required for key systems, a PBX can be a very large unit requiring special power and environmental control (i.e., computer room).

Although all of these systems were primarily designed to support voice transmission, there is an increasing amount of data traffic being sent through these systems today. Nonetheless, the components involved in these systems are relatively few and are very well standardized:

Station connection: Most voice systems are designed to use an RJ-11 or RJ-45 jack (SIO).

Wire: The most common form of voice wire in use today is two, three, or four unshielded twisted-pairs (usually 24- or 26AWG), although voice communication (analog) does not require high-quality or twisted cable pairs. Many voice systems installed before 1980 used quad or untwisted cable pairs and were still able to provide local connectivity throughout entire campuses (up to a few kilometers).

Cross-connects: Voice applications (analog) can operate over almost any type of cross-connect block available today. Line loss levels (insertion loss from a cross-connect) are not a problem for voice as they are for digital (data) applications.

13.3.2 Implementation Issues

The implementation of a Centrex system is very straightforward (there are no components other than the incoming telco circuits and the telephone handsets). For key systems and PBXs, however, the issue of system location can be quite important. A PBX or key system should always be housed in a centralized location near the building distribution frame (BDF). This is usually not a problem for PBXs since there is usually one per building. In multitenant buildings, however, individual tenants may have their own version of a BDF that serves their occupied floors, which would result in multiple PBX locations. For implementation purposes we can think of the proper PBX or key system location as being in the BDF of a particular organization. For large campus applications, the equipment should be located in the central campus distribution frame (CDF). See Fig. 13.2.

Key systems, being much smaller than PBX equipment, are often located in a distributed fashion on each floor. The principal argument

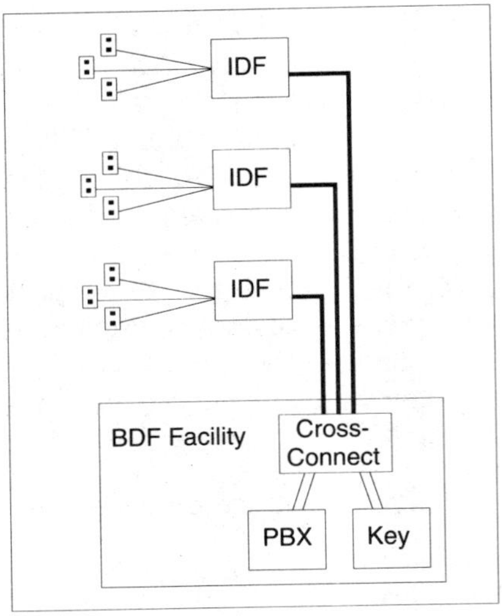

Figure 13.2 PBX and key locations.

for this placement is that it reduces the distance between the system and telephones and that it gives a greater level of control to the workgroup. This approach is common in some government offices where multiple government agencies may coexist within a single building (all using Centrex but seeking individual control of their units). Despite this control rationale, key systems should be treated as PBX equipment and located in a single BDF or CDF facility. This will help to ensure that the equipment is properly maintained in a controlled environment. It also helps to promote good system management, especially for cabling since the distributed approach creates mini-BDF facilities.

When planning the distribution or serving area of a voice system, it is important to take into consideration other nonvoice requirements. This is important from a cable distance perspective. Although most voice systems can easily support hundreds of meters between the central system and the individual telephone handsets, most data applications cannot support more than 100 m between centralized hubs and end-user devices.

13.4 LOCAL AREA NETWORKS

LANs are now the most commonly used systems for networking personal computers (PCs). They are used to connect PCs to minicomputers, mainframes, printers, file servers, modems, facsimile machines, and specialized communications devices. The rapid growth of LANs has been fueled by the rapid growth of the PC industry in the mid-1980s. During the 1980s the PC rapidly replaced the typewriter, accounting ledger, and file cabinet. It allowed the automation of many tasks previously handled manually.

As the number of PCs began to grow, so too did the number of peripheral devices, such as printers and modems. The cost of putting a letter-quality printer on every desk, coupled with a tremendous requirement to share information in a real-time basis with other computer systems, helped to bring about the growth of LANs. The use of shared peripheral devices such as printers and modems over a LAN became common to help reduce the cost of adding office automation. Through the use of a LAN, many PC users could access devices and information on other computers. The use of devices such as modems, bridges, routers, and gateways allowed users to share devices and information with remote users who were not located on the local LAN.

The growth of LAN technology was largely ad hoc in the 1980s. There were many different types of networking systems, each produced by a different vendor trying to gain market share. Eventually the market dictated a standardization process that resulted in several of the more popular LAN systems becoming industry standards. This standardization process allowed many different vendors to produce equipment that could interoperate, providing the end user with a variety of vendors and a more competitive market (lower cost).

Part of this standardization process involved the consolidation of LAN functions. This consolidation has allowed us to classify different types of LANs by their function into three basic categories of LANs:

Departmental LANs: Also referred to as workgroup LANs, a departmental LAN supports a single department or group of users. It often supports a limited geographical area, such as a floor or wing of a building.

Establishment LANs: An establishment LAN serves an organization within a single site or location. This type of LAN may be a backbone or campus LAN that links smaller departmental LANs together.

Enterprise LANs: The enterprise LAN is more of an internetwork than a LAN. It provides a backbone network facility for linking all networks within an entire organization. It may cover many different locations throughout a city through a metropolitan area network (MAN) or around the world through a combination of public or private wide area network (WAN) facilities (T1, X.25, FDDI, SONET, etc.).

Each of these network classifications can be supported by many different types of networks, with many different possible configurations.

13.4.1 Physical Media

The number of possible physical media types that can support LAN technology is as diverse as the number of LANs available today. Some of the common transmission media include the following:

Twisted-pair: This includes both shielded and unshielded cable for a variety of AWG sizes (22, 24, 26). While most LANs operate over solid conductor cable, some can operate over smaller AWG stranded cable.

Coaxial cable: One of the original networking transmission media, coaxial cable supports a wide range of LANs over many different types of cable (coax, twinax, triaxial, etc.).

Fiber optics: The use of fiber-optic cable has enabled the LAN to reach into the high-speed range where transmission is measured in hundreds of megabits per second. LANs can operate over many different types of fiber, such as multimode, single-mode, and even plastic fiber.

Microwave: Although uncommon in the office environment, microwave is used extensively to provide LAN-LAN connectivity in the campus and metropolitan market.

Infrared: Nonphysical transmission media, such as infrared, which uses open-air light signals, have been making big strides in the LAN market. Originally capable of only low-speed proprietary transmission, this technology can now support high-speed networks such as Ethernet and Token-Ring (although these implementations are still proprietary). The biggest constraint of infrared LANs is their line-of-sight requirement between transmitters and receivers.

Radio: Similar to microwave systems, there are LAN systems available today which use spread-spectrum and cellular phone technology. These types of wireless LANs are likely to become very popular in the next few years when the FCC determines how the frequency spectrum will be allocated. Some vendors are already prototyping next-generation wireless LANs that conform to networking standards, such as Ethernet/802.3, something that is lacking in most of today's current offerings.

One of the interesting aspects of LAN technology is the ability of individual LANs to operate over many different types of media. Ethernet is one example. Even the Fiber Distributed Data Interface (FDDI), based upon fiber-optic technology, has been shown in a lab environment to operate over high-performance twisted-pair.

13.4.2 Topology

There are three common topologies associated with LANs: bus, ring, and star. Each of these topologies is suited for different situations and may be associated with a particular type of network technology:

Bus networks: A bus network connects individual users together in a serial fashion. A single cable is connected point to point between each device on the network. One of the advantages of a bus network is that it can make it relatively easy to connect several devices located within the same room. One of the major disadvantages is that a break anywhere in the network cable results in users on either side of the break being isolated from each other. Another disadvantage is that bus networks can be difficult to install in office environments where there are little or no point-to-point cable facilities. This type of network also offers little control or management over individual users or portions of the network since there is no central intelligence in the network. See Fig. 13.3.

Ring networks: A ring network consists of a looped cable that connects multiple devices together. This is very similar to a bus network with the exception that the two ends of the network are joined together. Most of the disadvantages that apply to bus networks also apply to ring networks. There are, however, some exceptions. For example, some ring networks can continue to operate even after a cable failure. See Fig. 13.4.

Star networks: A star network, unlike a bus or ring, uses a centrally located hub. Individual cables connect the hub device to each networked device. One of the advantages of this configuration is that it follows the hub-oriented approach of most existing building wiring systems, allowing current distribution facilities to be used. Additionally, most star-shaped networks are designed such that a cable break between the hub and a networked device will not dis-

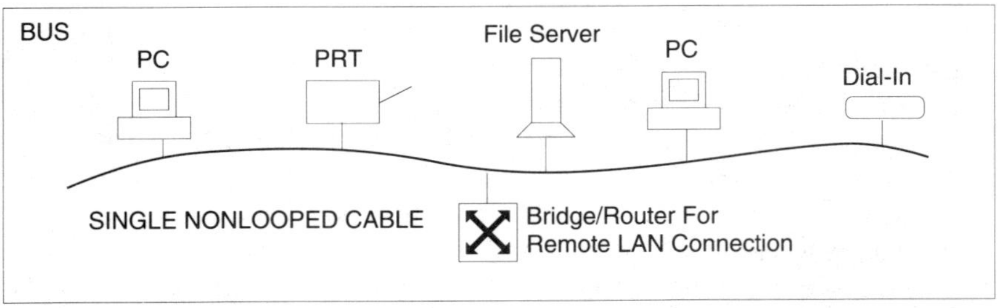

Figure 13.3 Bus LAN topology.

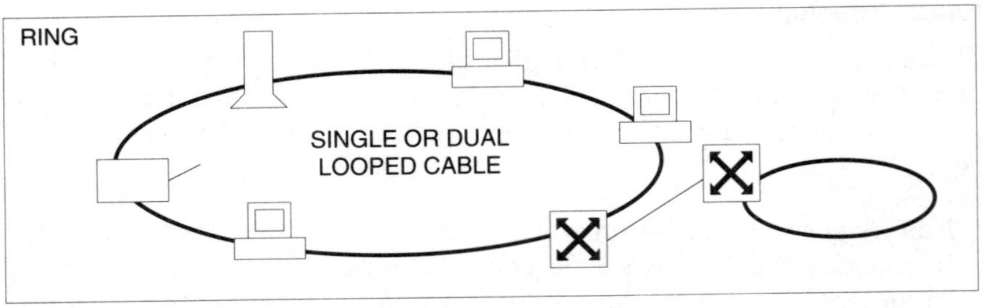

Figure 13.4 Ring LAN topology.

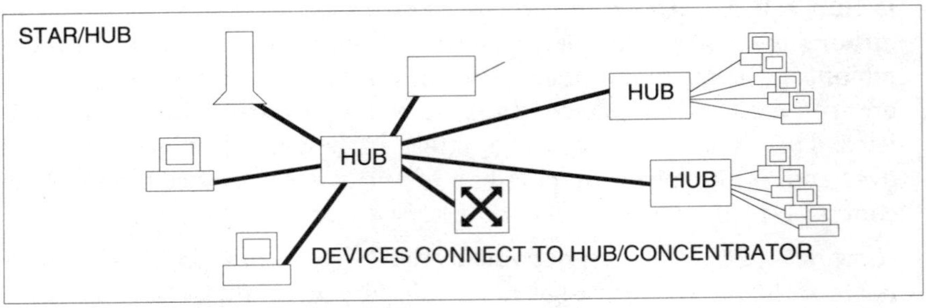

Figure 13.5 Star or hub LAN topology.

rupt the ability of other devices on the network to communicate. See Fig. 13.5.

As with transmission media, it is not uncommon to see a LAN that can support multiple physical topologies. This can enable a LAN designer to construct a LAN that uses the best topology to suit a particular requirement within a building, campus, or metropolitan area.

13.4.3 Access Methods

Within each networked device there exists a set of hardware and software that controls the input and output of the device onto the network. This set usually consists of a network interface card (NIC), which resides within the unit, and some form of LAN operating software. Together, these two units control all communications between networked devices. An important component of this control system is the access methodology. The access method used by a network device determines how, and with what frequency, a user may transmit or receive data. While this may not directly involve the selection of a particular wire or cable (the access method is part of the OSI Layer 2 protocol), there are certain access methods that are associated with LAN systems that operate over limited transmission media, which are relevant to the subject of building wiring.

The two most common access methods are carrier sense multiple access (CSMA) and token passing. These systems are very different in nature:

CSMA: The CSMA protocol is a random access protocol that allows any device on a network to transmit a message at any time, provided that no other device is currently transmitting (or has not transmitted within a fixed past-event window). This protocol also enables a device to be constantly listening to the network to receive information from other devices. Should two or more users transmit onto the network during the same general time frame, a collision between these multiple transmissions is likely to occur. When this happens, each device must redetermine that the network is "not busy" in order to retransmit its message. The CSMA protocol is commonly bundled with collision detection (CD) software that is used to determine if a network collision has taken place. CSMA is most commonly associated with Ethernet and Appletalk networks. One inherent disadvantage of the CSMA protocol is that the number of network collisions tends to increase rapidly as the number of users on the network increases. A large user base means that more devices are likely to attempt to transmit at the same time. This limitation is not serious when taken into account during the system design process. Devices such as bridges and routers can enable large numbers of users to share a single networking system. See Fig. 13.6.

Token passing: The token-passing protocol is a deterministic access protocol that uses a "talk-now" token that is passed from device to device in a preset sequence. A device can receive or transmit information only when the device controls the token. The use of a token allows a network designer to predetermine how frequently a user will have the token and to reduce the chances of network collisions (as occur with CSMA networks). To prevent a single user from monopolizing the network, a restriction is placed upon the size of the message that can be transmitted with any token. This protocol is used by many high-speed networks, such as Token-Ring and FDDI. One advantage of the token-passing access method is that it allows a higher number of users on the network without the extreme performance degradation (from network collisions) of CSMA networks. This reliability, however, can be a disadvantage on large networks when traffic volume is low. Since each device must wait for the token to pass sequentially through the network before the device can transmit, there is a built-in network delay factor (not a problem for small LANs).

Because of the differences in the access method, CSMA and token-passing networks differ greatly in the type of performance they can of-

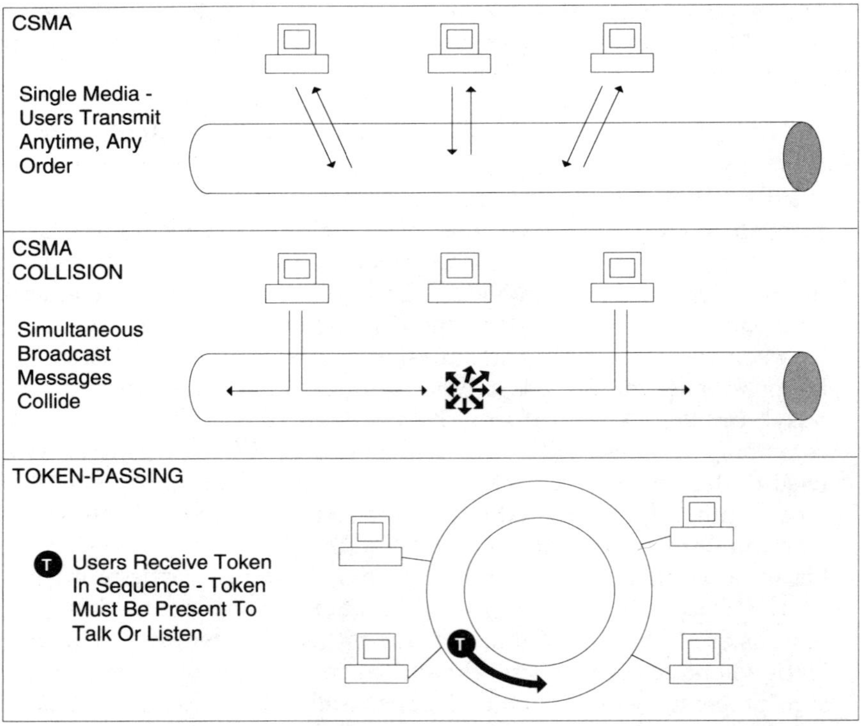

Figure 13.6 LAN access methods.

fer. CSMA is well suited to networks that are not heavily congested; it offers superior performance when only a portion of the network devices will be transmitting at any one time (burst-oriented network traffic). Under heavy network loads, the CSMA protocol tends to result in degraded performance. In contrast, token passing maintains a relatively even performance metric regardless of the number of users on a network.

13.4.4 Ethernet Background

The Ethernet networking system is one of the most popular LANs in use today. Originally developed by Xerox during the late 1970s, Ethernet uses the CSMA/CD access method. The Ethernet specification is very broad and allows the implementation of network speeds between 1 and 20 Mbps, although the 10-Mbps Ethernet system is by far the most common today. There are three different versions of Ethernet. Versions 1 and 2 belong to Xerox as proprietary LAN systems and are usually not installed any longer. The third version of Ethernet is the IEEE 802.3 network (developed by the 802.3 working group).

The Institute of Electrical and Electronics Engineers (IEEE) adopted a derivative of Ethernet Version 2 as its 802.3 LAN networking standard. Since they all use the same CSMA/CD access protocol, Ethernet (Versions 1 and 2) and the IEEE 802.3 network can coexist on the network cable. However, since they use different formats for their information packets (data transmitted between end devices), they are unable to communicate with each other. Since the Ethernet Versions 1 and 2 networks are no longer installed in great quantities, the term Ethernet is generally used to describe the IEEE 802.3 standardized networking system.

As part of the standardization process, the 802.3 committee is also responsible for the standardization of Ethernet transmission media, since Ethernet can operate over multiple types of Layer 1 systems. There are four popular IEEE groups involved in this standardization process, generally known as 10Base5, 10Base2, 10BaseT, and 10BaseF. The name of each working group is based upon the type of network system they are developing. The first set of numbers represents the transmission speed of the network in megabits per second (i.e., *10* Base5). The notation *Base* indicates that this is a baseband network. And the last characters indicate either the distance in hundreds of meters the network can span (i.e., 10Base *5*) or the type of transmission media (i.e., 10Base *T* operates over twisted-pair). See Fig. 13.7.

10Base5: This was the first popular IEEE 802.3 networking standard, also known as *thicknet Ethernet*. It uses an RG-8/9 coaxial cable in a bus topology to connect user devices together. User devices are connected through a NIC to the coaxial cable through the use of a 50-m attachment user interface (AUI) cable (a.k.a. transceiver cable) which connects to a vampire-tap transceiver attached to the co-

	10Base5	10Base2	10BaseT
Transmission Speed	10 Mb/s	10 Mb/s	10 Mb/s
# of Users w/o Bridge	1024	1024	1024
LAN Cable Length	500 m	200 m	100 m/user
Topology	Bus	Bus	Star/Hub
Cable Type	RG8 50-Ω Coax	RG9 50-Ω Coax	24AWG UTP
Cables Per LAN	3	3	Hub Specific
Cable Interface	DB15 AUI	BNC	RJ-45
Transceiver Cable	Yes: 50 m	Opt: 12 m	Opt: 12 m
Transceiver Type	External	Int/Ext	Int/Ext

Figure 13.7 Ethernet specifications.

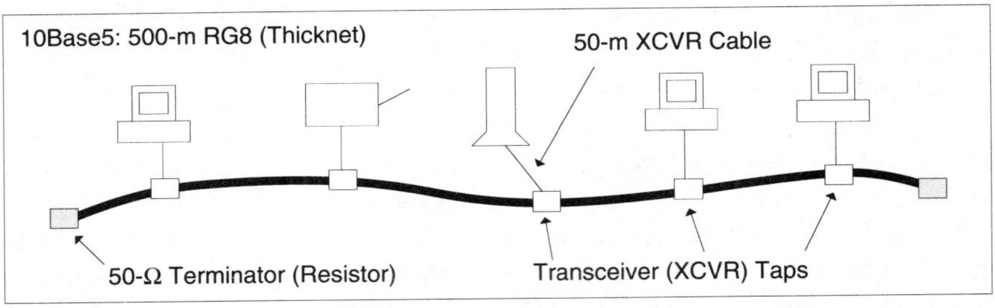

Figure 13.8 10Base5 connection.

axial cable. The term *vampire tap* refers to the fact that the network transceiver must connect to the coaxial network cable through a metal rod placed in a hole that is drilled into the core of the cable. The transceiver (sometimes written XCVR) is responsible for the actual listening and transmitting of data onto the network. To prevent data signals from reflecting from the ends of the network cable, a 50-Ω resistor is used at each end of the network cable as a *network terminator*. This type of network is very difficult to install because of the thickness of the coaxial cable and tapping process. It is mainly used today as a backbone network in a building or campus serving workgroup or establishment LANs. See Fig. 13.8.

10Base2: This version of the IEEE 802.3 networking standard was developed to provide an alternative to the thicknet configuration. Using a thin RG-58/9 coaxial cable, commonly known as *thinnet*, this standard was much easier to install (and less expensive). Still a bus topology network, the 10Base2 standard combined the NIC unit with the external vampire transceiver. The thinnet cable was now able to plug directly onto each Ethernet NIC through the use of a T-connector with a bayonet network connector (BNC) plug. This version of Ethernet also uses a network terminator at each end of the coaxial cable. See Fig. 13.9.

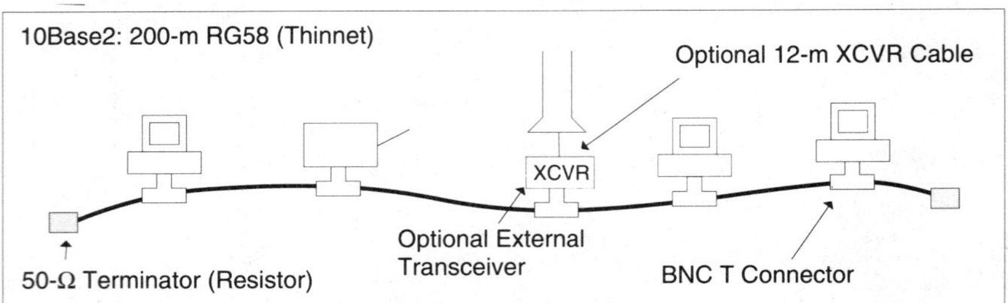

Figure 13.9 10Base2 connection.

10BaseT: This version of the IEEE 802.3 networking standard is a real departure from the previous Ethernet standards. First, it allows the network to operate over 24AWG twisted-pair (shielded or unshielded). Second, it allows network devices to be configured in a star shape through the use of an intelligent wiring hub. The distance between the hub and an individual user device is 100 m (provided a good quality cable with relatively few cross-connects). The 10BaseT standard uses two cable pairs terminated in an RJ-45 jack on pins 1, 2, 3, and 6. This networking system offers the ability to centrally manage Ethernet LANs through the intelligent wiring hub. Individual ports on the hub can be monitored and turned on or off (or have a cable severed) without affecting the rest of the network. This is the most common form of Ethernet being installed today. See Fig. 13.10.

10BaseF: This is an upcoming version of Ethernet designed to operate over fiber-optic cable. The 10BaseF is still years away from being finalized. There are currently several vendors who offer fiber-optic Ethernet solutions using the Fiber Optic Inter Repeater Link (FOIRL) protocol.

In addition to these groups, there are also groups responsible for other less-common types of Ethernet, such as 10Broad36 which operates at 10 Mbps over a broadband network for up to 3600 m in length (a broadband network uses frequency division multiplexing to allow multiple networks to share a single cable, unlike baseband which uses the entire bandwidth of a cable for a single network implementation).

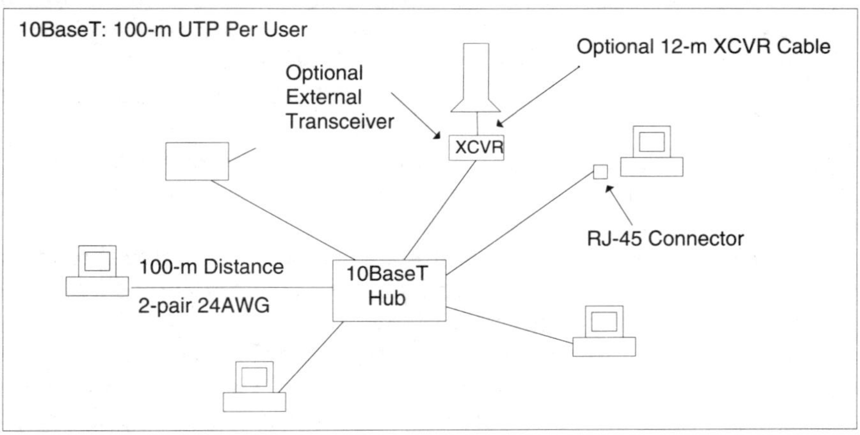

Figure 13.10 10BaseT connection.

13.4.5 Token-Ring Background

The IBM Token-Ring Network (TRN) was designed in the 1980s as a token-passing network operating at 4 Mbps over two pairs of shielded 22AWG twisted-pair cable. Adopted by the IEEE as the 802.5 LAN standard, the Token-Ring network now supports unshielded twisted-pair (24AWG). The use of unshielded cable effectively reduces the number of users allowed on a single network from 256 to 72. In addition, the latest version of the Token-Ring network provides 16 Mbps instead of 4 Mbps (part of the increased speed of this network comes from an "early release" capability which allows more than one token on a network at a time).

Although known as a ring network, the Token-Ring network uses a centralized wiring hub, called a multistation access unit (MAU), to connect devices together. Multiple MAU devices, not workstations, may be connected in a ring manner (the workstation ring is on the backplane of the MAU). Most MAU devices are unintelligent and operate through a physical relay connection which is made whenever a plug is inserted into a MAU receptacle. Ports on a MAU that have no plug inserted are bypassed by the ring backplane. Many vendors are now selling intelligent MAU devices that offer greater control over port communications and management. Like Ethernet, Token-Ring requires the use of a network interface card, commonly called a token interface card, or TIC, in each device. Many IBM devices, such as minicomputers, mainframe computers, and terminal controllers, now support TIC adapters and can communicate with each other over any IBM or IEEE 802.5 network. See Fig. 13.11.

13.4.6 Appletalk and Localtalk Background

Appletalk is the proprietary OSI Layer 3 networking protocol (based on XNS) used in all Macintosh computers. There are two versions of Appletalk, Phase I and Phase II. Appletalk is usually associated with Localtalk, a CSMA network operating at 230 kbps that is native to all Apple Macintosh computers (the Macintosh can also support other networks such as Ethernet through internal and external Ethernet NICs).

The Localtalk system was originally designed to operate in a bus configuration using a single shielded twisted-pair (24AWG) up to approximately 300 m in length. Each Localtalk device was connected through a special terminator box that provided connection between the special Localtalk port and the network bus cable. Recently, there have been third-party implementations of Localtalk that allow the network to be configured in a star topology through the use of an intelligent wiring hub. Using either one or two twisted-pairs, this star topology allows Localtalk to operate over many existing building wiring systems. See Fig. 13.12.

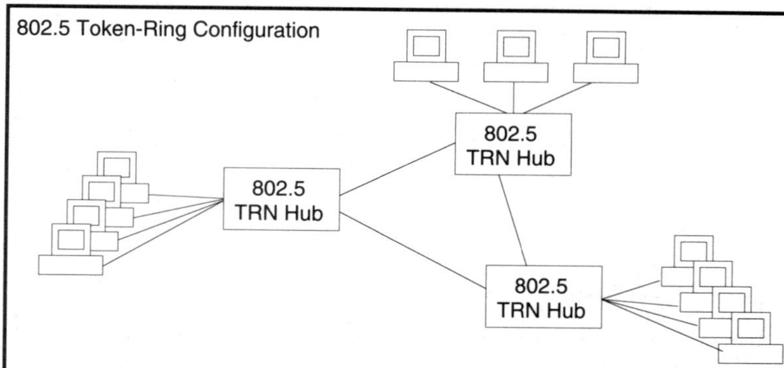

	Shielded TP	Unshielded TP
Transmission Speed	4/16 Mb/s	4/16 Mb/s
# of Users w/o Bridge	264	≤72
LAN Cable Length	100-300 m	≤100 m
Hub/Hub Cable	≤100 m	≤50 m
Topology	Star/Ring	Star/Ring
Cable Type	22AWG STP	24AWG UTP
Cable Interface	DB9/Data	DB9/Data/RJ45
Transceiver Cable	No	No

Figure 13.11 802.5 Token-Ring.

13.4.7 FDDI Background

The FDDI network is a 100-Mbps token-passing system that operates over multimode fiber-optic cable. FDDI is still in the development process with the American National Standards Institute (ANSI) X3T9.5 committee. A single FDDI network can span up to 100 km with a maximum distance between network devices of 2 km. There are four basic elements (protocols) of the FDDI standard:

- *Station management (SMT):* The SMT protocol is designed to support enhanced management features for FDDI devices, including fault isolation and bandwidth allocation.
- *Physical layer (PHY):* The protocol defines the physical frame and clocking structure for packets transmitted across an FDDI network.
- *Physical media dependent (PMD):* The PMD protocol specifies a

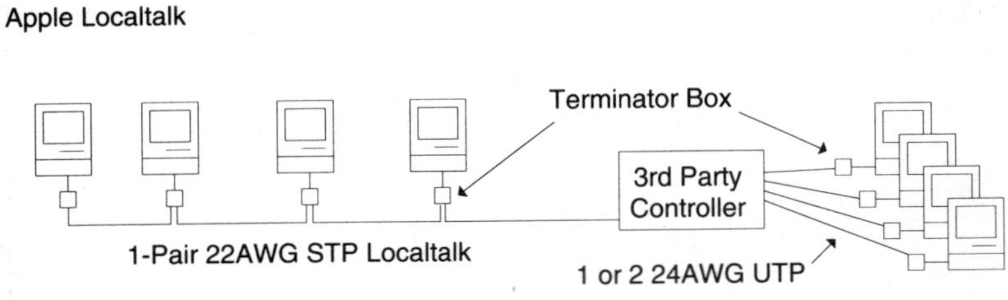

	Localtalk	3rd Party
Transmission Speed	230 kb/s	230 kb/s
Topology	Bus	Star
Cable Type	22AWG STP	22/24 AWG UTP
Cable Interface	DIN-3	DIN-3/RJ-11

Figure 13.12 Localtalk.

duplex fiber-optic connector supporting 62.5/125-μ multimode fiber (1300-nm signal at 125 MHz with 25 Mbps of network overhead).

- *Media access control (MAC):* This is an OSI Layer 2 protocol that controls how individual FDDI devices access the network.

The FDDI standard specifies two strands of fiber, between each network device, configured in a counterrotating ring. For example, the inner fiber sends signals in a clockwise fashion while the outer fiber sends signals counterclockwise. The FDDI standard is designed to automatically bypass faulty portions of the network (either cable or nodes) by looping back the two fibers on either side of the network failure. This allows an FDDI network to suffer a direct network failure and still operate without loss of user access (except for those directly attached to a faulty interface node). See Fig. 13.13.

An FDDI network can support two main types of devices, dual-attached stations (DAS) and single-attached stations (SAS), in addition to user devices. A DAS is capable of connecting directly to, and participating in, the FDDI ring (terminating four fiber-optic strands). An SAS can only connect to two fiber-optic strands and must attach to a dual-attached device to gain access to the FDDI ring (a single-attached device can connect to multiple user devices).

This network is ideally suited for backbone networking throughout a building, campus, or metropolitan area. An FDDI LAN will often be used to interconnect multiple lower-speed networks together (such as Ethernet and Token-Ring).

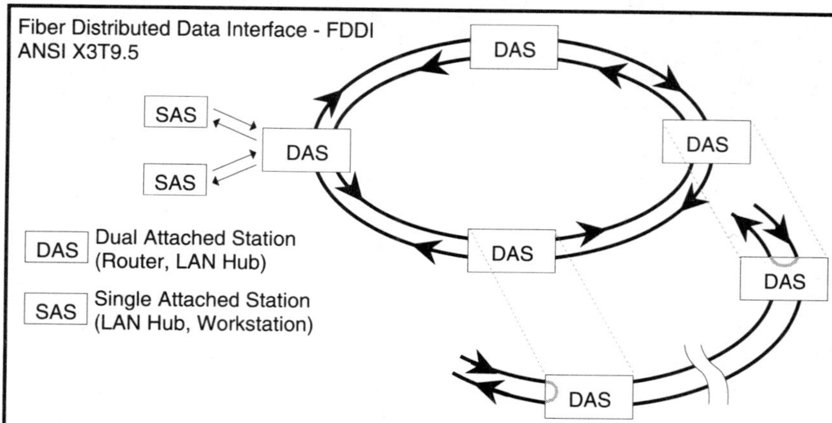

	FDDI
Transmission Speed	100 Mb/s
Topology	Dual Ring
Cable Type	62.5/125 μ
Cable Interface	Duplex Connector
Access Method	Token-Passing
Node-Node Distance	≤2 km
Total Ring Distance	≤100 km

Figure 13.13 FDDI.

There is a new version of FDDI, FDDI-2, under development. Possible enhancements to the FDDI include a single-mode fiber-optic interface (allowing greater distances), an increase in the performance of the network, and limited support for voice traffic or other communications systems that require dedicated "pass-through" data paths.

13.4.8 LAN Internetworking

There are four different types of networking devices that can be used to connect multiple LANs together (internetworking): repeaters, bridges, routers, and gateways. Each of these systems operates at a different OSI layer and differs greatly from the other units in complexity, functionality, performance, and cost. See Fig. 13.14.

Repeaters: A repeater operates at the OSI Layer 1. It sends all network signals from one LAN onto an adjacent network segment. Its

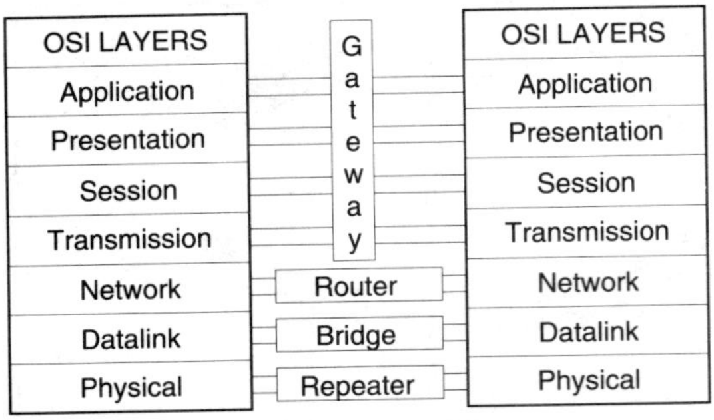

Figure 13.14 OSI internetworking.

primary use is to extend the physical distance of a network. Repeaters usually cost less than $1200.

Bridges: A bridge operates at the OSI Layer 2. It links two or more LAN segments into a single, larger, LAN (combining LANs A and B forms a single larger LAN C). A bridge differs from a repeater in that a bridge can learn where individual users are located on various network segments and make a forward or do not forward decision on each MAC Layer packet it receives. This function, called *filtering*, prevents traffic from being forwarded onto LAN segments that do not serve the destination device(s) of the packet. A bridge does not need to know the overall topology of a network in order to function. Bridges typically cost anywhere from $1500 to $30,000 depending upon their performance (the number of packets they can process per second) and the number of networks they can connect together.

Routers: A router operates at the OSI Layer 3. It provides a means of connecting multiple, independent networks together. Networks that are connected through routers maintain their individuality from a management and protocol perspective. Routers use protocols such as TCP/IP and XNS to pass traffic between user devices. Since the OSI Network Layer protocols use logical rather than physical addresses, traffic is never sent to a remote LAN unless it is destined for a user on that particular network. Routers are generally slower than bridges because of the extra processing required to interpret the logical addresses and create a map of the entire network (so they know where to send packets). Routers usually cost anywhere from $5000 to $75,000 depending upon their performance and the number of networks they can connect together.

Gateways: A gateway is a device that can translate between different types of protocol. As such, a gateway may operate at any or all of the OSI layers. It is common to see gateways combined with bridge or router functionality. Gateways, providing much more functionality than a repeater, bridge, or router, generally process information at a much slower rate. Gateway costs range from a few thousand dollars to over $100,000 for large systems.

A large internetwork (multiple LANs connected together) may often involve a mix of the above products.

13.4.9 LAN System Design Options

There are many different types of LAN networking equipment, such as bridges, wiring hubs, and file servers, that may be present within a building or campus environment. Deciding what type of network topology to implement, and where to place these components, can often be difficult. The following guidelines are intended to provide a suggested approach to LAN networking and system integration that helps to take advantage of a structured building wiring system:

Workgroup LAN topology: Star-shaped LANs should be used whenever possible. Bus and ring topologies can be much too difficult to install and maintain. The central hub of the LAN should be located within the first distribution frame (DF), such as an intermediate distribution frame (IDF), or computer facility (provided the computer facility is directly adjacent to the DF). In some cases, the use of a multitopology approach can be beneficial. For example, a small workgroup within a single room may connect together through a bus network, which in turn is connected to the remainder of the network via a centrally located wiring hub. See Fig. 13.15.

Building or campus LAN topology: Large building or campus LANs may require a ring or star topology. Where possible, however, a star shape is recommended. This includes installations of FDDI, which can be configured in a star through the use of double fiber runs. See Fig. 13.16. In the building, individual LAN hubs are placed in IDFs and bridged together in a star configuration to a central hub in the BDF. The same approach is used for connecting buildings through the CDF.

Another interesting option for building and campus LANs is the collapsed backbone, which uses a relatively small ring backbone with hub spokes connecting each remote system. This differs from a traditional backbone network where the backbone physically passes through each service site. Another option for backbone networks uses a single high-performance node as a backbone entity. For example, a single high-performance bridge or router, located centrally within a building, can serve as a backbone network (the backbone is actually the backplane of the unit). This reduces the number of internetworking devices required and also simplifies network management.

Hardware placement: The networking components of a LAN system (bridges, routers, modems, hubs, etc.) must be located in a cen-

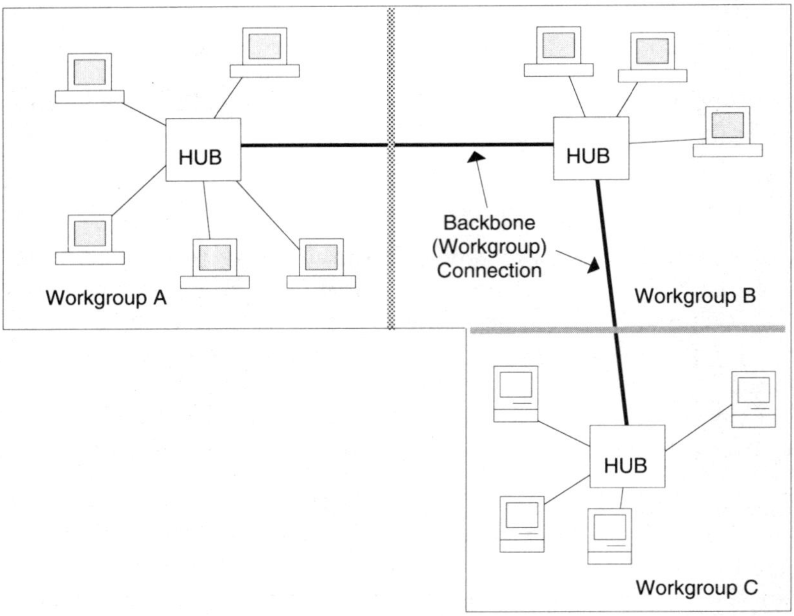

Figure 13.15 Workgroups.

tral facility that offers security and access control. Additionally, these devices should be located near the center of the wiring system since they will be required to serve as consolidation devices. It is recommended that any type of networking device (not file servers or other systems) be located in or near a DF facility.

13.5 HOST AND TERMINAL SYSTEMS

Most host and terminal networking systems are configured in a natural star topology. Individual workstations or terminals must be connected to either a computer or concentration device. When concentration devices are used, it is recommended that the cable path between the terminal and the concentrator follow the distribution path of the building wiring system. This results in a concentration node that is located within a DF facility (or in an adjacent computer room).

For systems that connect directly to a computer, it is recommended that the computer have a bulk cable installation between itself and the nearest DF. Terminals should connect to the computer through the building wiring system and be cross-connected to the bulk computer connection.

Terminals should never be connected in a straight-line fashion (i.e., separate from the heirarchical building wiring structure) un-

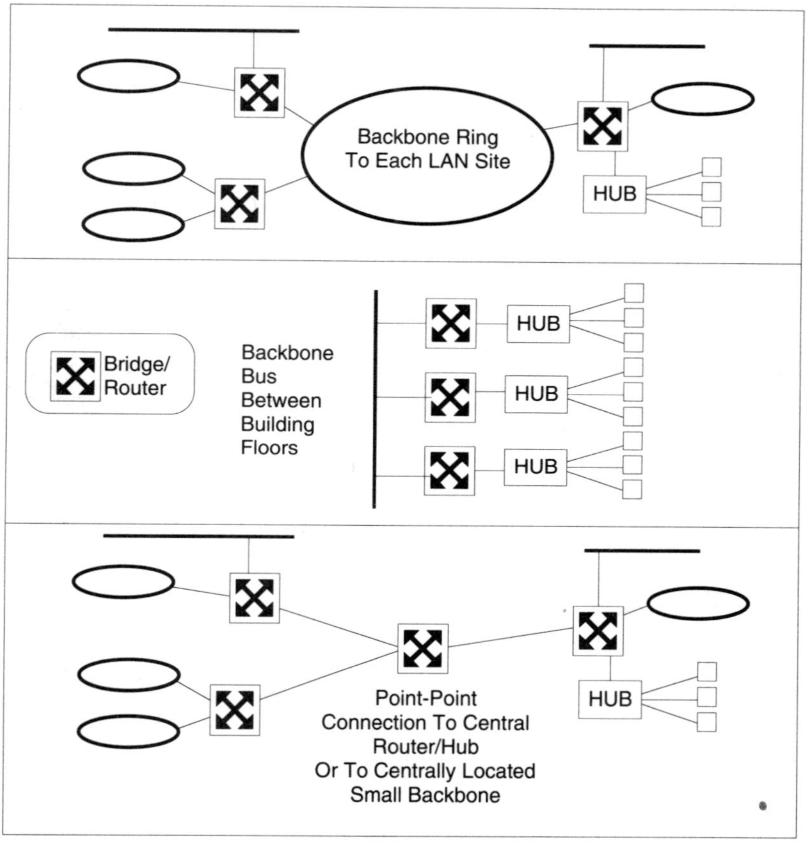

Figure 13.16 Building or campus LANs.

less they are located within the same area as the concentrator or computer.

13.6 SUMMARY

The following is a summary of the key points presented in this chapter:

- The system integration process involves the integration of voice, data, and video systems with a building wiring system.
- The system integration process should be part of the initial wiring requirements analysis, design, and planning.
- The implementation of networking hardware, such as wiring hubs, should coincide with DF locations in support of workgroup, establishment, and enterprise networks.

Chapter

14

Universal Wiring Standard

Definition: *Universal wiring standard* A collection of standardized guidelines used to shape the development of a site-specific building wiring plan.

While a building wiring plan addresses the site-specific requirements of an organization, there are situations in multisite organizations where coordination between each site is advantageous. This can be accomplished through the development of a universal wiring standard (UWS). A UWS presents a series of guidelines for the development of standardized site-specific wiring plans. This chapter takes a look at several aspects of a wiring standard, such as how a wiring standard differs from an individual wiring plan, how standards can be developed, and what are the limitations of standards. There is also information presented on existing industrywide standards that have been designed jointly by the vendor and user communities.

14.1 STANDARD GOALS

An important tool for shaping and managing building wiring resources is the UWS, which is a collection of guidelines designed to provide the guidance and direction for individual building wiring plans. It can be an invaluable tool for many large organizations not only for the shaping of building wiring plans but for the implementation of networking systems as well. A UWS specifies a standardized list of components and options that may be used during the development of a specific building wiring plan. The goal of a UWS is to provide a flexible architecture that can be both standardized and can allow custom tailoring to suit the needs of specific applications. This can be termed an *application-based configuration* (ABC). The standard provides a predetermined list of options that can be configured based upon the various requirements of the site-specific building and communications applications. Components covered by a standard include the following:

Standard information outlet (SIO) unit: A standard may list a predetermined type of SIO unit (with options for specific requirements). Different units may be specified for use with certain applications (wall mount versus inset, fiber versus coax, etc.).

Horizontal wiring: Information on the types of horizontal distribution cable that may be used are covered here.

Distribution frames (DFs): A standard will include specific types of cross-connect and patch-panel equipment that can be used.

Distribution types: Information such as the preferred type of distribution layouts and hardware (i.e., zone distribution using ducts in suspended ceilings), with options for specific buildings, is included.

Cable management: Similar to a building wiring plan, a wiring standard should set forth fixed guidelines to ensure that all wiring systems are managed in a consistent manner.

Configuration details: The plan may suggest configuration guidelines for individual components. For example, the placement of SIO units within an office or power pole may be predetermined. The configuration of a DF may be required to follow a preset structural organization to facilitate management between different facilities. Information on where to locate DF facilities, as well as their serving areas, is also included.

Applications support: There may be strict guidelines for the support of certain applications. For example, the standard may require that all Ethernet local area networks (LANs) installed be supported by unshielded twisted-pair only (no coaxial, fiber, or shielded cable allowed).

A building wiring standard can be considered a "super wiring plan" that deals with any foreseeable requirement, with the exception that the entire plan need not be implemented in any single building. Only those details of the standard that apply to the individual site installation are used. The actual configuration details, relating to sizing, exact equipment placing, or the building configuration and layout, are determined by the site-specific wiring plan.

The advantages of a complete wiring standard track closely to the benefits of a building wiring plan. Some of the additional benefits, above those associated with a wiring plan, include the following:

Simplified plan development: The guidelines set forth in an organizationwide UWS can be used to simplify the requirements analysis and design process by providing a clear set of options for plan implementation. The choices that must be made during the plan development focus on preset options for wiring components and the design configuration of the building/campus.

Component availability: The development of a wiring standard for an entire organization can result in the use of standardized components, such as station cables, SIO units, and cross-connect blocks. Standardizing allows bulk purchase and stocking of equipment. Additionally, if there are certain components, such as SIO units that must be custom built, a UWS can allow the components to be built for several sites at once and stocked at a central organization (provided the cost economies are present).

Reduced costs: Given that the components specified by the standard will be used throughout an entire organization, not just at a single site, reductions in vendor cost can be realized.

14.1.1 Standard versus Plan

An organization planning to build or renovate an office facility can benefit greatly from a building wiring plan. A well-structured plan can function well in two situations. The first is when there is a single implementation cycle taking place. A plan provides a one-stop, top-to-bottom configuration that is ideal for an implementation project. The second is when the organization is relatively small in scope. A good wiring plan is relatively simple and provides for the majority of end-user applications without custom installations.

There are, however, situations when a building wiring plan is not enough. For example, there are limitations to the complexity of a building wiring plan that become apparent in extremely large, multisite organizations. It is difficult to accommodate in a single plan every type of application that can occur, especially in organizations that have a large number of specialized groups that are physically isolated from other users. This can also be true for organizations that are composed of groups acquired through acquisition or corporate mergers. A single site or facility is often best served by a site-specific plan that addresses individual requirements. However, the coordination of plans between different facilities should take place to ensure consistency throughout the entire organization. The difficulty here is in implementing a process that helps to provide guidance to each site, providing standardization while also providing for unique site-only requirements. It is here that the implementation of a UWS that follows the application-based configuration concept can be extremely beneficial.

To understand the full benefits of a wiring standard versus a site-specific plan, it is important to address the differences between the two planning tools. A wiring standard can differ from a wiring plan in several key areas, including the following:

Site specifics: A wiring plan must include a great deal of information about the specific requirements of a building or campus. This includes a wealth of specifics that pertain to items unique to the implementation process (such as building size, office locations, cross-connect configurations, riser equipment, cable lengths, etc.). However, while a wiring standard is concerned with shaping individual configurations, it is not concerned with specifying the implementation details of a single site.

Duration: A building wiring plan has a life cycle limited to the planning and implementation of the plan (although the wiring plan can be used for ongoing management). A wiring standard, however, is designed to shape the wiring planning process for a much longer time period (at least 5 to 10 years).

Scope: A wiring plan is designed to support an installation of fixed proportions. The wiring plan, using blueprint techniques, dictates a finite set of design parameters, custom-tailored to a single building or campus, which do not apply to future installations or installations at other facilities. In contrast, a wiring standard is intended to be used to set generic guidelines for multiple sites. Although a standard helps to shape a plan, the standard itself is not concerned with the blueprint specifics of a single building. Since it applies to multiple sites, and organizations, the wiring standard is much broader in scope than a wiring plan.

Complexity of planning: Both a wiring plan and a wiring standard can be extremely complex. However, they differ in the planning complexities. The development of a wiring plan requires that a great deal of information on current requirements be evaluated. This information must be very exact (i.e., the exact number of twisted-pair cables required for each SIO) or the plan cannot be implemented. In contrast, the development of a wiring standard requires a great deal of information about applications (and cable support requirements) that could occur within the organization. The standard then sets guidelines for supporting these potential requirements. For example, a wiring plan may state that "four unshielded twisted-pairs must be installed to each data jack within all SIO units." A wiring standard, however, may state that "four unshielded twisted-pairs for data use are required in facilities that must support any of the following data applications...."

Sphere of influence: While a wiring plan is directly shaped by the requirements of the immediate user community, a wiring standard tries to influence user requirements. As an example, an organization may wish to have a standardized type of cable used for the majority of data applications across all facilities. Thus, they may state in a wiring standard that all future data applications implemented within the organization must be capable of operating over a certain type of cable. A common example of this is twisted-pair for data. Many organizations are only installing data applications that can be supported with twisted-pair. This means that 3270 terminals, which could use baluns and twisted-pair rather than coaxial cable, would be permitted. Other applications that could not use twisted-pair, such as a broadband network, would not be permitted (or at least not supported fully). Another example involves the number of conductors required for user-device access. It is common to restrict user devices to three or four twisted-pairs within an RJ jack (SIO unit). New applications that would require a larger number of conductors would not be supported. While this may seem a bit harsh, most data systems have evolved to the point where they can be supported in a fairly standardized fashion, common to many systems (just look at the high number of data applications, previously only supported by coaxial cable, that are now supported by fiber-optic or twisted-pair cable).

Controversy: The development of a building wiring plan is based upon real and immediate requirements. There is typically a set time frame for the design and implementation of the plan, based upon the occupancy requirement (this is true for both new construction and renovation projects). The set of requirements can be easily obtained by looking at the current and near-term requirements of the user community. While this process may result in some dissention, it usually can be overcome by the impending sense of urgency brought about by the user occupancy date. The development of a wiring standard, however, is typically not driven by any firm date, such as a building occupation date. By removing the sense of urgency, and increasing the scope of the activity, the level of controversy can rise rapidly. Given that a standard must be designed to support communications requirements throughout an entire organization (including multiple site locations), end users can be quite vocal concerning applications that they feel may be implemented in the coming decade. This often results in heated arguments over topics like single-mode versus multimode fiber and unshielded versus shielded twisted-pair for data applications. The only way to prevent this from taking place is to have high-level management support with imposed deadlines and a project manager who can focus the effort and can also act as a mediator between the various groups within the organization.

It is important that these differences be completely understood before undertaking the development of a wiring standard. Failure to completely understand the differences, or the additional complexity, can result in a wiring standard that is either incomplete or does not fully support the building wiring requirements.

14.1.2 Standardization Process

The development of a building wiring standard can be beneficial to most mid- to large-sized organizations involved in the expansion of their communications systems. However, the least desirable time to develop a wiring standard is when pressured by an impending requirement. Such plans are often rushed and lack the proper insight to support user requirements for longer than a few years without major revisions or changes. Organizations that are not currently involved in building projects can also benefit from the development of a wiring standard to help ensure that future projects follow a consistent and unified implementation structure.

A building wiring standard can be developed by any organization that is responsible for the decisions made concerning wiring support systems. Many wiring standards are outgrowths of existing standardization processes taking place within management information systems (MIS) and telecommunications departments. In fact, given that the role of a wiring standard is to accommodate future as well as current applications, it is advisable that the development of a wiring standard involve any organization responsible for network-related service or support (this includes end users to a certain extent).

The support for the standardization process must begin at the top of the organization. The development of a standard can take close to a year depending upon the size of the organization and the diversity of the communications requirements. For example, the development of a standard for a small organization occupying only a single building could be accomplished within a single month. In contrast, the development of a wiring standard for a large organization, such as a state government where multiple diverse applications are the rule, can easily take up to a year. Top-level management support is clearly needed to sustain a lengthy project.

14.1.2.1 Outside support. The development of a wiring standard often requires the services of an outside agency, an industry expert familiar with telecommunications trends and developing wiring standards and plans. Finding such a consultant can be more difficult than finding a consultant to help design a single-site wiring plan. While there are many consultants who have designed wiring plans, there are a much smaller number who have actually been involved in the development of a UWS. Fewer still are consultants who have developed a UWS for large national, international, or governmental organizations. Nonetheless, there are experienced personnel available within the consulting industry who can provide the wiring experience (and time resources) that may be lacking within an organization.

Many user organizations solicit the assistance of vendors when developing building wiring plans and standards. The problem with this is that the vendor is approaching the process from a financial perspective. A wiring system vendor is going to direct the process toward the type of wiring products they would like to sell. The equipment vendor (computer, PBX, LAN, etc.) is going to be interested in structuring a plan or standard to support their own product line, possibly resulting in an inability of the wiring standard to support other vendors of similar equipment products (i.e., suggesting an SIO unit that is configured to support a proprietary jack interface or promoting the use of a certain cable, such as two shielded twisted-pairs for data applications, that may restrict the use of data applications that require two, three, or four unshielded twisted-pairs).

The best way to deal with vendor organizations during the standardization process is to use them as an information resource rather than the driving force. Most vendors are extremely willing to present information on their current and future cabling architectures, perhaps even suggesting components for inclusion in the standard. This may result in the selection of components offered by particular vendors; however, the choice will be made based upon the organization's requirements and not the vendor's overstock inventory or networking architecture.

14.1.2.2 User's guide. A mandatory component of any building wiring standard is the user's guide. While a standard should present a component-by-component listing of wiring-related items, a user's guide provides information on how a wiring standard should be implemented depending upon site specifics. As an example, the implementation of a wiring system may vary considerably between an office building and health-care building because of the diverse function of occupants. Likewise, the implementation of horizontal distribution systems would be quite different in a new office building with suspended ceilings than it would in an older office building with concrete walls and a plaster ceiling.

The user's guide should also present information on options to the standard. For example, while the standard for an entire organization may specify that drop-tile ceilings should be used for distribution of unshielded twisted-pair, a particular facility that has high EMI levels may require the use of shielded cable. It is the responsibility of the user's guide to present alternatives to the standard for nonstandard installations.

14.1.2.3 User education. Once the user's guide and wiring standard have been developed, the education process can begin. A high-quality standard will be ineffective within any organization if it is not properly implemented. Even worse than an ineffective UWS is a UWS that is not well known in the organization. If the user organization does not know about the UWS, it will not be utilized during the design process. Likewise, if the UWS is not fully understood by the user base, it will be either improperly utilized or ignored altogether. A full education process on any UWS must include several key elements:

Distribution: The UWS, and any associated documentation, must be distributed liberally to all individuals involved in the building wiring, computer networking, or MIS support systems.

User training: For large organizations, such as state governments, hands-on training can be performed for each major group within the organization. This will help to familiarize each staff member with the plan and offer the chance to ask questions directly of the UWS development team.

Selling the UWS: Not every person in an organization will feel good about the standard. Many may feel that they should not be constrained by a set of standards developed by somebody else (this is very common in extremely large organizations). In addition, many people may feel that the standard does not offer them the flexibility that they need during the development of their own site-specific wiring plan. To counter this, it is necessary to "sell" the standard to the appropriate personnel. This sell involves information on the goals and benefits of the standard as well as where a building wiring designer can go to seek additional information or assistance with implementation issues.

14.1.2.4 Ongoing development. The development of a standard does not stop once the standard has been published and distributed. It continues throughout the life of the standard through any number of installations. It is impossible for a wiring standard to accommodate 100 percent of every possible requirement (because of changing and unforeseen evolutions in both technology and user requirements). As a result, it is inevitable that changes to the standard will be required.

There should be a mechanism in place to make modifications to the UWS and its associated user's guide. Changes to the standard should be based upon suggestions from users of the standard and from information obtained during a "review" of the standard which should take place after every major implementation. The review process should evaluate the effectiveness of the UWS. Areas of concern include the flexibility of the UWS to accommodate special cases and the hit or miss ratio between UWS standards and the requirements of the site-specific user base.

14.2 STANDARDS LIMITATIONS

The process of standardizing components used in the workplace can be extremely beneficial. Equipment conforming to the standard can be purchased from different sources, often at more competitive pricing than would be possible given the use of proprietary components. The use of standardized equipment can also help to ensure compatibility between different systems, such as computers or networks. However, there are certain limitations inherent in any standardization process. This is true for almost any item being standardized, computers, mechanical equipment, or car parts. These limitations can be extremely important in a fast-moving market, such as telecommunications, where technology is evolving at an extremely rapid rate.

There are several limitations to the average building wiring standard. While these limitations are minor compared to the advantages of standardization, it is important to know and understand them:

Lack of completeness: It can be difficult to take into account all possible requirements, especially in extremely large and diverse organizations.

Rigidity: When not designed properly, wiring standards can lack the flexibility to respond to diverse building environments or requirements.

Obsolescence: The telecommunications industry is evolving at an incredible pace. New technology, such as plastic fiber, may become a viable alternative within the next 5 to 10 years. There are new data applications, such as Token-Ring and Fiber Distributed Data Interface (FDDI) over unshielded twisted-pair cable, that require new high-capacity cable that differs in electrical characteristics from most installed cable today. This trend will continue. Multimode fiber-optic cable is currently very popular for networks such as Ethernet and FDDI. However, single-mode fiber may replace multimode for in-building installation as the new FDDI specification is developed and prices for single-mode fiber optics drop to the range of multimode equipment. Unfortunately, it is easy to select a type of cable, or any component, that may not be able to support applications in the future.

Knowing these limitations can be extremely helpful during the development of a building wiring standard. Proper research and planning (involving an in-depth analysis of trends in the networking industry) can help to avoid these problems.

14.3 EXISTING STANDARDS

The number of organizations that have developed building wiring standards is rapidly growing. Organizations such as energy and utility companies, defense industries, state governments, investment and financial institutions, and even political lobbying groups have all developed wiring standards. The scope of their particular standards ranges from the very basic (covering an organization with only 200 staff members) to the incredibly complex (covering all government buildings within an entire state).

In addition to end-user organizations, there are standards-development groups that have also developed standards that address the building wiring market. These standards are generally generic in nature and can be applied to almost any situation since they are not overly specific. They are, more than anything else, an excellent source of guidance on implementation guidelines. This section discusses four established wiring standards available today from the following groups:

The Electronics Industry Association (EIA)

Building Industry Consulting Services International (BICSI)

The State of Connecticut

The State of California

Each of these standards is considered by many to be a role model in the development of end-user wiring standards.

14.3.1 EIA Wiring Standard

The EIA has developed a complete set of standards that addresses the issues of building wiring. The purpose of the EIA standards is to present a complete set of guidelines to benefit those involved in the development of building wiring plans. The standards address three areas:

EIA T568 addresses building wiring for the commercial use sector, such as office buildings and large multiuse facilities.

EIA T569 addresses the architectural issues of building wiring systems.

EIA T570 addresses building wiring for both residential and light industry facilities.

The EIA, along with the Canadian Standards Association (CSA) and the Telecommunications Industry Assoc. (TIA), has spent over 3 years developing these standards (CSA T530 in Canada). Specific areas addressed within these standards include the following: horizontal wiring, backbone wiring, distribution frames (telecommunications closets), equipment rooms, and network access points. There is currently a working group (TR41.8.4) working on a specific standard for backbone (FIS) and riser cables.

The standards themselves are very complete, detailing a wealth of information on most aspects of building wiring. However, they tend to be overly generic in that they allow just about every wiring option available under the standards umbrella (such as the cabling and equipment offered by AT&T's PDS, the IBM Cabling System, and DEC's DECconnect). There is very little from a component aspect that is not permitted. This is more a group of installation guidelines than a true wiring standard that might be expected based upon other EIA standards, which are fairly fixed and inflexible (such as cable connector specifications—RS-232—which are very well-defined with very few options).

As a tool, however, the EIA standards are highly recommended for review prior to the development of any building wiring plan or standard. As with any prepackaged standard, it will require customization for implementation in most organizations.

14.3.2 BICSI Wiring Standard

BICSI, a consulting organization primarily concerned with the development of building wiring systems, has developed one of the finer reference documents available today. The Telecommunications Distribution Methods Manual (TDMM), originally presented by BICSI and

GTE Telecom, Inc., is a complete document that provides in-depth information on practically every technical aspect of building wiring. Specific areas highlighted within the document include the following:

Service entrance and termination

Equipment rooms

Telecommunications closets

Riser systems

Grounding systems

Line protection

Horizontal distribution systems

Fire protection

Telecommunications technology

Codes and standards

One of the areas that the BICSI TDMM does not specify is the actual configurations or type of components for implementation. For example, while the manual covers the various implementations of SIO units, there are no specific recommendations for what type of SIO unit should be used or what type of jack configuration is required. Instead, the manual offers a broad array of guidelines for the placement and installation of SIO units. Likewise, the manual does not specify the type of cable or cross-connect equipment that should be installed. However, there is no shortage of information on cable design (including information on horizontal distribution such as conduit and duct sizing and design parameters) or cross-connect placement. Detailed and valuable information on design and construction techniques for most any type of cable distribution system is included.

The structure of the manual is very stiff. There is little text but lots of very good technical charts and diagrams. This format, and the lack of specific information on individual components, keeps the TDMM from being a true building wiring standard. However, it is highly recommended as a technical resource for anybody involved in the building wiring industry. Complete knowledge of the TDMM is currently used as the basis for certification as a Registered Communications Distribution Designer (RCDD). The RCDD is a certification offered by BICSI that ensures a minimum (although very complete) level of knowledge with building wiring services.

14.3.3 State Standards

Many state governments have undertaken the task of developing comprehensive building wiring standards. These standards are intended for use in all state facilities. The advantage to a state government is that any person involved in new building construction or renovation activities can have a single source of wiring recommendations on which to base a site-specific building wiring plan. This results in a great deal of consistency between wiring systems in different organizations or facilities (helping to promote cross-management). Additionally, given that organizations within a state government tend to change offices on a regular basis, there is now a good chance that whoever is the current occupant, the building wiring system will be able to support the voice and data telecommunications requirements. In fact, it is not uncommon to see a state government use compliance with a building wiring standard as a requirement for communications purchases.

The two most widely known standards within state governments were designed by the State of California and the State of Connecticut. Each of these standards provides a great deal of information on specific components, such as SIO units and cross-connect blocks. This contrasts to the more generic industry standards available today which place a greater emphasis on implementation options.

14.3.3.1 State of California. California was one of the first states to actively develop a true building wiring standard for implementation in all state buildings. Developed in 1986, the standard uses a combination of unshielded and shielded twisted-pair and fiber-optic cable for telecommunications system support. Designed with an original life cycle of 7 to 10 years, the standard uses four 24AWG unshielded twisted-pairs for voice and two 22AWG shielded twisted-pairs for data, possibly based upon the IBM Cabling System. However, specifics on the type of SIO unit or cross-connect equipment to be utilized are lacking. Items such as data jacks and SIO pin configurations are incomplete in the original standard.

The structure and specific areas addressed by the California standard include the following:

Horizontal and vertical cable specifications

Distribution hardware specifications

Cable management systems

System documentation specifications

System planning guidelines

Equipment evaluation guidelines

There are several strengths to the California wiring standard. Although it is incomplete in some areas (primarily because it was one of the first of its kind), there are some areas of the standard that are well thought out and complete. For example, the cable management aspect is covered in great detail. Similarly, the standard includes a great deal of information on system management (including the integration of computer and wiring system management). There is also a great deal of information on planning and evaluation including equipment depreciation guidelines and project planning guidelines, something lacking in many other building wiring standards.

14.3.3.2 State of Connecticut. Connecticut followed the path of California in 1987 and began the design of a building wiring standard. Completed in 1988, the standard was one of the first to include a separate user's guide to assist individual state agencies with the implementation process. Following a combination of four 24AWG unshielded twisted-pairs for both voice and data with 110-blocks supplemented by fiber-optic cable (showing its foundation in AT&T PDS), the standard includes information on the following areas:

Station termination (SIO units)

Horizontal distribution (cabling)

Intermediate distribution frame (floor DF)

Interframe backbone (riser)

Building main distribution frame (building DF)

Campus systems

Cable management

In addition to the standard, there is an accompanying user's guide that presents a series of guidelines that address development planning, system implementation, and bid specifications.

The Connecticut standard is very well suited for the new-building environment. It places a great deal of emphasis on the large office building environment and includes specific information on SIO unit configurations, cable types, and cross-connect equipment. Detailed options for the different types of office environments are also included.

There are, however, some weak areas of the standard. One of the major deficiencies of the standard is the lack of support for specialized nonoffice structures. For example, information on diverse installations, such as medical facilities or renovations, is lacking. Additionally, although the standard allows the use of nonstandard facilities in special cases (cable, jacks, cross-connects, etc.), it does not provide in-depth information on supporting these nonstandard components (both design and management are lacking here).

Overall, the Connecticut standard is well-structured and provides a complete foundation for any person involved in the network or building wiring system design process. It is a good example of how a wiring standard can be designed to provide a comprehensive wiring infrastructure to support a diverse user base, provided the user base does not have a nonstandard supported communications requirement.

14.4 SUMMARY

The following is a summary of the key points presented in this chapter:

- The development of a UWS using the application-based configuration concept can benefit most mid- to large-sized organizations.
- A UWS is fundamentally different from a site-specific wiring plan in terms of complexity and functionality.
- A UWS must be supported by extensive education for those involved in telecommunications or it will not be properly utilized.
- Ongoing revision and development of a wiring standard is required to maintain support for user applications and take into account emerging technology.
- Existing standards, such as the BICSI and EIA offerings, are a good source of information of the wiring planning and implementation process.

Chapter

15

Looking to the Future

Definition: *Looking to the future* An analysis of emerging trends that may affect all or part of your telecommunications system. Something that must be performed on an ongoing basis for a telecommunications wiring system to be a success.

The typical building facility can have a useful life expectancy of at least 50 years. This can hold true for most types of buildings, such as offices and medical, maintenance, and industrial facilities. Planning today for a building and end-user environment that will last 50 years is almost an impossible task. Building renovations and occupancy changes are likely to occur on an almost yearly basis. Even the best building wiring plans will require modifications every 5 to 10 years to accommodate new communications requirements and new building wiring technology.

Nonetheless, taking a look at the current and emerging trends can help to plan for those 5- to 10-year adjustments in advance. Understanding the direction of telecommunications, computer, global business, and wiring trends can help you to plan for a building wiring system that is guaranteed to require modification as user requirements change during the coming decades. This chapter presents a list of emerging trends that could influence the building wiring environment. Accompanying each trend is a possible impact that this trend may have upon your wiring system.

15.1 TELECOMMUNICATIONS TRENDS

The telecommunications, or networking, industry is evolving very rapidly. Formerly concerned with devices such as voice private branch exchanges (PBXs, or phone switches) and analog modems, this technological area has helped to redefine the communications patterns of our entire planet. The following trends are reshaping our networking systems, and thus our telecommunications wiring requirements:

- *TREND: Wireless local area networks (LANs):* The development of wireless LAN technology has taken root with major LAN and networking vendors.

  Analysis: Several different types of wireless LANs are available today. Using technology such as infrared or spread-spectrum radio, these systems are gaining in popularity. New technological applications involving cellular and microwave technology should push these types of products into the 10- to 16-Mbps throughput range by the mid-1990s at cost-effective pricing. The IEEE has established a working group (802.11) to evaluate potential wireless LAN solutions. Providing distances of up to 100 m between central transmitters and workstations, wireless LANs could seriously affect the requirements for cable-based LANs by the end of the late 1990s. This trend will probably expand to include wireless voice networks as well through the personal communications system (PCS).

- *TREND: Wireless voice networks:* Voice networking vendors (PBXs) have recently begun working with wireless phone systems.

  Analysis: Using systems similar to cordless telephones and limited cellular systems, these new devices offer the ability to provide a personal communications device (PCD) that relates to a single person and not a fixed phone location. The implementation of this technology is much easier (except for power supplies) than wireless LANs since the communications link between devices need only be 64 kbps. These systems may become popular in the near future, displacing some wire-based systems.

- *TREND: Unshielded twisted-pair networks:* Data applications are beginning to center around the most commonly available transmission media: unshielded twisted-pair.

  Analysis: Most local networking systems either currently support, or are likely to in the near future, the use of unshielded twisted-pair as a primary transmission media. Systems such as Ethernet, Appletalk, ARCnet, Token-Ring, and even 3270-terminal applications are all available on unshielded twisted-pair. Even high-speed networking systems, such as 16-Mbps Token-Ring and 100-Mbps Fiber Distributed Data Interface (FDDI), have all been tested over this type of cable. There is a strong likelihood that this transmission media will continue to be the dominant networking cable for the next 5 to 10 years.

- *TREND: Internetworking:* The 1990s are on target to be the decade of internetworking, where the networking of networks (voice, data, video) takes place on a building, campus, metropolitan, and international arena.

  Analysis: The internetworking of networks requires a hierarchical design structure, similar to that of a building wiring system. Individual workgroup networks are grouped together through floor or building backbones. Entire buildings are linked together through campus networks. And a complete organization can be tied together through the use of an enterprisewide network (actually a LAN/MAN/WAN architecture). The use of networking devices, such as bridges, routers, gateways, and switches (i.e., frame relay switches), will help to build these internet systems. However, their foundations must be in a structured wiring approach since the location of networking wiring hubs and internetworking devices will probably be in computer or distribution frame rooms. Hierarchical networks require hierarchical wiring systems.

- *TREND: Collapsed backbones:* The collapsed backbone is becoming increasingly popular in building and campus networking.

  Analysis: A collapsed backbone uses a small backbone that ties buildings or networks together through a star-shaped hub (where the hub is a small backbone network) rather than the traditional looping backbone that connects remote networks in a serial fashion. This trend will force a greater reliance upon star-shaped hierarchical building and campus wiring systems.

- *TREND: Integrated Services Digital Network (ISDN):* While ISDN has yet to become popular as a service, there are areas where it has made inroads into end-user networks.

  Analysis: ISDN is being used to provide frame relay networks, terminal adaption, and switched digital services. The integration of voice and data onto a single networking platform as provided by ISDN (2BQ) could help to standardize or reduce the number of cables required to each standard information outlet (SIO) unit (ISDN jacks use four unshielded twisted-pairs). Unfortunately, ISDN does not have a large user base and is unlikely to replace conventional LANs (Ethernet, Token-Ring, etc.) during the 1990s. A potential role for ISDN-type services is likely to be for switched-circuit access allowing data networks to link together over dial-up lines. This will require ISDN devices both on the desktop and in the DF and computer rooms, requiring cable and power supplies.

- *TREND: Voice, data, and video integration:* The integration of voice, data, and video onto a single networking system is beginning to take place in the business and residential environments.

  Analysis: While ISDN offers the hope of multiapplication integration for the business community, there are other areas where this is taking place on a larger scale. The cable television (CATV) industry is carefully developing the ability to offer integral voice and video services. Likewise, the telephone industry is trying to determine if they will be able to offer integrated video services along with their existing voice and data offerings (this is a regulatory as well as a technical issue). The integration of these services into a single offering, such as a broadband CATV system, could shape the way that network services are provided to the building and desktop, most likely over fiber-optic cable.

15.2 COMPUTING TRENDS

Increased computer usage within all sectors of the business community has dramatically altered the manner in which we conduct our lives. While this change has been extremely visible in the way that we build networks, it has its roots on the desktop. It is important to understand how this trend will continue to change the way people work and the wiring systems needed to support such change.

- *TREND: Increased desktop performance:* The processing power available in personal computer (PC) platforms is continuing to increase at an unchecked rate, accompanied by an equal reduction in price per unit of performance.

  Analysis: The advent of PCs that have performance measured in millions of instructions per second (MIPS) has served to increase the number of PCs on desktops while decreasing the number of centralized mini- and mainframe computers. This type of a decentralized computing system has increased the data networking requirements to the desktop. This process will continue, resulting in more data devices on the desktop that must be supported by building wiring systems.

- *TREND: Increased bandwidth requirements:* As the number of PCs, and their processing power, in the workplace has increased, so too has their requirement for networking bandwidth.

Analysis: The use of 64 kbps to the desktop (the equivalent of an analog voice channel) is no longer sufficient for the majority of networking applications. Even networks such as Ethernet that provide 10 Mbps or higher are not capable of supporting large computer-aided design (CAD) or graphics applications. The use of 16-Mbps Token-Ring and FDDI to the desktop will continue to increase for the next decade, as will the use of hierarchical networks that use Ethernet and Token-Ring (with a small number of users per network) feeding into much faster networks, such as FDDI (1 and 2) and 802.6 Switched Multimegabit Data Service (SMDS). These high-speed networks, currently operating only on fiber-optic cable, will slowly move into the building campus environment as costs decrease and user demand increases.

15.3 GLOBAL NETWORKING

ISDN and the Global Network are on their way. How can the buildings of today be designed or modified to support the communications technology of tomorrow? The advent of a global marketplace and worldwide networking capabilities have resulted in increased demands being placed upon the building wiring system.

- *TREND: Worldwide market:* Many organizations have an increasingly large number of networks that are global in nature.

  Analysis: A networking system that spans several continents must be operational for a greater period of time than the average metropolitan or countrywide network. Given the time differences between countries, organizations must be able to transfer information 24 hours a day (for example, the Tokyo stock exchange is active long before the New York exchange opens; however, many organizations in the United States must act on information from Tokyo in a real-time fashion). This requires that both networks and building wiring systems have support organizations that can react at any time to correct network problems. This can be quite a strain on organizations that relegate building wiring maintenance to daily office hours only.

- *TREND: Global standardization:* As organizations expand into the world market, they are beginning to establish global standards for networking, and thus wiring, systems.

  Analysis: The creation of networking standards across a global organization can pose serious problems if the networking systems cannot operate over the available building wiring system. This will force the adoption of global wiring standards which, although they might not be identical between all countries, will ensure universal networking support.

15.4 WIRING TRENDS

While computer and networking trends will certainly shape the types of building wiring systems designed and installed, the underlying element supporting all of these trends is the development of the building wiring products themselves.

- *TREND: High-speed unshielded twisted-pair:* Vendors are starting to develop new high-speed versions of unshielded twisted-pair for use with high-speed data networks.

  Analysis: This new type of cable has improved electrical characteristics and can provide higher speeds and greater distances than traditional cable. While this may allow networks such as FDDI to be supported over twisted-pair, it is expensive and can add significant cost to wiring installations. However, in the long run, it may be a viable option over fiber-optic cable. An important point with high-speed twisted-pair is that it may force a rethinking of older cross-connect architectures such as the 66-block. As throughput requirements continue to increase over twisted-pair, there will be a point at which the cross-connect block becomes the weakest link (because of high signal loss).

- *TREND: Plastic fiber:* New plastic fiber technology can support local networking systems up to 2 Mbps for distances up to 100 m.

  Analysis: The use of plastic fiber-optic cable is extremely limited today. The cable, while easier to work with and less expensive than optical fiber, does not yet provide sufficient distance and performance qualities for large-scale implementation. However, some vendors have indicated that they are developing high-speed (10 Mbps) versions of this fiber that could support Token-Ring or Ethernet LANs. The IEEE 802.8 committee is currently evaluating plastic fiber for inclusion into existing LAN standards. This development trend is likely to continue in the future and may be a viable alternative only if the costs of glass fiber-optic cable and splicing systems remain high or if the cost of plastic fiber can approach the cost of twisted-pair systems.

- *TREND: Single-mode fiber:* Communications vendors appear to be heading toward a switch from multimode to single-mode fiber-optic cable.

Analysis: Single-mode fiber offers much greater distances between repeater devices than multimode fiber. Its major detraction, high cost, appears to be slowly fading. New digital networking systems, such as Synchronous Optical Network (SONET), Asynchronous Transfer Mode (ATM), and 802.6 Metropolitan Area Networking (MAN), all require the use of high-speed single-mode fiber. Users who are planning to install fiber today should carefully evaluate their intended networking systems to determine if single-mode fiber can be a viable alternative for their application (this means an in-building shift from multimode fiber between networking node sites).

- *TREND: Modularity:* Vendors are developing extremely low-loss modular devices for terminating and cross-connecting building wiring.

  Analysis: The use of modular equipment in the building wiring system is increasing because of the high-quality product and the requirement for low-loss systems to support high-speed data applications. Users may want to opt for modular approaches over standard equipment provided the modular equipment has long-term viability and network equipment vendor support.

- *TREND: Technology:* The cost of computer-based technology is decreasing at such a rate that the implementation of a private digital cross-connect (PDX) system may become a reality.

  Analysis: Similar to large digital cross-connect switches used by the public telephone company, the PDX would automate the cross-connect process by performing connections through a digital switch rather than on a physical block. Incoming cables would be terminated on bulk connectors attached to the PDX. Any cable pair (or conductor) could be connected to any other pair (conductor) through electronic circuitry (much like a PBX only quite a bit slower since the connections would be nailed-up like a standard block system). The advantages of such a system are: (1) All cross-connects are guaranteed to be solid (no cable or connection failures) barring electrical component failure, (2) reconfigurations can take place from any remote computer terminal (perhaps even through a LAN or similar network), and (3) pair sequencing can be quickly rearranged to suit any jack application. Such a system could become an economic reality provided the cost could be balanced against the typical cable maintenance costs (adds, moves, and changes).

Appendix

Abbreviations and Acronyms

ABBREVIATIONS AND ACRONYMS COMMONLY USED WITHIN THE TELECOMMUNICATIONS INDUSTRY

μ	Micro (micron)
Ω	Ohm

A	Amp
ac	Alternating current
A/C	Air conditioning
A/D	Analog-to-digital
ACK	Acknowledgment
AM	Amplitude modulation
ANSI	American National Standards Institute
ASCII	American Standard Code for Information Interchange
AT&T	American Telephone and Telegraph
AUI	Attachment user interface
AWG	American Wire Gauge

BDF	Building distribution frame (also BMDF)
BERT	Bit error rate test
BICSI	Building Industry Consulting Service International
BMDF	Building main distribution frame (also BDF)
bps	Bits per second

CA	Collision avoidance
CAD/CAM	Computer-aided design and computer-aided manufacturing
CATV	Community antenna television
CCITT	Consultative Committee International Telegraph and Telephone
CD	Collision detection

CDF	Campus distribution frame (also CMDF)
cm	Centimeter
CMDF	Campus main distribution frame (also CDF)
CO	Central office
CPU	Central processing unit
CSA	Canadian Standards Association
CSMA	Carrier sense multiple access
CSU	Channel service unit
CTS	Clear to send

D/A	Digital-to-analog
DACS	Digital access cross-connect system
dB	Decibel
dc	Direct current
DCE	Data communications equipment
DCO	Digital central office (opposite of analog central office)
DDS	Dataphone digital service
DEMARC	Demarcation point
DF	Distribution frame
DOV	Data over voice
DSR	Data set ready
DSU	Data service unit
DTE	Data terminal equipment

EIA	Electronic Industries Association
EMI	Electromagnetic interference
ESS	Electronic switching system

FAX	Facsimile machine
FCC	Federal Communications Commission
FDM	Frequency division multiplexing (as found in broadband LANs)
FDR	Feeder system
FM	Frequency modulation
ft	Feet
ft^2	Square feet
ft/lb/sq	Foot/pounds per square inch

GHz	Gigahertz

HDLC High-Level Data Link Control
HVAC Heating, ventilation, and air conditioning
Hz Hertz

IDF Intermediate distribution frame
in Inch
in^2 Square inch
ISDN Integrated Services Digital Network

kbps Kilobit per second
kVA Kilovolt-amperes

LAN Local area network
LAP Link access protocol
LATA Local access transport area
lb Pound
LEC Local exchange carrier
LED Light emitting diode

m Meter
m^2 Square meter
MAC Media access control
MAN Metropolitan area network
MAU Media access unit
Mbps Megabit per second
MDF Main distribution frame
MH Manhole
MHz Mega (million) hertz

NAP Network access point
NCC Network control center
NCP Network control program
NIC Network interface card

OSI Open systems interconnection

PAD Packet assembler/disassembler
PBX Private branch exchange
PC Personal computer

PCM	Pulse code modulation
PLC	Power line communications
POP	Point of presence
POTS	Plain old telephone service
PR	Pair (twisted)
PUC	Public Utility Commission
RBHC	Regional Bell Holding Company
RBOC	Regional Bell Operating Company
RCDD	Registered Communications Distribution Designer (BICSI)
RG	Radio government (MIL-C-17)
RJ	Registered jack
RTS	Request to send
SDF	Satellite distribution frame
SMA	Subminiature type A
SNA	System network architecture
SNR	Signal/noise ratio
ST	Straight-tip
SWB	Switchboard
TDM	Time division multiplexing
TIA	Telecommunications Industry Association
UL	Underwriters Laboratories
UDG	Underground
UPS	Uninterrupted power supply
WAN	Wide area network
XCVR	Transceiver

Glossary

10Base5 The IEEE 802.3 Ethernet standard for Ethernet transmission over thick (RG-8/9) coaxial cable as defined by the IEEE 802.3 10Base5 committee. Each 10Base5 network, configured in a bus topology, may be 500 m in length.

10Base2 The IEEE 802.3 Ethernet standard for Ethernet transmission over thin (RG-58/59) coaxial cable as defined by the IEEE 802.3 10Base2 committee. Each 10Base2 network, configured in a bus topology, may be 200 m in length.

10BaseT The IEEE 802.3 Ethernet standard for Ethernet transmission over unshielded twisted-pair cable as defined by the IEEE 802.3 10BaseT committee. The 10BaseT network is configured in a star topology with two twisted-pairs running from a centralized hub to each end-user workstation. Individual cable runs over twisted-pair can be up to 100 m in length.

aerial cable A cable suspended above the ground, such as those cables strung between buildings or telephone poles.

air duct A duct used to circulate air in a building. Air ducts are considered plenum environments in most buildings. Air ducts may be either square or round in shape and are generally not used to house communications or power cables.

alternate access point An access point through which backup cable systems may enter a building or room. Typically used as a redundant backup in case of primary system failure.

alternate path An alternate path through a network or cabling system through which network traffic may pass if the primary path has been disrupted.

American Wire Gauge (AWG) A measurement of cable diameter used to describe copper communications or power cables. The larger the number, the smaller the cable diameter (e.g., 24AWG is smaller in diameter than 22AWG).

amplifier A device used to regenerate a communications signal, common with broadband networks. An amplifier takes a weakened signal and adds gain, or power.

analog transmission A communications transmission of analog signals, such as analog voice transmission.

apparatus closet An alternate name for a telephone closet (i.e., BMDF, IDF).

applications-based configuration The concept of using a generic or universal system (either wiring or networking) that can be custom-tailored to meet specific end-user applications without additional hardware development or the use of nonstandard equipment.

asynchronous A data transmission characterized by the presence of Start and Stop indicators at the beginning and end of each message for clocking purposes. The devices communicating only contact each other when a message is being sent. The opposite of synchronous.

attenuation The dissipation, or weakening, of a signal as it passes down a communications line, usually measured in decibels (dB) lost over a given distance.

backboard A board secured on a wall in a telecommunications room on which cross-connect equipment is mounted. Backboards are usually made of fireproofed plywood.

backbone A main network segment used to provide interconnection between remote computers or other networks. Also called a trunk.

balanced signal A signal that is impedance balanced (the same) over both leads of a cable to promote better transmission capabilities and reduce potential crosstalk or interference between cable pairs. Twisted-pair cables are typically balanced while coaxial cables are not (the outer and inner conductors have different electrical characteristics).

balun (balanced/unbalanced) A device used to convert an unbalanced signal (typically coaxial cable) to a balanced signal (typically twisted-pair). Baluns are used in pairs and known as coax-replacers.

bandwidth The range between the highest and lowest frequencies available for transmission over a cable. This is often mistakenly referred to as the transmission speed of a network.

baseband A transmission that is frequency, or bandwidth, independent where the entire bandwidth of a cable is used for a single signal, such as Ethernet and Token-Ring.

bearing wall A wall that is responsible for supporting a portion of the building directly above the wall or directly adjacent to it (i.e., adjacent wall or floor structure). This differs from a wall that does not support any building weight (such as a typical drywall wall used in many buildings) that can be removed without any structural damage to the building.

bipolar A digital signal consisting of two opposing electrical pulses, representing 1 and 0.

bit A single binary digit (1 or 0); 8 bits form 1 byte of data. *See bits per second.*

bit error The condition that occurs when a bit is erroneously altered from a 1 to 0 or a 0 to 1. Bit errors typically occur as a result of EMI or other electrical interference on a cable. Can be monitored with a bit error rate test (BERT) device.

bits per second (bps) The number of bits (binary information, either a 1 or 0) that are transmitted per second. Most transmission speeds are measured in thousands or millions of bits per second (i.e., 1.2 kbps, 56 kbps, 1.544 Mbps).

bonding The process of permanently securing multiple ground wires together to form a single common ground.

braided conductor A metallic conductor composed of many small strands braided together. Commonly used for shielding and as the outer conductor in coaxial cable.

bridge tap A tap on a circuit, usually made in a cross-connect block, that splits the circuit into multiple destinations from a single source.

broadband A transmission process that is frequency dependent and allows multiple signals to be simultaneously transmitted over a single cable (i.e., CATV). Each signal is assigned a specific frequency range over which to transmit a data signal.

building core The central location of a building that typically houses the stairways, elevator shafts, and vertical building wiring distribution system. A building may have more than one building core if the building is particularly tall or long.

building distribution frame (BDF) The primary distribution frame (cross-connect) within a single building. The BDF is the central distribution point from which all cables, either directly or through riser systems, originate. Also referred to as a building main distribution frame (BMDF).

Building Industry Consulting Service International (BICSI) An organization promoting proper wiring and telecommunications practices.

buried cable A cable that has been buried permanently in the ground between two endpoints. This differs from cable placed in underground conduits, ducts, or tunnels.

bus A network or signal path consisting of a single data path that connects multiple devices together in a serial fashion (i.e., Ethernet over coaxial cable).

byte A unit of data consisting of 8 bits.

cable Any material, such as copper or optical fibers, used to transmit a communications signal between two or more devices. Often referred to as a wire.

cable duct A tray or duct used to house telecommunications cable in the ceiling, ground, or floor.

cable insert An opening in a conduit or cable duct through which cables can be pulled to a local termination.

cable shield A metallic shield placed within a communications cable designed to provide EMI protection to the inner conductors. A cable shield may also be a rigid metallic casing placed around a cable for physical protection. Also referred to as cable armor.

cable tray A tray placed within a ceiling designed to transport communications cable (a form of raceway).

cable trough A cable path (raceway) placed into either concrete flooring or soil. Access to cable troughs is typically through movable steel plates placed over access points (inserts).

CAD/CAM (computer-aided design and computer-aided manufacturing) CAD/CAM systems are computer systems used to help design intricate systems, such as computer boards, computer chips, automobiles, aircraft, etc. CAD systems rely upon a graphical interface that allows detailed design drawings of an object to be made. They are becoming popular for use in designing and managing building wiring systems.

campus distribution frame (CDF) The primary distribution frame (wiring center) for an entire campus. The CDF is the central facility that connects all BDF systems within a campus together. Also referred to as a campus main distribution frame (CMDF).

carrier detect An electric signal that indicates an active carrier on a communications circuit.

C-band A group of frequencies used for microwave and satellite communications.

cellular phone A phone that uses the radio phone network (UHF range) and is not physically connected to any transmission media. Commonly referred to as a mobile telephone. This is different from a cordless phone which uses radio transmission to connect the receiver unit with the main phone system (across a very limited distance ≤50 ft) which is physically connected to an access line or circuit.

central office equipment The equipment used by the telephone company to connect and switch circuits connecting the central office (CO) to customer sites.

Centrex Telephone service provided by the telephone company in which all switching or call routing (i.e., PBX function) is performed by central office equipment.

Centronics cable A cable using a 36-pin interface to provide parallel communications between computers and printer devices.

channel attached The connection of two devices directly and locally between the input and output ports (e.g., not using remote or network communications).

circuit Any transmission or communications system between two or more locations. May be cable or free-transmission (microwave, infrared, satellite) based.

circuit switch A switch that provides a dedicated communications path between two devices. An example of a circuit switch is the data PBX.

cladding The outer fiber-optic material that surrounds the inner core of a fiber-optic cable.

coaxial cable An unbalanced communications cable that consists of an inner conductor, encased by an insulating material, that is surrounded by a mesh outer conductor.

CODEC (coder/decoder) A device that converts analog signals to digital signals, typically to transmit analog video signals over digital circuits.

community antenna television (CATV) A broadband communications system based on radio frequency (RF) signals.

communications cabinet A metal cabinet or rack (19 or 23 in wide) designed to house communications equipment, such as patch panels, cross-connects, and networking devices.

communications processor Any device used to process, route, or manipulate communications transmissions.

communications room/computer room A room specially designed to house communications equipment such as computers and PBXs. May include special power, A/C equipment, raised floor, and fire system.

composite link A circuit carrying multiple signals between two multiplexer devices, such as might be used within the backbone or riser of a building wiring system.

concentrator A device that is used to concentrate, or group, many terminal devices for connection to a processor via a single communications circuit (i.e., IBM 3270 or 3174 controllers).

concrete fill A mass of concrete placed around an underfloor conduit, underground building entrance, or buried cable for protective purposes. Also referred to as a concrete pour.

conduit Metallic pipe used to house electrical and communications cables. Used to provide protection for cable exposed to harsh environments and to house non-plenum-rated cables within a plenum-rated environment.

Consultative Committee International Telegraph and Telephone (CCITT) An international group established by the United Nations to promote the development of worldwide communications standards.

control cable A cable used to transport control information associated with HVAC, mechanical, or security systems.

core drill A hole drilled in the floor or ceiling of a telephone closet to allow riser cables to pass between floors. Part of the vertical riser system.

core walls The walls surrounding the building core. These walls are typically concrete and provide both structural and fire protection benefits.

cross-connect A hardware device used to provide physical connection between two or more different communications cables (i.e., AT&T 110- or 66-block).

crossover cable A cable designed to connect two communications devices that typically require intermediate equipment without the use of intermediate equipment. The crossover cable connects opposing cable leads, such as Data Transmit and Data Receive.

crosstalk The interference, or bleeding over, of one communications signal onto another, between copper cables because of the electrical characteristics and relative signal strengths.

data communications equipment (DCE) The equipment used to provide a data communications circuit, such as a modem, in contrast to end-user data termination equipment (DTE).

data over voice (DOV) The transmission of low-speed data and voice simultaneously over a single voice circuit.

data PBX A private branch exchange used to switch digital data only (no analog transmissions).

data terminal equipment (DTE) The user equipment used to originate a data transmission, such as a computer or terminal.

decibel (dB) A measurement of signal strength, usually used to represent signal loss over distance.

dedicated line A communications line that provides a link between two locations and cannot be used to establish connections to other outside locations.

de facto standard A specification or idea that becomes standardized through widespread use rather than a formal standardization process.

demarcation point The location at which a telco (or other) network cable is terminated within a building or campus. This is the boundary between the telco and end-user ownership of cable.

dial-up circuit A communications circuit that can be accessed by dialing the telephone number of the circuit with a modem.

digital circuit A communications circuit used to transmit a digital (binary) communications signal (opposite of analog circuit). May serve digital voice, data, or digital video.

digital data Any data or information represented in a digital (binary 0, 1) format.

direct current (dc) An electrical signal consisting of a constant flow from positive to negative.

direct-in dial (DID) The ability to access a PBX-attached telephone (dial directly) without going through the main console of a PBX.

direct-out dial (DOD) The ability to place a call from a PBX-attached telephone without using the main console.

distribution frame (DF) A collection of cross-connect equipment used to bridge two or more sets of communications circuits. *See also intermediate distribution frame (IDF) and building main distribution frame (BMDF).*

diverse routing The ability to route a communications signal over multiple paths, depending upon the availability of communications facilities.

duct *See cable duct or air duct.*

earth ground A ground connected directly to the earth (such as a buried rod or copper coil).

earth station or ground station A transmitter (satellite dish) used to transmit or receive data from a satellite.

echo Any reflection of an electrical signal (analog or digital) that occurs as the result of cable faults or cable characteristics. Also the duplicate transmission signal that sometimes appears on voice circuits.

electrical ground A cable or metal system that provides connection between the ground wire of an electrical device to the earth for purposes of establishing an electrical ground. A ground may consist of many different items, such as a copper wire, metallic water pipe, or metallic plates (referred to as grounding electrodes). All ground systems must terminate in the soil.

electromagnetic interference (EMI) A signal disruptive to communications signals created as a result of electrical or magnetic power.

Electronic Industries Association (EIA) An organization that develops and promotes communications standards.

electronic switching system (ESS) An electronic telephone switch capable of providing enhanced services (more than simple circuit switching) such as call hold, transfer, and call waiting.

emergency power A backup power supply that automatically becomes operational in the event of a main system power failure. *See uninterrupted power supply (UPS).*

equipment room *See communications room.*

Ethernet A 10-Mbps local area network developed by Xerox, Intel, and DEC in the late 1970s. A variation of Ethernet has been designated as the IEEE 802.3 standard. Ethernet uses a random access method (CSMA). Variable media (coaxial and twisted-pair) are permissible under the 10Base5, 10Base2, and 10BaseT specifications.

expansion joint A joint placed in a cable distribution system to allow for system expansion. Such a joint is primarily used between different segments of a building, especially in areas that are prone to frequent earthquakes, although they are also used in many large conduit systems.

facility A common term used to describe or refer to communications equipment or transmission media.

facsimile machine (FAX) A machine capable of scanning a piece of paper, digitizing its text or images, and sending the resulting data to a remote location for printing.

Federal Communications Commission (FCC) A federal agency established in 1934 to regulate all telephone, television, and radio communications.

feeder cable A short cable that connects to a larger cable or a cable that feeds a given area, such as a building. *See also tail circuit.*

feeder duct A small duct or conduit system that feeds into a larger (main) duct system.

fiber-optic system Any system that uses optical fiber for communications purposes. *See optical fiber.*

file server A special computer used to store common access files on a local area network (LAN), wide area network (WAN), or multiuser computer system.

fireproof Any material that is nonflammable or resistant to flames and high heat.

fire rating A rating given to a material based upon the amount of time that the material can resist flames or extreme heat without burning. Any material that is nonflammable or resistant to flames and high heat and has been assigned a specific fire rating, such as a 3-hour firewall, is considered to be fire rated.

firestop The placing of a material that is nonflammable or resistant to flames and high heat in such a location as to prevent the spread of a fire through openings in walls or floors.

firewall A wall designed to prevent a fire from spreading from one area of a building or room to another adjacent area.

foam dielectric A dielectric material made of solidified foam (such as polyethylene). Also referred to as a cellular dielectric.

four-wire A telephone or communications circuit that uses two twisted-pairs (four conductors), one pair for transmit and one pair for receive. *See full duplex.*

frequency The rate of occurrence of a signal wave (how often the signal wave is repeated in a set period of time).

full duplex Being able to communicate in both directions (simultaneously send and receive) over the same circuit.

gain The increase of a signal (amplification) measured in decibels (dB).

gateway A communications device used to connect two or more unique networks, usually involving protocol translation (i.e., an X.25/LAN gateway translates between X.25 Layer 2 packets and LAN Layer 2 packets).

gel-filled cable A cable that has a gel placed around the inner conductors to prevent water or moisture from reaching the metallic conductors.

ground terminal A section of metal placed in a telecommunications facility that serves as a common access point for multiple ground wires. Individual ground wires from electrical systems are connected to this terminal, which in turn provides access to an earth ground.

ground wire A copper wire that serves as the electrical ground for a system, commonly 6AWG.

half duplex A communications circuit (typically two wires) that is only able to either transmit or receive at one time. The opposite of full duplex.

hardware The physical components of a computer, cable, or communications device, such as processor boards, cross-connects, patch panels, etc.

hard-wired A communications circuit between two devices that is permanently or directly attached, such as an RS-232 cable between a printer and computer.

harmonic frequency A frequency that appears as the result of a strong signal at a lower frequency. For example, a 4-MHz signal transmission produces 16- and 32-MHz harmonic frequencies.

head end The main or "top" transmitting device in a broadband network through which all transmissions are routed.

hertz (Hz) A measure of frequency cycles per second.

hierarchical database A database designed in a hierarchical or tree structure with distinct relationships between elements on different branches of the structure.

hierarchical network A network based on a hierarchical structure with network responsibilities broken down into a layered approach with each layer providing an independent and interchangeable network function.

High Level Data Link Control (HDLC) A Layer 2 communications protocol used to maintain data link connections, typically associated with synchronous transmission between LAN bridges and routers.

host computer A computer (mini, micro, or mainframe) accessible through local or remote terminals.

hub A central point in a network structure where all network segments terminate (typically a star configuration). A single network may have multiple hub locations in a distributed fashion with minor hubs connected to a main central hub.

impedance The total electrical resistance to an electrical signal transmission. The impedance of a cable determines the propagation delay of a cable (the amount of time it takes an electrical signal to travel a given length of cable).

independent telephone company A local telephone company not affiliated with AT&T or any of the seven Regional Bell Operating Companies.

infrared signal A high-energy light signal, in the infrared range, transmitted through the open air to link two devices together. Open line of sight is required for this type of communications.

inner core The inner fiber-optic material in the center of a fiber-optic cable through which a lightwave is transmitted.

inner duct A small cable duct or conduit that is located within a much larger duct system. Such systems allow protection and eased installation of cable in crowded ducts.

insertion loss The degradation or decrease of a communications signal due to devices being attached to the circuit. Attached devices may include terminal equipment as well as connectors.

Integrated Services Digital Network (ISDN) An all-digital network structure being developed and implemented worldwide to promote connectivity of voice, data, and video signals over any switched telephone network.

intelligent network A communications network capable of sophisticated functions such as dynamic routing or in-depth network control. This concept is currently being applied to the public switched network (PSN) by AT&T and the Regional Bell Holding Companies (RBHCs).

interconnect Any system that connects communications lines together. May also refer to a vendor that provides interconnection services. *See also cross-connect.*

interexchange carrier (IXC) A communications carrier that provides connection between different LATAs or BOC facilities, such as AT&T, MCI, and US Sprint.

interference Any disruption of a communications signal due to other disruptive signals or power noise.

intermediate distribution frame (IDF) The primary distribution frame (cross-connect) on a single floor of a building.

International Standards Organization (ISO) An international organization that promotes the standardization of communications systems.

ISDN *See Integrated Services Digital Network.*

laser A system for generating a high-frequency light beam commonly used in single-mode fiber optics (*l*ight *a*mplification by *s*timulated *e*mission of *r*adiation).

leased line A dedicated communications line leased from the local telephone company.

light emitting diode (LED) An electric diode that emits a light signal. Commonly used to drive multimode fiber.

line driver A device that modifies or powers a communications signal to increase the distance allowable between two communications devices.

line loss The degradation or decrease of a signal on a communications line (dB).

line-of-sight system A communications system that transmits nonterrestrial signals, such as microwave or infrared, and requires a direct and unblocked sight path between send and receive devices.

local area network (LAN) A network designed to serve a small geographic region, such as an office, building, or campus.

main distribution frame (MDF) The primary distribution frame for a single building. In a campus environment, referred to as the building main distribution frame (BMDF).

main terminal The point at which cables coming into a facility are connected to the building interior cable distribution system (often involving electrical protection). *See also demarcation point and network access point (NAP).*

major distribution point A large distribution point, such as a campus, intermediate, main, or satellite distribution frame.

media access control (MAC) A subcomponent of the OSI Data Link Layer 2 for the IEEE 802 specification that controls access to networks.

metropolitan area network (MAN) A network designed to serve a single metropolitan area, often by bridging multiple LANs together.

microwave system A system used to transmit nonterrestrial radio communications signals between two points. Requires direct line of sight.

minor distribution point A small distribution point between major distribution points.

mobile telephone *See cellular phone.*

modem (modulate/demodulate) A device which modulates a digital communications signal to an analog communications signal, usually for transmission over the public switched network.

modem eliminator A device that can be used in place of a modem over limited distances (2 km). This device "fakes out" the end DTE/DCE devices, through crossed circuits, and causes them to think that a modem, and the appropriate signal conversion, is in fact being used. Also referred to as a null modem.

multidrop line A communications line that has one origin and several destinations.

multimode fiber An optical fiber, commonly used for local communications under 3 miles, that is capable of transmitting a signal containing multiple frequencies. Typically driven by LEDs.

network A system of communications facilities connected together to provide communications between end-user devices that are connected to the communications facilities.

network access point (NAP) The demarcation point within a building or campus where an external network or cable is terminated for internal communications. The interface point between an end-user facility and the local telco network and central office.

network architecture The design and layout of a network, involving physical cable, software, and protocol configuration.

network control center (NCC) The central facility through which a network is controlled and managed.

node A device on a network. A node may be either a user device (such as a terminal or printer) or a network device (such as a switch or PAD).

noise Electromagnetic interference (EMI) that disrupts a communications signal. Noise may be the result of power signals, other circuits, or natural interference. Usually expressed in decibels (dB).

nonterrestrial A circuit that is not bound by physical land-based systems, such as cable or fiber optics, but is instead transmitted through open air as a radio wave signal.

null modem *See modem eliminator.*

off-hook When a telephone set or line is active (i.e., the receiver is up).

on-hook When a telephone set or line is inactive (i.e., the receiver is down).

open A section of cable that has not been terminated properly (e.g., a cut in a cable).

Open Systems Interconnection (OSI) A seven-layer protocol architecture designed by ISO that defines individual layers each responsible for a different networking function.

optical fiber A thin fiber strand made from optical glass or plastic and capable of transmitting lightwaves representing communications signals.

parallel interface A cable interface in which all information regarding a single character or word is transmitted at the same time (parallel transmission). The opposite of serial interface.

patch cable Any cable used to connect two sides of a cross-connect or interconnect system.

personal communications system (PCS) A communications device that serves a single user with voice or data capabilities (often mobile, i.e., wireless).

plastic fiber A fiber-optic cable made of plastic instead of glass. Usually much fatter in diameter than glass fiber (900 to 100 μ) and capable of only limited distances (<100 m). Advantages of plastic fiber include ease of splicing and troubleshooting. The fiber is very thick and does not require exact core matches for splicing. Additionally, the light source is a simple LED that can be visually detected (to determine if a circuit is connected and unbroken).

poke through A hole that has been poked or drilled through a wall or floor to provide cable access between different floors or rooms. Smaller than a core drill.

power line communications (PLC) Communications using analog signals over power lines for low-speed communications.

power pole or utility pole A hollow pole between a floor and ceiling designed to provide a space for running electrical or communications cables. Power poles typically provide electrical or station outlets and are used in large open rooms that must provide power or communications cables to a large number of desks or workstations.

pressurized cable A cable that is air pressurized within the outer sheath to prevent water from passing through to the inner conductors.

prewire The installation of communications wire prior to the selection or implementation of a communications system. It is common to prewire a building during construction to provide for telephone or computer systems that may not be installed for years to come.

private branch exchange (PBX) A privately owned telephone switch (non-telco) serving a limited geographical area, such as a building or campus.

private line or network A communications circuit or network that belongs to, or is leased to, a private organization and is dedicated to their own use.

public network A network owned by a telco or interexchange carrier (IXC) for use by any customer, such as the networks owned by AT&T or the Bell Operating Companies.

Public Utility Commission (PUC) An organization that regulates the communications environment within a given state.

pull A segment of cable that has been "pulled," or installed, from one point to another, typically within a building environment. The act of installing cable. *See run.*

pull box A box placed in the middle of a conduit segment which allows the installer access to the cable during the installation process.

pull strength *See tension strength.*

pull wire A thin wire used to pull communications cable through a conduit or duct. This is similar in function to a "fish-tape" used to fish (or pull) wire through a wall or ceiling area. Pull wires are often placed in the conduit during initial installation.

quad cable A type of telephone wire that has four solid copper conductors (red, green, black, and yellow) sealed in PVC. Not acceptable for digital data communications.

raceway A generic term referring to a conduit, duct, or similar system used for the distribution of communications or power cable. Can support horizontal or vertical distribution.

raised floor An artificial floor that is supported above the original floor. Raised floors are designed to provide an open space between the original and artificial floors that can be used to run cables or serve as an airway for air conditioning systems. Raised floors are commonly used in computer rooms. Sometimes referred to as an access floor.

reflection loss The loss in signal strength (dB) because of signal reflection that occurs when a communications signal encounters a change in the electrical or physical characteristics of a cable (such as encountered at a splice or open termination).

Registered Communications Distribution Designer (RCDD) A BICSI-certified designer of wiring systems.

registered jack (RJ) A common jack interface specification used by most phone systems providing connectivity for between one and four twisted-pairs (i.e., RJ-11, RJ-14, RJ-45, RJ-48).

repeater A device that regenerates and amplifies a signal for transmission over extended distances.

rigid cable A cable that uses a rigid or hard outer sheath for protective purposes. Rigid cables are commonly used outdoors. A rigid sheath may be either hardened plastic or metal.

riser A vertical raceway system used for the vertical distribution of communications cables between floors of a building, usually aligned with a building core and phone closets.

run A segment of cable that connects two points, typically within a building environment. *See pull.*

satellite distribution frame (SDF) A distribution frame used to connect remote or satellite locations on a single floor to the primary floor distribution frame (IDF). Also referred to as a satellite room or closet.

serial interface A cable interface in which all information regarding a single character is transmitted 1 bit at a time in series. The opposite of a parallel interface.

serving area (or serving region) An area served by a given telecommunications device or network (such as a single PBX or LAN).

shielding Any metallic component used to protect a device or cable from EMI.

short When two or more segments of cable that are supposed to be physically separate come into direct contact.

short-haul communications Communications within a local campus or metropolitan area.

signal beam A beamed transmission, such as between microwave stations, that passes between two predetermined points and is not broadcast to other areas.

signal/noise ratio (SNR) The ratio of a communications signal to the background noise on a communications circuit.

simplex A communications media that can only transmit a signal in one direction (e.g., fiber-optic cable).

single-mode fiber A laser-driven optical fiber, commonly used for long-distance communications in excess of 3 miles, that only carries a single light signal at a given frequency.

skinning The passing of electrical signals along the outer edge of a solid conductor rather than through the center of the conductor.

slab A section of poured concrete that serves as the foundation for a building, walkway, or floor or ceiling.

sleeve A metal or plastic conduit, usually a minimum of 4 in in diameter, that passes from one floor to another to allow cables to pass between floors.

slip sleeve A short segment of conduit into which two smaller conduits are placed for connection. The slip sleeve can be slipped to one side to provide access to the inner conduits.

slot A hole cut into a floor to allow cables to pass through from one floor to another.

sneak current A gradual rise in the voltage present on a communications cable.

solid conductor An inner conductor in an electrical or communications cable constructed of a single solid wire.

spread spectrum A radio transmission technique that spreads a single signal over a broad range of radio spectrum for transmission purposes. Common with emerging LAN technology used to augment and/or replace wire-based systems.

standard A set of specifications that defines a specific method or technology for use in a fixed set of applications. A standard may be defined by any organization for any purpose.

star configuration A network in which devices are served by a central hub or computing device located in the middle or center of the geographic serving area (i.e., a hub and spoke system).

station An end-user device on a network.

stranded conductor A copper conductor that is composed of many individual small strands (similar to braided conductors) as opposed to a single large copper conductor.

support cable, strand, or wire A strong cable running between two points (i.e., buildings or telephone poles) to which a communications cable is lashed or attached.

surge A rapid rise in the voltage of a communications cable, often caused by lightning or faulty electrical systems.

suspended ceiling An artificial ceiling made of lightweight tiles, typically for appearance only, suspended from a structural ceiling. Suspended ceilings do not offer load-bearing support.

switch A device that can connect multiple communications lines together, generally with the ability to route communications traffic between different combinations of lines.

switchboard A central device or console through which all incoming calls are routed in a PBX or key system. An operator's console.

synchronous A form of transmission that requires a constant timing signal be passed between communicating devices, even if data are not transmitted. Opposite of asynchronous.

tail circuit The local loop portion of a circuit connecting a user location to a central office or network node. Also, a short connection cable linking a user device to a main circuit.

Teflon A material developed by Du Pont (polytetrafluoroethylene) and widely used in plenum-rated communications areas as a fireproof coating on cable and wire.

telecommunications The communications of voice, data, or video systems over any type of transmission circuit (fiber, copper, microwave, infrared, etc.).

telephone closet A room designed to house telephone and/or communications cable and cross-connect equipment. Also referred to as an equipment closet, distribution frame, or wiring closet.

telephony Communications involving telephone systems.

tension strength The amount of tension that can be placed on a cable during installation without damage to the cable or conductor. Also called pull strength.

terminal (1) A device used to provide a user interface for communications with a computer or network service. (2) A device used to terminate a communications or electrical ground cable (such as a cross-connect block).

terrestrial Any communications circuit that uses a land-based physical media, such as copper or fiber optics.

throughput The amount of information (traffic, bits, etc.) that can be passed over a circuit for a given period of time. *See also bandwidth.*

token bus A token-passing network using a single bus configuration to connect terminal devices together. The IEEE 802.4 standard defines a token-bus network.

token passing A network access method in which a "talk-now" token is passed between users. Users cannot transmit any messages unless they possess the token.

token-ring A token-passing network using a ring configuration to connect terminal devices together, such as the IBM Token-Ring (IEEE 802.5) and the ANSI FDDI standard.

topology The actual configuration, either logical or physical, of a communications network (such as tree, star, bus, mesh, loop, etc.).

transceiver A communications device, capable of both transmitting and receiving messages, that serves as the interface between a user device and a network.

transmission speed The rate at which data (a transmission of data traffic) are being sent over a communications circuit, usually expressed in the number of bits per second (bps).

tray An open raceway that is placed in ceilings for cable transport. Trays provide cable support while also allowing easy cable access through the open top.

tree topology A network topology structured like a tree with a single source leading to many different branches or network segments, such as CATV. May be either physical or logical.

trunk Any major circuit or network link between two or more locations.

twinaxial cable A variation of a coaxial cable that has two stranded inner conductors.

twisted-pair cable A cable consisting of two solid copper conductors, each coated in an insulating material, that are twisted together. Commonly used for telephone and data applications.

two-wire circuit A circuit consisting of one twisted-pair used to provide one-way, or half-duplex, communications. Full-duplex communications may be possible at low data speeds (300 to 1200 bps).

underfloor duct A cable raceway system installed in a concrete floor slab. Underfloor ducts that operate on two axes (i.e., x and y) with one axis above or below the other are called bi- or two-level systems.

uninterrupted power supply (UPS) A backup (redundant) power supply that can supply emergency power (temporary <30 min) to a system in the advent of main system power failure.

voice band or grade Any circuit that is capable of passing an analog voice signal (300 to 3300 Hz).

wall stud A vertical beam, either steel or wood, which supports a wall.

wide area network (WAN) A network that spans a large geographic area, such as an entire state or country.

wire Any copper cable or strand installed between any two points and used for communications or power purposes.

wiring closet Any closet or room used to house wiring equipment, such as cross-connect blocks or patch panels. *See also distribution frame or telecommunications closet.*

workstation A device on a network that provides access to a user, such as a PC or terminal.

X.25 A CCITT standard for packet-switched network communications (OSI 2/3).

zone of protection A protective zone created by a lightning rod within which a lightning strike will have limited impact.

ABOUT THE AUTHOR

Fred J. McClimans is the director of strategic marketing at Newbridge Networks, Inc., in Herndon, Virginia. Prior to joining Newbridge, he served as a telecommunications consultant and analyst specializing in the design and implementation of local communications systems, such as LANs, packet networks, and building wiring systems. He has been involved in the development of wiring plans and standards for a wide range of environments and has provided consulting services to commercial, state, and federal clients. Mr. McClimans is the author of numerous articles and enjoys speaking and teaching at industry conferences and training seminars on topics such as building wiring, enterprise networking, local area networks, bridging/routing internetworks, and emerging technologies, such as wireless voice and data communications, and cell/frame relay networking.

Index

Access point [*see* Demarc(ation point)]
Aerial cable, 207, 209, 212–214, 218, 222, 225, 331
Air duct, 331
Alternate access point, 331
Alternate path, 331
American National Standards Institute (*see* ANSI)
American Wire Gauge (AWG), 17, 331
AMP, 93
Amplifier, 331
Analog transmission, 331
ANSI (American National Standards Institute), 76, 263
  ANSI X3.131, 22
  ANSI X3T9.5 FDDI, 280, 295–297
Apparatus closet, 332
Appletalk, 281
Applications based configuration, 304, 331
AT&T wiring, 19, 93, 96, 97
Attenuation, 14, 332
AWG (American Wire Gauge), 17, 331

Backboard, 179, 198, 332
Backbone, collapsed, 299, 321
Backhoe fade, 210
Balanced signal, 332
Baluns, 85
Band-marked color code, 12
Baseband, 332
Bayonet connector, 45, 63, 64
BDF (Building distribution frame), 171, 189, 201, 333
Bearing wall, 332
Belden Cable, 135
Bell Atlantic, 93
Biconic connector, 63
BICSI (Building Industry Consulting Service International), 261, 312–314, 332, 343
Bits per second (bps), 333
BNC/TNC connector, 45
Bonding, 76, 333
Braided conductor, 17, 333
Brand-Rex, 94
Bridge tap, 333
Bridges, 298
British Telecom, 93
Building core, 190, 192, 206, 265, 333
Building distribution frame (BDF), 171, 189, 201, 333
Building Industry Consulting Service International (BICSI), 261, 312–314, 332, 343
Building segmentation, 162
Buried cable, 61, 140, 211, 222, 226, 333

C/FM (Configuration/facilities management), 100
Cable:
  buried (*see* Buried cable)
  Centronics, 334
  color coding of, 232
  fault isolation, 243, 245
  fault resolution process, 246
  gel filled, 76, 207, 338
  installation, 163, 270
  marking, 225, 228, 230, 232
  procurement, 269
  protection, 207
  selection, 133, 202
  shielding, 206
  sizing, 204
  testing, 272
  types:
    coaxial (*see* Coaxial cable)
    composite, 19
    cross-connect, 181
    fiber optic, 53, 58, 60
    multimedia, 61
    quad, 15
    speciality, 19
    stranded, 15, 345
    twisted-pair, 11, 347
Cable armor, 207
Cable, composite, 19
Cable duct, 142
Cable insert, 138, 140, 333
Cable management, 94
  administration of, 241, 242
  automation of, 236, 237
  definition of, 227
  documentation, 228, 229, 234

Cable management (*Cont.*):
  expert systems, 249
  implementation issues, 252
  problem escalation, 247
  selection, 238
  work orders, 235
Cable shield, 333
Cable tray, 333
Cable troughs, 136, 138, 334
Campus distribution frame (CDF), 171, 187, 201, 214, 334
Canadian Standards Association (CSA), 263
Capacitance, 14
Carpet, 175
Carrier sense multiple access (CSMA), 289
CATV (Community Antenna Tevevision), 335
CCITT (Consultative Committee on International Telegraph and Telephone), 335
CDF (*see* Campus distribution frame)
Ceiling raceway, 144
Cellular phone, 334
CENTREX, 33, 37, 277, 282, 283, 334
Centronics cable, 334
Cladding, 54, 56, 59, 62, 334
Clamps, 209
Coax-replacers, 86
Coaxial cable, 42, 286, 334
  advantages of, 49
  applications of, 47
    cost of, 48
  definition, 41
  dielectric, 42
  disadvantages of, 50
  impedance of, 42
  RG-rating, 43
  types of, 42
Collapsed backbone, 299, 321
Color coding, 12, 232
Common access phones, 129
Community Antenna Television (CATV), 335
Competitive bidding, 269
Composite cable, 19
Concentricity imperfections, 62
Conduit, 50, 86, 95, 104, 136, 143, 335
  cable insert, 333
  capacity, 161
  ceiling distribution, 141
  distribution frame (DF), 184
  distribution frame access, 165
  EMI with, 158
  firestopping, 167
  in freestyle distribution, 147
  in plenum areas, 135
  installation timing, 164, 219
Conduit (*Cont.*):
  levels, 139
  marking of, 226, 228
  power, 126
  types:
    ceiling, 143
    feeder, 338
    inner duct, 339
    molding, 151
    sleeve, 345
    underfloor, 138
    underground, 210
Configuration/facilities management (C/FM), 100
Connectionless network service, 281
Connections, home run, 146
Connectors, 83, 108
  bayonet, 45, 63, 64
  biconic, 63
  BNG/TNC, 45
  coaxial, 44, 75, 112
  EIA 550 (FD), 45
  FDDI, 64
  fiber-optic, 61, 63, 113
    difficulties with, 63
    splicing, 64, 70
  genderless, 23
  M34, 22
  pin, 21
  RS-232, 108
  SCSI, 22
  SMA, 63
  ST, 63
  twisted-pair, 20, 21, 109
    keyed or shifted, 21, 111
    RJ, 20, 21, 109, 111, 283, 344
  V.35, 22
Construction Specifications Institute (CSI), 263
Consultants, 260
Consultative Committee on International Telegraph and Telephone (CCITT), 335
Control cable, 4, 34, 134, 335
Control system, 4, 5, 172, 181
Core (cable), 54, 56, 59, 62
Core drill, 183, 335
Costs, 48, 77, 81, 197
  estimation of, 240, 266
  of fiber-optic systems, 67
Cross-connect blocks, 23, 178, 283
  adapters, 84
  cable, 23
  configurations of, 180
  marking of, 233
  placement of, 179
  selection of, 196
  types of, 23
    110-block, 27

Cross-connect blocks, types of (*Cont.*):
66-block, 25
Bix-block, 30
coaxial, 46
cutting displacement, 24
modular, 31
pressure displacement, 24
Crosstalk, 14, 17, 42, 88, 110, 170, 336
CSA (Canadian Standards Association), 263
CSI (Construction Specifications Institute), 263
CSMA (carrier sense multiple access), 289

Dark fiber, 203
Data switch, 34
DEC wiring, 94, 98, 100
DECnet, 281
Demarc(ation point), 182, 336
Design programming, 267
DF (*see* Distribution frame)
Digital PBX, 33
Distribution, freestyle, 146
Distribution frame, 336
center-located approach, 190
characteristics of, 174
definition of, 171
distribution impact, 214
equipment, 173
functions of, 172
installation of, 198
layered, 198
serving areas, 193
space requirements, 265
types of, 171
building, 189
campus, 187
intermediate, 191
satellite, 194
secondary, 177
shared, 176
Distribution systems, 136–157
ceiling, 141
molding, 150
poke-through, 157
power pole, 156
undercarpet, 154
underfloor, 136
Duct (*see* Conduit)

Education, of users, 310
EIA (Electronic Industries Association), 263
EIA 550 (FD) connector, 45
Electrical systems:
and telecommunications, 87
distribution frame power, 174, 181, 185
Electrical systems (*Cont.*):
noise, 87
placement of, 88
rule of 6, 88, 159
Electromagnetic interference (EMI), 15, 17, 38, 42, 49, 58, 69, 82, 88, 135, 158, 174, 187, 206, 337
Electronic Industries Association (EIA), 263
EIA 550 (FD) connector, 45
EMI (*see* Electromagnetic Interference)
Entropy factor, 227
Environmental control, 175
Expansion joint, 141, 337
Expert systems:
definition of, 249
management integration of, 251

F-Series connector, 45
FCC (Federal Communications Commission), 74
FDDI connector, 64
Federal Communications Commission (FCC), 74
Feeder cable, 337
Fiber-optic cable, 102, 286
characteristics of, 56
definition of, 53
elements of, 54
graded index, 60
multimode, 58, 203
single-mode, 59
step index, 60
Fiber-optics:
advantages of, 68
applications of, 66
disadvantages of, 69
Fill, protective, 211, 212, 335
Fire protection, 175
Firestopping, 166, 199
Flag markers, 226
Floor support, 175
Frame interconnection system:
cable options, 202–207
design goals, 217
design of, 214
distribution, 208, 209
existing, 215
installation timing, 218
troubles with, 221
Freestyle distribution, 146

Garage phones, 130
Gateways, 298
Gel-filled cable, 76, 207, 338
Gel-filled sheaths, 207
Grounding, 17, 165, 166, 174, 176, 207, 314

Grounds, 336, 338
  types of:
    building steel, 176
    buried, 176
    cold water pipes, 176

Half-duplex transmission, 56
HDLC (High-Level Data Link Control), 339
Hierarchical network, 339
High Level Data Link Control (HDLC), 339
Home run connections, 146
Host systems, 300
Hubble, 94

IBI (Intelligent Building Institute), 263
IBM wiring, 19, 94, 101, 102
IDF (intermediate distribution frame), 171, 177, 201
IEEE (Institute of Electrical and Electronics Engineers), 263
  IEEE 802.3 Standard, 21, 34, 66, 280, 286, 290–293, 331
  IEEE 802.4 Standard, 280, 346
  IEEE 802.5 Standard, 23, 34, 44, 258, 280, 294, 346
  IEEE 802.6 Standard, 280, 323, 325
  IEEE 802.8 Standard, 324
  IEEE 802.11 Standard, 320
Impedance, 14, 339
Imperfections:
  concentricity, 62
  ovality, 62
Inductance, 15
Infrared, 79, 286
  advantages of, 81
  applications of, 80
  disadvantages of, 82
Insert, 140
Insertion loss, 340
Installation, timing of, 265
Institute of Electrical and Electronics Engineers (*see* IEEE)
Integrated Services Digital Network (ISDN), 321
Intelligent building, 3
Intelligent Building Institute (IBI), 263
Intercom systems, 34
Intermediate distribution frame (IDF), 171, 177, 201
ISDN (Integrated Services Digital Network), 321

Jack, phone, 20
Junction box, 140

Key system, 33, 37, 95, 128, 282–283

Kilovolt ampere (kVA), 88

Ladders, 144
LAN (*see* Local area networks)
Lasers, 55
LCS (local communications system), 5
Lighting:
  ballast, 160
  in the DF, 174, 198
Lightning, 129, 174, 185, 207, 213, 262, 347
Line of sight, 79
Local area networks (LANs), 66, 76, 284
  10Base5, 2, T, 331
  Appletalk, 294
  design of, 299
  Ethernet, 290, 331, 337
  FDDI, 295
  history of, 285
  integration of, 299
  media used in, 286
  Token-Ring, 294, 346
  topology of, 287
  twisted-pair, 34
  types of, 285
  wireless, 286, 320
Local communications system (LCS), 5

MAC (media access control), 341
Main distribution frame (MDF), 340
Marker plate, 225
MDF (main distribution frame), 340
Mechanical shop phones, 129
Media access control (MAC), 341
Mesh riser support, 209
Microwave, 48, 73, 75, 286
  advantages of, 77
  disadvantages of, 78
Mod-Tap, 94
Molding, 150
Multimode fiber, 341
Multitenant buildings, 176

N-series connector, 45
NAP (*see* Network access point)
National Electrical Code (NEC), 262
National Fire Protection Association (NFPA), 262
NEC (National Electrical Code), 262
Network access point (NAP), 182, 186, 190, 201, 219, 265
Network Services:
  connectionless, 281
  hierarchical, 339
NFPA (National Fire Protection Association), 262
Noise (*see* Electromagnetic interference)

Open Systems Interconnection (OSI) Model, 278–282
Outdoor phones, 129
Outlet cables, 119
  cross-over, 120
  straight through, 120

Ovality imperfections, 62

Patch cable, 23
Patch panels, 83
  twisted-pair, 84
PBX (private branch exchange), 282, 343
PCN (Personal Communications Networks), 76
Phone jack, 20
Phones:
  garage, 130
  mechanical shops, 129
  outdoor, 129
  public pay, 128
  security access, 130
Planning:
  project, 263
  space, 264
Plastic fiber, 324, 342
Plenum rating, 18, 35, 43–44, 48, 55, 68, 135, 142, 156, 205, 262, 346
Poke through, 136, 157, 199, 342
Polyvinylchloride (PVC), 18
Power pole, 126, 156
Pressurized sheaths, 207
Private branch exchange (PBX), 282, 343
Procurement issues, 269
Protective fill, 211, 212, 335
Public pay phones, 128
PVC (polyvinylchloride), 18

Quad cable, 15

Radio, 74, 286
Raised floor, 155
RCDD (Registered Communications Distribution Designer), 261, 314, 330
Recreational areas, 130
Registered Communications Distribution Designer (RCDD), 261, 314, 330
Requirements definition, 257
Resistance, 14
Riser support, mesh, 209
Riser system, 183, 203, 205, 208
Rodents, 211, 212
Routers, 298
  protocols in use of, 280
Rule of 6, 88, 159

Satellite distribution frame (SDF), 171, 177, 201, 344
Satellite equipment room (SER), 100
SDF (satellite distribution frame), 171, 177, 201, 344
Secondary distribution frame, 177
Secondary transmission systems, 73
Security access phones, 130
Security systems, 34
Segmentation, building, 162
SER (satellite equipment room), 100
Sheaths:
  gel-filled, 207
  pressurized, 207
Siecor, 94
Signal interference:
  crosstalk, 88
  electrical noise, 87
SIO (*see* standard information outlet)
Slotted zone distribution, 149
SMA connector, 63
Specialty cables, 19
Splicing, 64
ST connector, 63
Standard information outlet (SIO), 268
  configuration of, 114
  definition of, 107
  installation of, 131
  location of, 123–125, 127–130
  mounting methods, 116, 118, 119
  multiuser, 115
  power pole integration, 126
  surge protection, 130
  testing of, 132
Steel strands, 207
Stranded cable, 15, 345
Structural support, 175
Support staff:
  integration with MIS, 249
  organization of, 248
Surge protection, 130, 185
Switchboard facilities, 128
System installation, 270
System integration 238, 275

T568(A/B) standard, 110
T568, T569, T570 standards, 313
Tape marker, 225
TCP/IP (Transmission Control Protocol/ Internet Protocol), 280
Teflon, 55, 346
Telecommunications Distribution Methods Manual, 313
Telecommunications Industry Association (TIA), 313
Tensile strength, 70
Terminal systems, 300
Termination box, 141
Testing, 85, 135, 165, 245, 247
  distribution components, 199

Testing (*Cont.*):
  frame interconnection system, 224
  horizontal distribution, 168
  SIO units, 132
Testing and acceptance procedures, 272
TIA (Telecommunications Industry Association), 331
Time domain reflectometers, 245
TNC connector, 45
Token passing, 289
Transmission, half-duplex, 56
Transmission Control Protocol/Internet Protocol (TCP/IP), 280
Twinax connector, 45
Twisted-pair, 7, 286, 320
  advantages of, 36
  applications, 33
  characteristics of, 14
  construction of, 12
  cost of, 34
  definition of, 11
  disadvantages of, 37
  marking of, 12
  physical options, 16
  shielding of, 17, 134
  sizing of, 134
  trends of, 324
  types of:
    composite, 19
    high performance, 19
    imitations, 15
    quad, 15
    specialty, 19
    stranded, 15
    untwisted, 15
Two-wire circuit, 3, 47

UL (Underwriters Labratories), 262
Undercarpet cable, 19
Undercarpet distribution, 154
Underwriters Labratories (UL), 262
Universal service order code, 109
Universal Wiring Standard (*see* Wiring Standards)
User's Guide, 310
UWS (*see* Wiring Standards)

Velocity of Propagation (Vp), 14
Vendor plans:
  selection of, 94
  services, 93
  types of:
    AT&T, 96
    DEC, 98
    global, 92
    hardware-based, 92
    IBM, 101
    modular, 103
    spec-based, 92
Vendor selection, 269
Voice systems:
  CENTREX, 33, 37, 277, 282, 283, 334
  integration of, 282
  key system, 33, 37, 95, 128, 282–283
  PBX, 282, 343

Wiring codes, 9, 262
Wiring standards:
  definition of, 303
  development of, 308
  existing, 312
    BICSI, 313
    EIA/TIA, 313
    State of California, 315
    State of Connecticut, 316
  limitations of, 311

Xerox Network System, 281

Zone distribution, slotted, 149
Zone of Protection, 347